Environmental Concerns with Energy

Environmental Concerns with Energy

Edited by Cody Long

SYRAWOOD
PUBLISHING HOUSE

New York

Published by Syrawood Publishing House,
750 Third Avenue, 9th Floor,
New York, NY 10017, USA
www.syrawoodpublishinghouse.com

Environmental Concerns with Energy
Edited by Cody Long

© 2017 Syrawood Publishing House

International Standard Book Number: 978-1-68286-494-4 (Hardback)

Cataloging-in-Publication Data

Environmental concerns with energy / edited by Cody Long.
    p. cm.
Includes bibliographical references and index.
ISBN 978-1-68286-494-4
1. Power resources--Environmental aspects. 2. Clean energy. 3. Renewable energy sources.
4. Biomass energy. I. Long, Cody.
TD195.E49 E58 2017
333.794--dc23

Printed in the United States of America.

TABLE OF CONTENTS

PREFACE

This book aims to highlight the current researches and provides a platform to further the scope of innovations in this area. This book is a product of the combined efforts of many researchers and scientists from different parts of the world. The objective of this book is to provide the readers with the latest information in the field.

This book on environmental concerns with energy discusses the energy generation practices and the level of pollution that such industries cause. The major sources of energy generation on earth is still coal, oil and natural gas and the burning of fossil fuels causes considerable damage to the earth's atmosphere. Efficient energy practices aim to reduce emissions while providing sustainable energy. Technological innovation and progress are the pillars that will facilitate efficient energy production. This book is a compilation of chapters that discuss the most vital concepts and emerging trends in this field. It elucidates new techniques and their applications in a multidisciplinary approach. For someone with an interest and eye for detail, this book covers the most significant topics in this field. It will help new researchers by foregrounding their knowledge in this branch.

I would like to express my sincere thanks to the authors for their dedicated efforts in the completion of this book. I acknowledge the efforts of the publisher for providing constant support. Lastly, I would like to thank my family for their support in all academic endeavors.

Editor

Using climate-FVS to project landscape-level forest carbon stores for 100 years from field and LiDAR measures of initial conditions

Fabián B Gálvez[1†], Andrew T Hudak[1*†], John C Byrne[1], Nicholas L Crookston[1] and Robert F Keefe[2]

Abstract

Background: Forest resources supply a wide range of environmental services like mitigation of increasing levels of atmospheric carbon dioxide (CO2). As climate is changing, forest managers have added pressure to obtain forest resources by following stand management alternatives that are biologically sustainable and economically profitable. The goal of this study is to project the effect of typical forest management actions on forest C levels, given a changing climate, in the Moscow Mountain area of north-central Idaho, USA. Harvest and prescribed fire management treatments followed by plantings of one of four regionally important commercial tree species were simulated, using the climate-sensitive version of the Forest Vegetation Simulator, to estimate the biomass of four different planted species and their C sequestration response to three climate change scenarios.

Results: Results show that anticipated climate change induces a substantial decrease in C sequestration potential regardless of which of the four tree species tested are planted. It was also found that *Pinus monticola* has the highest capacity to sequester C by 2110, followed by *Pinus ponderosa*, then *Pseudotsuga menziesii*, and lastly *Larix occidentalis*.

Conclusions: Variability in the growth responses to climate change exhibited by the four planted species considered in this study points to the importance to forest managers of considering how well adapted seedlings may be to predicted climate change, before the seedlings are planted, and particularly if maximizing C sequestration is the management goal.

Keywords: Carbon sequestration; Climate change; Forest vegetation simulator; General circulation model; Growth and yield; LiDAR

Background

Forests cover about one third of the Earth's terrestrial surface and have great capacity to store and cycle carbon (C). Living and dead wood, litter, detritus, and soil exceed the amount of C present in the atmosphere [1,2]. Forest resources supply a wide range of environmental services like mitigation of increasing levels of atmospheric carbon dioxide (CO_2). Recent research shows how changes in forest cover and land use affect CO_2 emissions to the atmosphere [3]. Evolution of new plant associations [4], shifts in the spatial distribution in tree species [5], redistribution of populations to local climates [6],

and changes in site index [7] are the effects that climate change are having and are expected to have on forest ecosystems now and in the future. Tree growth, mortality and regeneration potential are typically adversely affected by the climate changing from the "normal" conditions to which tree species have adapted [8-10]. Conversely, climate change may lead to increased growth in other species [11] or other positive effects.

Ecosystem process-based models take the approach of simulating underlying biogeochemical processes, such as photosynthesis and respiration, using mathematical equations that determine the allocation of C from atmospheric CO_2 into biomass. These models require parameterization for vegetation type, climate, and site conditions that constrain net primary productivity and ecosystem C balance. Forest-BGC (Biogeochemical

* Correspondence: ahudak@fs.fed.us
†Equal contributors
[1]USDA Forest Service, Rocky Mountain Research Station, 1221 South Main St., Moscow, ID 83843, USA
Full list of author information is available at the end of the article

Cycles) [12,13] and the Terrestrial Ecosystem Model (TEM) [14] partition C based on water and nitrogen limitations. The 3PG (Physiological Principles Predicting Growth) model has been linked to satellite image-derived estimates of canopy photosynthetic capacity to estimate forest growth [15,16]. Another alternative approach for assessing climate change impacts is to merge a state and transition model (STM) with the outputs from a dynamic global vegetation model such as MC1 [17] that predicts plant communities under equilibrium conditions. The MC1 model is so named because it combines biogeographic rules defined in the MAPSS [18] model with the CENTURY [19] biogeochemical model, which focuses on soil organic matter dynamics. The STANDCARB [20] model simulates both living and dead C pool dynamics at the forest stand level, which comes closer to what foresters expect as a measure of growth and yield. LANDIS-II (Forest Landscape Disturbance and Succession) [21,22] simulates landscape-level forest succession and disturbance processes, as well as forest management. All of the aforementioned models have the capacity to explore climate change effects on forest C sequestration, and most can operate in a spatially-explicit manner.

As opposed to the suite of process-based models favored by ecologists, forest managers traditionally use empirical models for predicting forest growth and yield. As climate is changing, forest managers have added pressure to obtain forest resources by following stand management alternatives that are biologically sustainable and economically profitable [23]. An empirical growth and yield model extensively used in the United States, the Forest Vegetation Simulator (FVS), is an approved quantification tool by the American Carbon Registry and is used broadly to predict forest stand dynamics. FVS operates at the individual tree level, simulating growth, mortality, and regeneration based on empirical studies. Forest managers use FVS to summarize and predict current and future forest stand conditions under different management alternatives, where outputs obtained from the model are used as inputs to forest planning models and other uses [24]. Other uses of FVS take into account how management and forest practices affect stand structure and composition, determine suitability for wildlife habitat, estimate hazard ratings for insect disease outbreaks or potential fires, and calculate consequent losses from these events. FVS is a powerful suite of models which has been linked to Forest Service forest inventory data bases and geographic information systems, evolving into a useful suite of tools for forest managers [25].

Climate-FVS is a recent improvement upon FVS that includes functions which take climate change and species-climate relationships into account when predicting tree growth, mortality, and regeneration establishment [8]. General Circulation Models (GCM) are specified within Climate-FVS since they are key to understanding future climates [26]. The variability in GCM outputs resulting from different model formulations and emissions scenarios are accounted for by running Climate-FVS such that different Climate-FVS runs are each informed by different GCM outputs. Neither FVS nor Climate-FVS currently have a spatial analysis capability.

Forest biomass and C stores and fluxes can be quantified at synoptic scales using remote sensing technologies, especially Light Detection And Ranging (LiDAR) [27]. Current commercial airborne LiDAR systems emit laser pulses of near-infrared light and measure the time elapsed until the light reflects off of the vegetation or ground and returns to the aircraft. Upwards of 100,000 laser pulses per second can be recorded, along with simultaneous inertial measurement unit (IMU) and global positioning system (GPS) measures of the aircraft position, to return a 3-dimensional point cloud characterizing at high resolution the x,y,z position of the ground and vegetation surfaces. LiDAR canopy height measures can be related to tree measures from forest inventory plots to map forest structure attributes of utility to forest managers [28]. Studies have applied LiDAR to extrapolate plot-level measures of forest biomass and C across forest landscapes [27,29], demonstrating the utility of area-based modeling methods for predicting (and mapping) current conditions.

While LiDAR and other remotely sensed data provide a snapshot of forest conditions in time, growth models such as FVS are commonly used to update the interval years between inventories, be they traditional field surveys or surveys that use both field and LiDAR or other remotely sensed data. Forest managers use FVS for planning purposes and for updating inventories under the assumption of an unchanging climate. This assumption of unchanging climate may be practical for predicting forest growth over the next 10 years or so. However, given the consensus among scientists that climate is changing, it is a difficult assumption to defend at longer time scales, such as a century, or the 50–80 year rotation length of managed, even-aged stands in the U.S. Northwest. In one previous study, Climate-FVS was employed to study the efficacy of active management alternatives applied in the aftermath of disturbances likely induced by climate change in the Rocky Mountains of Colorado, USA [30]. They concluded that adaptation-oriented management was necessary to provide for forest cover and accompanying C stocks during the 21st century.

The primary objective of this study is to project the effect of typical forest management actions on forest C levels, given a changing climate, in the Moscow Mountain area of north-central Idaho, USA (Figure 1). The secondary objective is to upscale plot-level projections from a map of initial biomass conditions as mapped from 2009 LiDAR data across the 20,000 ha study area. Thus, results are summarized at two scales: At the plot level, the trees

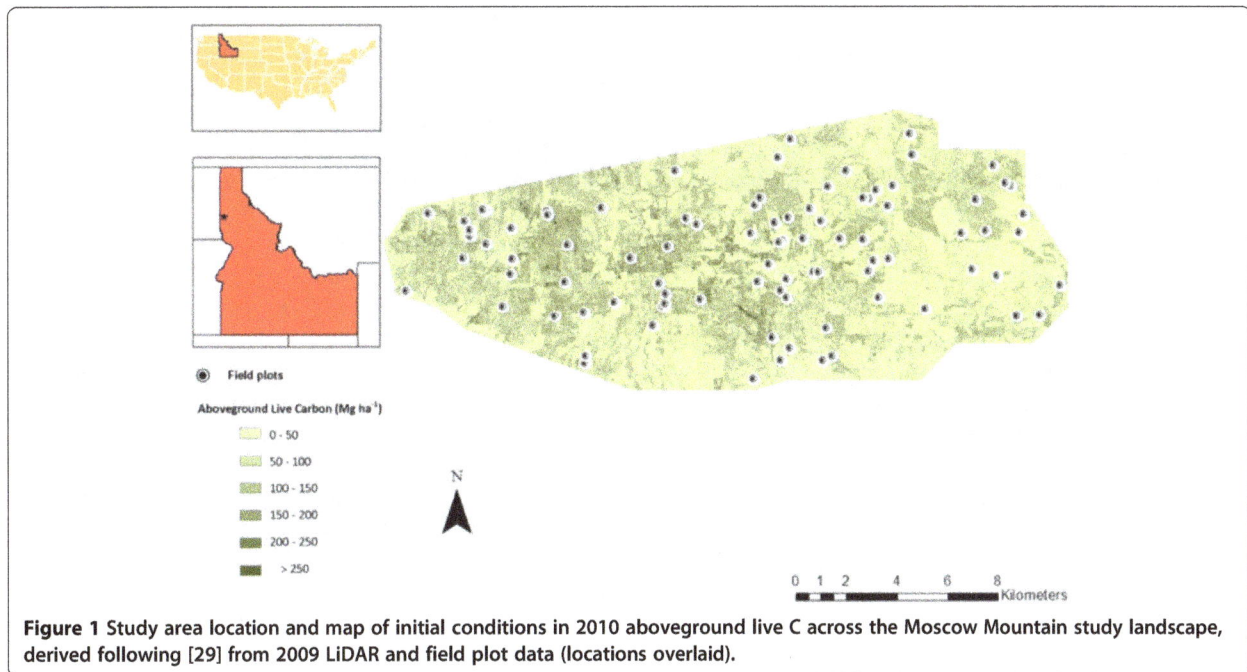

Figure 1 Study area location and map of initial conditions in 2010 aboveground live C across the Moscow Mountain study landscape, derived following [29] from 2009 LiDAR and field plot data (locations overlaid).

inventoried in 2009 are grown and summarized at decadal intervals for one century (2010–2110) using Climate-FVS. At the landscape level, the plot-level, decadal C projections are linked to a map of forest aboveground biomass predicted from the same forest inventory plot data and a 2009 LiDAR survey. Since forest management decisions are often made at the landscape level, the landscape-level projections may better inform a forest manager of the consequences of management alternatives on forest growth in the context of climate change. An implicit assumption in this study is that managers will want to sustain as productive a forest as possible by maximizing C sequestration on site.

Methods
Study area
Moscow Mountain is a western extension of the mixed-conifer forest type that dominates north-central Idaho, with agricultural lands more prevalent to the north, south, and especially Washington to the immediate west (Figure 1). Elevations range from 786 m to 1517 m, and annual rainfall from 630 to 1015 mm, with most precipitation falling during the winter and spring, while the summer and fall are dry. The common tree species, listed in order of decreasing drought tolerance [31] are: ponderosa pine [PP] (*Pinus ponderosa*), Douglas-fir [DF] (*Pseudotsuga menziesii*), western larch [WL] (*Larix occidentalis*), grand fir [GF] (*Abies grandis*), and western red cedar [RC] (*Thuja plicata*). Typical habitat types are the PP series at xeric sites on southern and western aspects, the DF and GF series on respectively moister sites, and the cedar/hemlock series on the most mesic sites on northern and eastern

aspects [32]. A volcanic ash cap layer is thicker on northeast aspects and increases soil water holding capacity [33], augmenting the important influence of aspect on forest composition in this topographically complex landscape (Figure 1). Moscow Mountain is the setting of four large University of Idaho Experimental Forest management areas, extensive landholdings by private timber companies, as well as many private and some public land inholdings. The landscape is actively managed, with 26% of the 20,000 ha study area harvested between 2003 and 2009 alone [29].

2009 Forest inventory
LiDAR data
LiDAR data were collected 30 June 2009 at a mean density of 8.52 points/m^2, including 50% overlap between adjacent flight lines limited to a scan angle of ±14° from nadir. A 4.3 cm vertical accuracy was achieved. Ground returns were classified using multiscale curvature classification [34], from which a 1 m resolution digital terrain model (DTM) was interpolated. LiDAR return elevations (Z) were normalized for topography by subtracting the DTM elevation from the LiDAR points, resulting in canopy height measures at every X, Y location sampled by the LiDAR. Canopy height, intensity, and density metrics characterizing the canopy structure were calculated at a cell resolution of 20 m across the entire LiDAR collection, along with topographic metrics from the DTM resampled to the same 20-m x 20-m (400 m^2) cells [29]. The same suite of LiDAR metrics were also calculated from the normalized point cloud data and the DTM surface located within 89 fixed-radius field plots sampled across the study area.

Field data

Field plots for forest inventory measurements in 2008 (4 plots) or 2009 (85 plots) were distributed following a random stratified design based on topographic elevation, slope, aspect, and a Landsat satellite image-derived map of percent canopy cover [29]. The 89 plots were 400 m^2 in size, within which all trees >10 cm were tallied. Saplings (≤10 cm and >1.37 m height) were tallied across the entire plot and seedlings (≤1.37 m height) within a 20 m^2 subplot situated at plot center. Multiple plot center positions were logged using global positioning system (GPS) units with differential correction capability and averaged for an estimated plot location uncertainty of about one meter.

Forest aboveground biomass C map

Tree biomass estimates aggregated at the plot level were associated with the plot-level LiDAR metrics in an imputation model, with species-level plot biomass forming the response variables and the LiDAR metrics forming the explanatory variables. These plot-level field and LiDAR data comprised the reference plots for imputing forest aboveground biomass across the Moscow Mountain landscape, with the gridded LiDAR metrics representing the target cells where LiDAR data were available but field data were not. Imputation was used to assign the ground-measured attributes of the 89 plots to similarly sized target cells where no ground-measures were taken. The 89 plots are called reference observations and the gridded cells are called target observations [35]. In this case, the 400 m^2 target cells were each assigned one of the 89 possible 400 m^2 reference observations of total aboveground tree biomass. The act of making this assignment is an imputation. The assignments depended on the similarity of the LiDAR metrics in the target cells to those in the reference plots. The closest reference observation in a multivariate space defined by LiDAR metrics is its nearest neighbor. The exact definition of the multivariate space used in computing the distances is conditioned by the relationships between the LiDAR metrics and the biomass metrics that are evident in the 89 sample plots where both sets of metrics are known. An advantage of imputation is that it maintains the co-variance relationships between all plot attributes, meaning any measurements taken on a plot can also be imputed, even if they play no part in nearest neighbor selection. Therefore, the plot-ID corresponding to the imputed aboveground tree biomass reference observations mapped by [29] was itself mapped as an ancillary variable for this study.

Climate-FVS

Like the standard FVS model, Climate-FVS reads initial stand or plot inventory information and uses it as starting values. In addition, Climate-FVS reads an additional input file that contains climate metrics (measures of temperature and precipitation as projected by down-scaled GCM outputs) and species viability information that are specific to the location and elevation of the site being simulated. This additional file is generated using the "Get Climate-FVS Ready Data" webpage on the Rocky Mountain Research Station (RMRS) website [36] and requires a text file containing longitude, latitude, and elevation for each plot location. The mortality submodel increases mortality rates when tree species viability scores, ranging from 0–1, drop below 0.5 [8]. High-mortality rates may lead to the loss of some currently existing tree species in future years; Climate-FVS estimates that if viability scores decrease to <0.2, then the species is absent, with a chance of survival equal to zero. This study used Version 1 of Climate-FVS, which does not take into account genetically different populations of the same tree species for predicting mortality rates.

In this study, each plot was projected using a standard forest management option (clearcut harvesting initiated when stocking reaches 65% of normal, followed by prescribed burning and tree planting at a density of 200 trees/acre) under one of twelve treatments. The twelve treatments were combinations of four planted species (PP, DF, WL, WP (western white pine (*Pinus monticola*)) and three climate model outputs which form our climate change scenarios (Figure 2). The three GCMs used in this study are from the Canadian Center for Climate Modeling and Analysis Global Coupled Model (CGM), the Geophysical Fluid Dynamics (GFD) Laboratory at Princeton University, and the Met Office Hadley Centre (HAD) in the United Kingdom. Each climate change scenario corresponds to one GCM run according to the A2 emission scenarios [37] as described by [36]. We used the A2 emission scenarios assuming the highest levels rather than the lower B levels because current greenhouse gas levels are already higher than those contemplated when the climate model projections were made. The plot data were projected considering the four management treatments (plus one control) and three GCM scenarios (plus one control), or 5 x 4 = 20 projections per plot. In addition, a single control without management and without climate change was run for each plot. In this study, the Climate-FVS C projections for a given treatment and GCM combination varied solely as a result of the variability in initial conditions as measured across the 89 sample plots.

Tree growth from 2008–2009 tree diameter measures was projected to 2010 as a starting point, then projected for 100 years while summarizing at decadal intervals. Besides the standard FVS outputs (stem density, basal area, volume, etc.), two additional outputs were requested: The first is called the Carbon Report and provides the

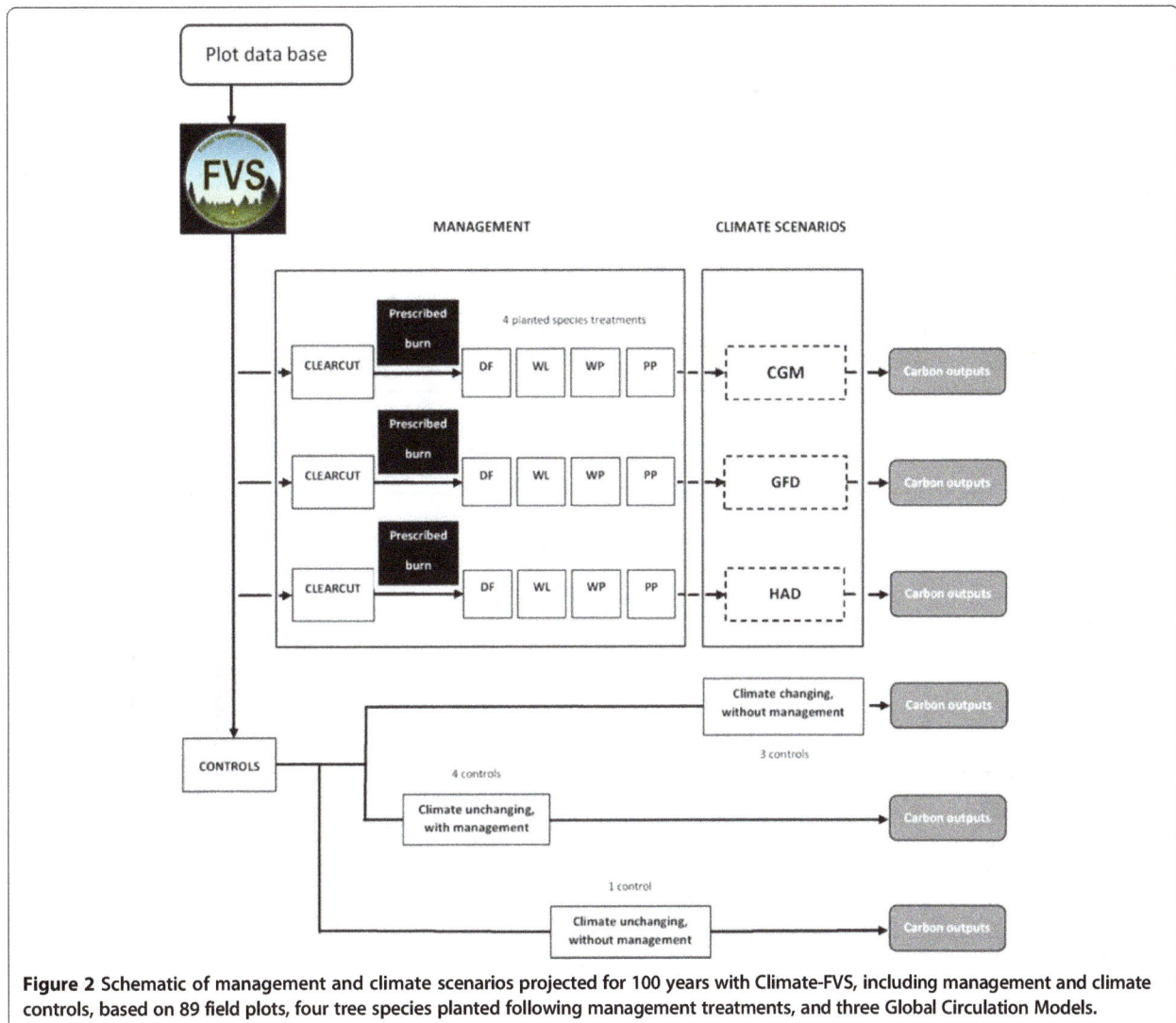

Figure 2 Schematic of management and climate scenarios projected for 100 years with Climate-FVS, including management and climate controls, based on 89 field plots, four tree species planted following management treatments, and three Global Circulation Models.

total aboveground live tree C as well as belowground live C, aboveground dead tree C, and total stand C. Also, it provides the C amount in the forest floor, forest shrubs, forest down dead wood, and the C that is removed by harvesting. The second special report is called the TREE-BIO report and it provides biomass of live and dead trees, standing and removed trees, and breaks down the biomass into stem, crown, or live foliage. It returns estimated biomass in dry weight tons per acre, which is then multiplied by 0.5 to convert to C units [38].

Computing area-wide totals

In a previous study [29], plot-level aboveground tree biomass at the 89 sample plots was measured and then those measurements were imputed to similarly sized map cells derived from the LiDAR, to form a study area-wide biomass map. Indeed, as stated above, those ground-based measurements are the same as the initial inventory data used here. A byproduct of the work by [29] was a data table

that relates each of the 89 inventory plot projections to counts of the number of map cells in the study area to which each plot was imputed; these map cell counts served as weights in computing area-wide C projections. To map C projections across the study area (Figure 1), the plot-level C projections were joined to the map of imputed plot-IDs.

Results

Plot-level C sequestration

Treatments and controls show a general tendency to increase C pools starting in 2010. Relative to the no management controls, management treatments always result in lower standing live tree C storage by 2110 regardless of tree species planted (Figures 3 and 4). Among these four species, PP plantings sequester C at the fastest rate in the first 50 years, but then C storage declines to 2110. DF and WL show the same trends but with less magnitude; DF peaks approximately a decade later, while WL trajectories are the least dynamic. Only WP steadily

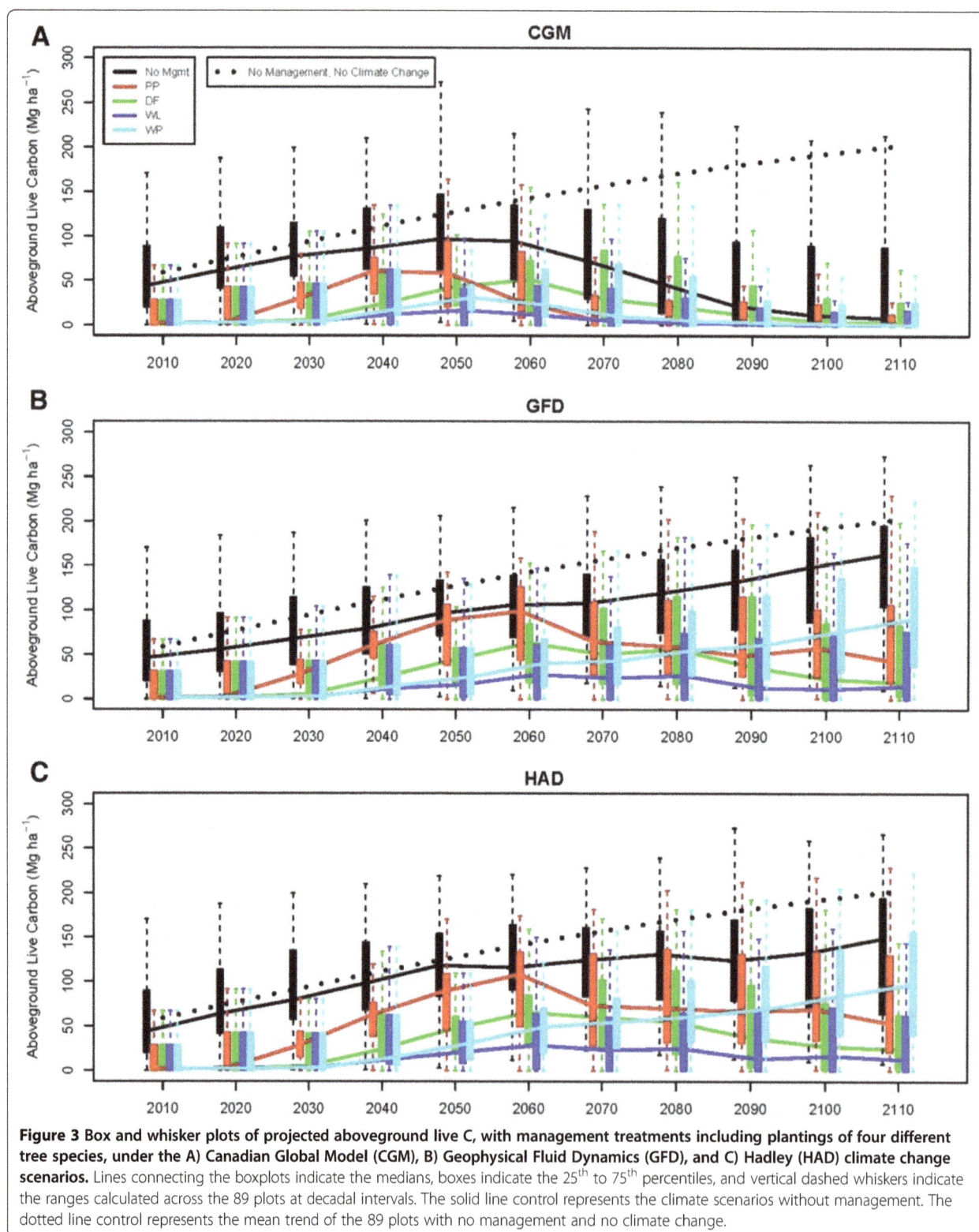

Figure 3 Box and whisker plots of projected aboveground live C, with management treatments including plantings of four different tree species, under the A) Canadian Global Model (CGM), B) Geophysical Fluid Dynamics (GFD), and C) Hadley (HAD) climate change scenarios. Lines connecting the boxplots indicate the medians, boxes indicate the 25th to 75th percentiles, and vertical dashed whiskers indicate the ranges calculated across the 89 plots at decadal intervals. The solid line control represents the climate scenarios without management. The dotted line control represents the mean trend of the 89 plots with no management and no climate change.

accumulates C to 2110, except under the CGM climate change scenario (Figures 3 and 4).

Similarly, C sequestration is invariably lower under all three climate change scenarios compared to the no

climate change control. Among the three climate change scenarios, the CGM scenario results in the lowest C storage, while C levels are intermediate under the GFD and HAD scenarios (Figures 3 and 4). The mean of the

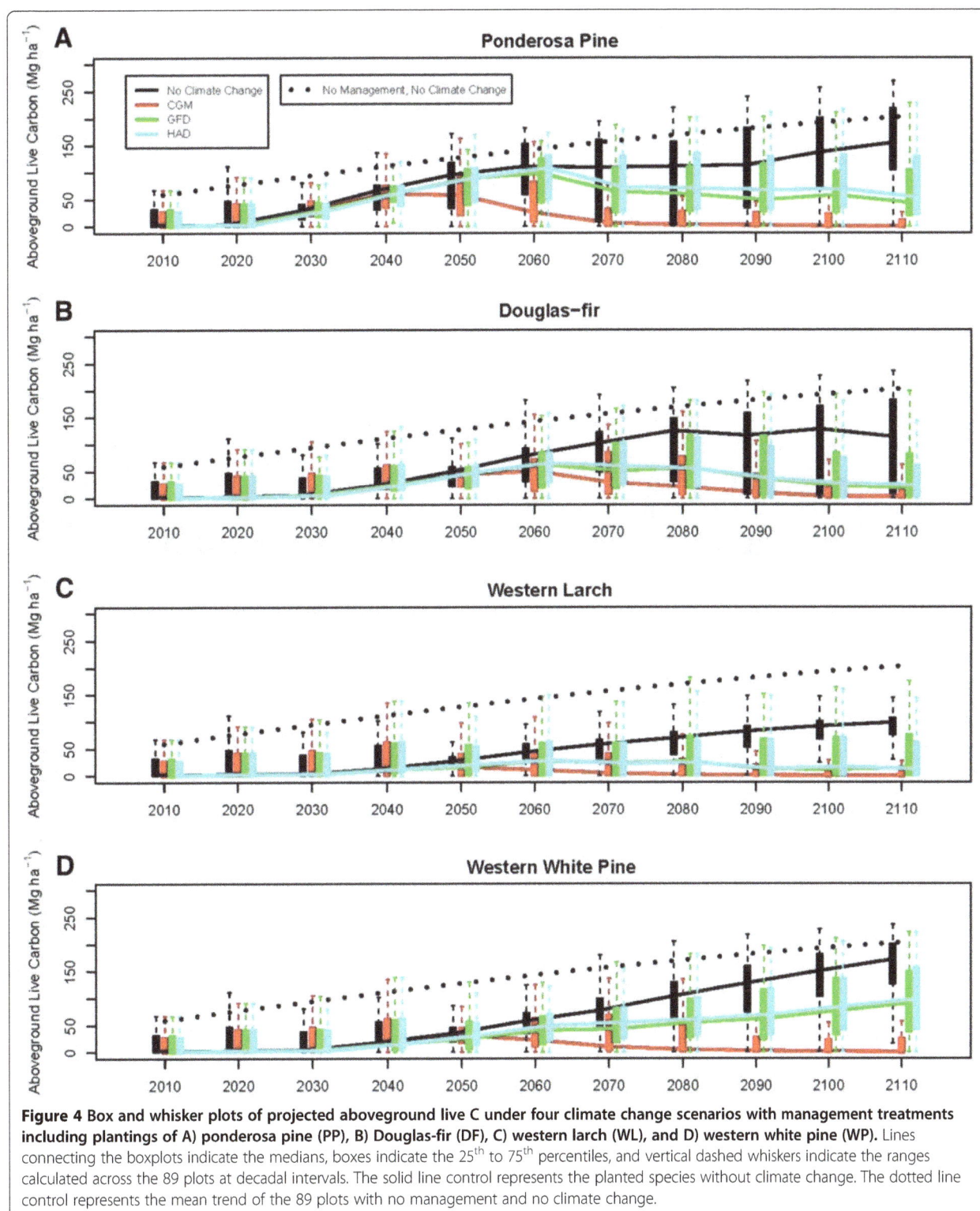

Figure 4 Box and whisker plots of projected aboveground live C under four climate change scenarios with management treatments including plantings of A) ponderosa pine (PP), B) Douglas-fir (DF), C) western larch (WL), and D) western white pine (WP). Lines connecting the boxplots indicate the medians, boxes indicate the 25th to 75th percentiles, and vertical dashed whiskers indicate the ranges calculated across the 89 plots at decadal intervals. The solid line control represents the planted species without climate change. The dotted line control represents the mean trend of the 89 plots with no management and no climate change.

full control projections that exclude both climate change and management is presented as a dotted line for reference in Figures 3 and 4.

Figures 3 and 4 illustrate plot-level projections for the aboveground live C pool, which is the largest of the C pools reported by Climate-FVS, but trends are similar for the other reported C components: belowground live, aboveground dead, and harvest removal (Figure 5). Among the four alternative plantings considered, planting WP sequesters the most C by 2110; it therefore comes closest to

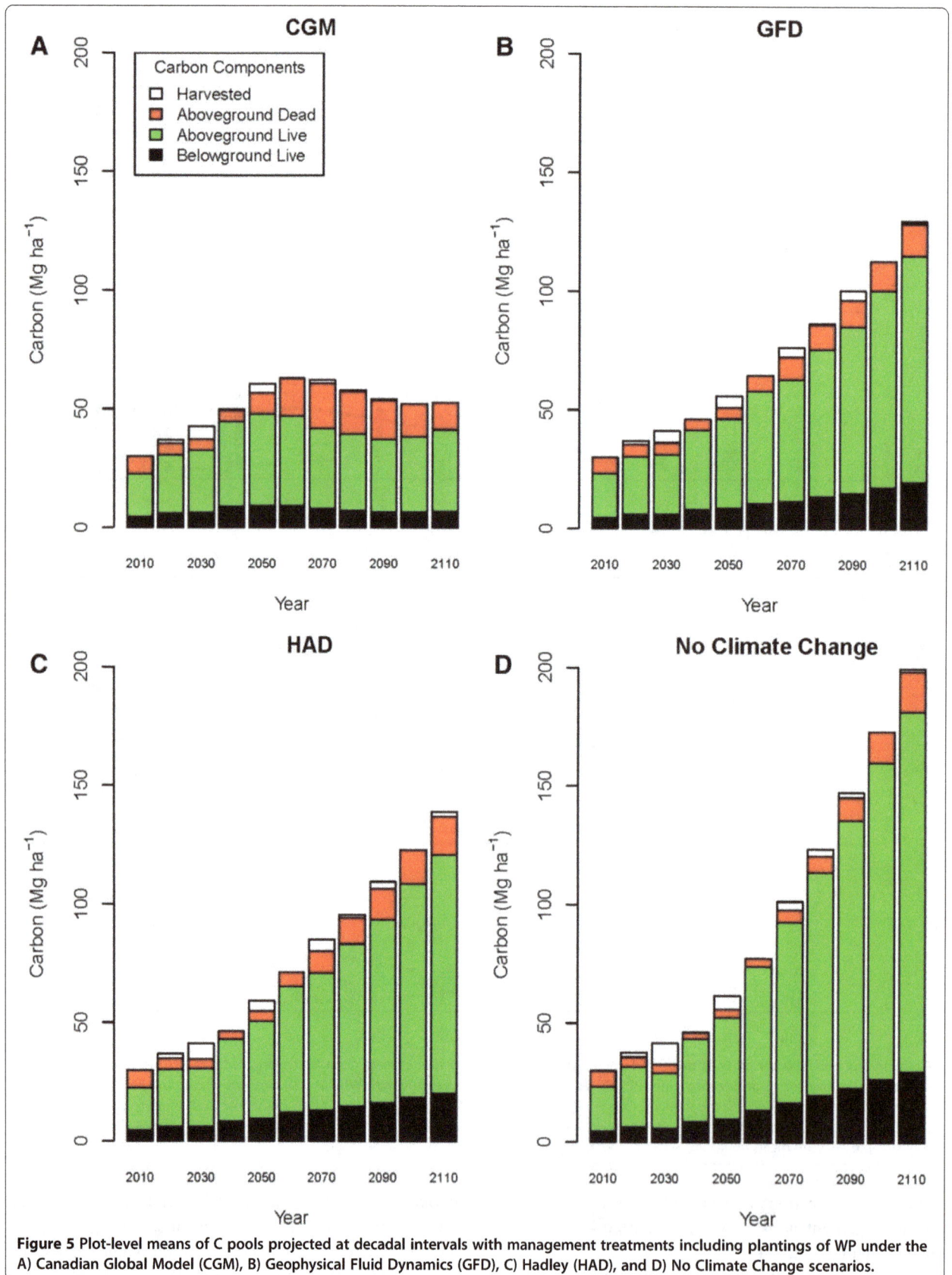

Figure 5 Plot-level means of C pools projected at decadal intervals with management treatments including plantings of WP under the A) Canadian Global Model (CGM), B) Geophysical Fluid Dynamics (GFD), C) Hadley (HAD), and D) No Climate Change scenarios.

the control scenario that maximizes C sequestration, which is why we selected it as an illustrative example (Figure 5). Even if harvest removals are added to the other C pools that comprise total C on site, total C under every alternative climate scenario or management treatment is always less than the total C projected under the control scenario of no climate change. Because the C component pool trends remain consistent, the results reported in this paper focus on the aboveground live C, the most relevant C pool for managing forest C stores on site.

Landscape-level C sequestration

Anticipated climate change induces a significant decrease in C sequestration potential regardless of which of the four tree species tested are planted (Figure 6). No climate change results in the greatest C storage regardless of tree species planted in the management treatments. C sequestration of PP plantings peaks earliest (2050–2060); DF plantings peak slightly later (2060–2080); WL plantings (besides the control) show the least amount of change; WP plantings produce the highest total C by 2110 of all four planted species in all three climate scenarios and the no climate change control; only the CGM scenario shows a decline, after peaking in 2050. Among the climate change scenarios (excepting the control scenario), HAD usually produces the largest C pools, followed closely by GFD, while CGM results in the lowest C storage.

Figure 7 shows an example of the landscape-level maps summarized in Figure 6, with projected 2110 aboveground live C after management with WP plantings, under the scenarios of no climate change or the three GCMs. The spatial patterns in C as depicted in Figure 7 are invariant because no attempt was made to predict the location of management treatments. However, depending on which climate change scenario is considered, the overall magnitude of C sequestration is dramatically affected by 2110. Similar differences are evident if maps of the different tree plantings are compared, but are not included here for brevity, as the cumulative magnitude of landscape-level differences in C sequestration are already indicated in Figure 6.

Discussion

For brevity, only the results for aboveground live C are presented in this paper, but the trends in the belowground live, aboveground dead, total, and other C pools reported in the FVS C report are similar, in that they are linked to the tree growth projections that comprise the growth engine that drives FVS and Climate-FVS [8,25]. Productivity rates drop dramatically when climate change is involved in the projections. Moreover, total production of wood volume at each simulated time step is the current standing volume of trees plus harvest removals for that timestep, which are ongoing over the simulation period.

Tracking total production in this manner shows that a decrease in total production is due to the increased tree mortality projected by Climate-FVS, due to declining tree species viability scores. The Climate-FVS output shows that climate change is responsible for the detriment of the tree species initially present in the stands. The CGM climate scenario had the greatest impact among the three GCMs tested in this study.

Much of the following discussion considers elements of uncertainty. We highlight some specific issues emerging from this analysis that we feel are noteworthy, rather than several other potential sources of error – such as those related to the underlying FVS growth model as it is already broadly applied [24]. A few things are firmly understood and one of them is that climate is changing. While the GCM predictions differ, none predict that climate is not changing; furthermore, the magnitude of change is substantial. Another firm idea is that management decisions need to be made that depend on predictions about the future. Uncertainty, therefore, cannot be avoided. Climate-FVS is intended as a decision-making tool for forest managers despite the uncertainties implicit with climate change.

We caution that the uncertainties in these C projections are at least as high as the uncertainties in the GCMs themselves. This uncertainty would be expected to increase with time since the initial conditions were specified. Thus, it is more likely that PP may show the most favorable growth response to climate change in the next 50 years, than that WP may show the most favorable response in the next 100 years (Figures 3 and 4). More important is the trend that all four planted species show a depressed ability to sequester C if forecasted warming and drying occurs, than if it does not (control scenarios). Of the four planted species considered, WP shows the most resilience to long-term climate change effects on C sequestration, while WL shows the least resilience to climate change. This agrees with the particularly deleterious effects of climate change on WL as noted by [39], who predicted dramatic latitudinal and altitudinal shifts in the climate space suitable for WL. Forecasted rates of climate change are expected to exceed the rates that trees, especially WL, can migrate to their shifted climate space, bolstering calls for assisted migration [40] of seeds from distal sources. Seeds from different sources have different capacities to store C, but the expression of these differences depends on the environment [9,40,41]. Variability in the growth responses to climate change exhibited by the four planted species considered in this study points to the importance to forest managers of considering how well adapted seedlings may be to predicted climate change, before the seedlings are planted. These decisions may involve not just the tree species growing on Moscow Mountain, but also

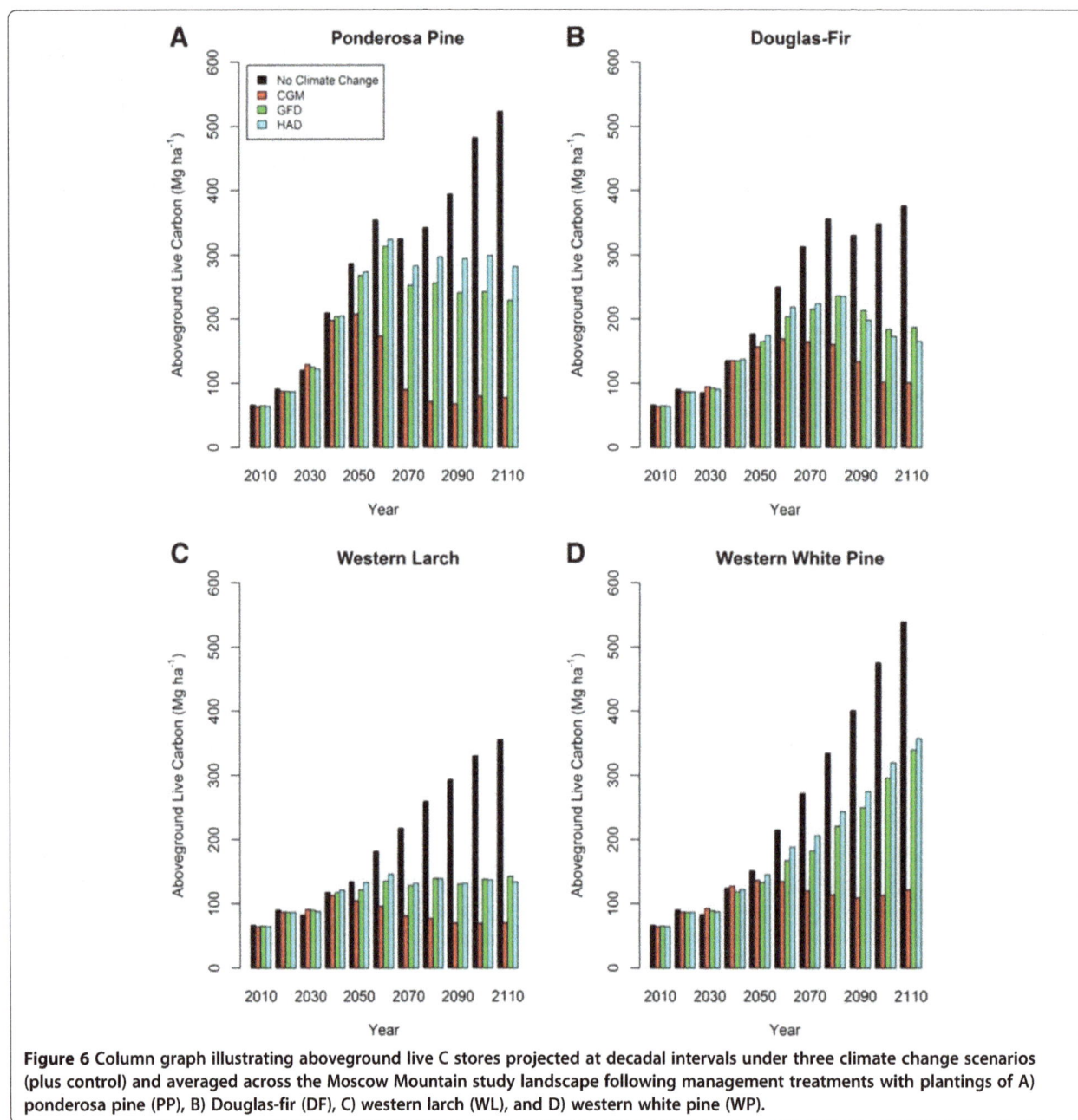

Figure 6 Column graph illustrating aboveground live C stores projected at decadal intervals under three climate change scenarios (plus control) and averaged across the Moscow Mountain study landscape following management treatments with plantings of A) ponderosa pine (PP), B) Douglas-fir (DF), C) western larch (WL), and D) western white pine (WP).

other species and spatially disjunct seed sources used to regenerate them.

There are additional sources of potential error in these simulations. As pointed out by [8], the Climate-FVS model uses empirically calibrated species-climate relationships based on the observed presence and absence of species. These data capture the realized niche of species that are due to competitive relationships between trees of different species as well as climate. The potential niche space is, by definition, larger than the observed, or realized, niche. Climate-FVS does not attempt to capture the potential niche effect and therefore it may overstate

the effect of climate change on composition. On the other hand, there is a dearth of data that can be used to measure the largely unobservable potential niche.

In managed forests in the U.S. Northwest, investment in initial species establishment success following harvesting is required by law under state forest practices legislation. So, in general, managers have no choice but to conduct silvicultural treatments such as prescribed burning, planting, spraying, and subsequent replanting as needed, in order to successfully regenerate stands with target species composition. However, our results show that subsequent interactions among climate and

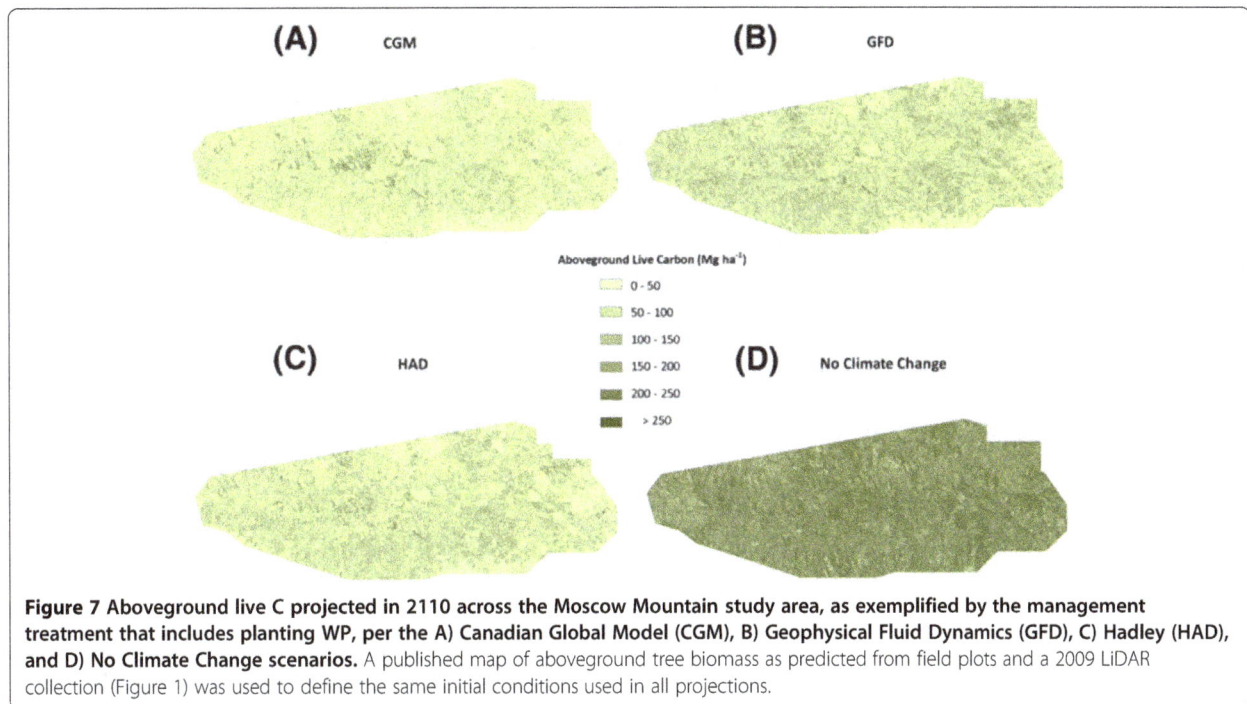

Figure 7 Aboveground live C projected in 2110 across the Moscow Mountain study area, as exemplified by the management treatment that includes planting WP, per the A) Canadian Global Model (CGM), B) Geophysical Fluid Dynamics (GFD), C) Hadley (HAD), and D) No Climate Change scenarios. A published map of aboveground tree biomass as predicted from field plots and a 2009 LiDAR collection (Figure 1) was used to define the same initial conditions used in all projections.

productivity that affect intermediate stand development may still be evident. It is unclear how species-specific interactions of climate and stand productivity may manifest themselves in mixed stands where interspecific competition is also at play.

Other modeling approaches to addressing climate change arguably could be used instead of Climate-FVS. The principle shortcoming of FVS and Climate-FVS currently is that it does not ingest spatially explicit inputs or operate in a spatially explicit analysis framework, as do many ecosystem process-based models [e.g., 14,15,17, 21,22]. Furthermore, forest-BGC [12,13], 3PG [15], and STANDCARB [20] were initially developed in and hence are well parameterized for U.S. Northwest forests. An application of MC1 in the U.S. Northwest [42] is particularly relevant in that it considers altered disturbance regimes due to climate change, and climate change effects on species assemblages. We view the fact that Climate-FVS operates at the level of individual tree species as an advantage for our study, because adaptation to climate change acts at the species level [10]. Landscapes like Moscow Mountain will likely remain a temperate coniferous forest for the next 100 years; it is the composition of that forest and rates of change in composition within that forest that are likely to change. Climate-FVS is arguably the best model to use in this context. We also chose to use Climate-FVS in deference to local managers' desire to manage forests on Moscow Mountain for timber production. The four species we selected for planting following treatment are the most marketable tree species.

Our field plots sampled the Moscow Mountain landscape following a random stratified design, and simulated management treatments were distributed across the landscape according to their imputed plot-ID; this is why the plot-level and landscape-level projections in this study show similar trends and lead to the same conclusions. It would be more realistic and useful to allow the user to target specific locations (i.e., stands) for treatment, such as topographic positions with habitat types and tree species that may be more vulnerable to mortality from a warmer and/or drier climate. For instance, it is reasonable to assume that currently wetter NE aspects will remain wetter than SW aspects even as climate changes, because topographic variables are practically static even as climate is dynamic. Although we have linked forest growth projections to existing forest structure as characterized with LiDAR across the landscape, further research is needed to consider landscape context in future growth projections and management treatment alternatives. Indeed, the application of FVS and Climate-FVS within a spatially explicit modeling framework is an object of current development work.

Conclusions

Different tree species sequester C at different rates, as constrained by genetics, site characteristics, and their interaction. Tree growth, total incremental production, and species viability are projected to be negatively affected by climate change in this mixed conifer forest in north-central Idaho, USA. While there are uncertainties in the GCMs, and differences between their outputs, there is little doubt that climate is changing. The projected declines in forest productivity and C sequestration potential under widely accepted climate scenarios deleteriously affect the four major commercial tree species most commonly planted in this environment. However, subsequent analyses to further understand potential climate impacts should distinguish among climate impacts on stand productivity, timing of individual stand harvests over the landscape, and optimally replanting individual stands to a variety of possible species or mixed-species alternatives over time in the context of landscape-level forest planning. Climate-FVS provides a powerful tool to forest managers regarding which trees to plant for mitigating the effects of global warming.

Competing interests
The authors declare that they have no competing interests.

Authors' contributions
FG conducted analysis, summarized results, wrote initial draft. AH conceptualized and supervised the study, wrote most of the manuscript in its current form. JB trained FG to use Climate-FVS, graphed results, helped write. NC led development of Climate-FVS, selected climate change scenarios, helped write. RK selected management treatment scenarios, provided critical feedback, helped write. All authors read and approved the final manuscript.

Acknowledgements
Fabián B. Gálvez expresses his sincere gratitude to Experience International, U.S. Forest Service (USFS) International Programs, the USFS Rocky Mountain Research Station, and all the helpful personnel at the Moscow Forestry Sciences Lab for their support.

Author details
[1]USDA Forest Service, Rocky Mountain Research Station, 1221 South Main St., Moscow, ID 83843, USA. [2]Department of Forest, Rangeland, and Fire Sciences, University of Idaho, 975 West 6th St., Moscow, ID 83844-1133, USA.

References

1. Food Agriculture Organization of the United Nations: *(FAO) Global Forest Resource Assessment. Progress Towards Sustainable Forest Management.* 147, VialedelleTerme di Caracalla, Rome, Italy: Food and Agriculture Organization of the United Nations Forestry; 2005.
2. Heath LS, Smith JE, Skog KE, Nowak DJ, Woodall CW: **Managed forest carbon estimates for the US greenhouse gas inventory 1990–2008.** *J For* 2010, **109:**167–173.
3. McKinley DC, Ryan MG, Birdsey RA, Giardina CP, Harmon ME, Heath LS, Houghton RA, Jackson RB, Morrison JF, Murray BC, Pataki DE, Skog KE: **A synthesis of current knowledge on forests and carbon storage in the United States.** *Ecol Appl* 2011, **21:**1902–1924.
4. Jackson ST, Overpeck JT: **Responses of plant populations and communities to environmental changes of the late quaternary.** *Paleobio* 2000, **26:**194–220.
5. Iverson LR, Prasad AM: **Predicting abundance of 80 tree species following climate change in the eastern United States.** *Ecol Monogr* 1998, **68:**465–485.
6. Tchebakova NM, Rehfeldt GE, Parfenova E: **Redistribution of vegetation zones and populations of *Larix sibirica* Ledeb and *Pinus sylvestris* L. in Central Siberia in a winning climate.** *Sib J Ecol* 2003, **10:**677–686. in Russian.
7. Monserud R, Yang Y, Huang S, Tchebakova N: **Potential change in lodge pole pine site index and distribution under climatic change in Alberta.** *Can J For Res* 2008, **38:**343–352.
8. Crookston N, Rehfeldt G, Dixon G, Weiskittel A: **Addressing climate change in the forest vegetation simulator to assess impacts on landscape forest dynamics.** *For Ecol Mgmt* 2010, **260:**1198–1211.
9. Rehfeldt GE, Ying CC, Spittlehouse DL, Hamilton DA: **Genetic responses to climate in *Pinus contorta*: niche breadth, climate change, and reforestation.** *Ecol Monogr* 1999, **69:**375–407.
10. Rehfeldt GE, Crookston NL, Warwell MV, Evans JS: **Empirical analyses of plant–climate relationships for the western United States.** *Int J Plant Sci* 2006, **167:**1123–1150.
11. Messaoud Y, Chen HYH: **The influence of recent climate change on tree height growth differs with species and spatial environment.** *PLoS ONE* 2011, **6:**e14691. Doi: 10.1371/journal.pone.0014691.
12. Running SW, Coughlan JC: **A general model of forest ecosystem processes for regional applications I. Hydrologic balance, canopy gas exchange and primary production processes.** *Ecol Mod* 1988, **42:**125–154.
13. Running SW, Gower ST: **FOREST-BGC, A general model of forest ecosystem processes for regional applications. II. Dynamic carbon allocation and nitrogen budgets.** *Tree Physiol* 1991, **9:**147–160.
14. Chen M, Zhuang Q: **Spatially explicit parameterization of a terrestrial ecosystem model and its application to the quantification of carbon dynamics of forest ecosystems in the conterminous United States.** *Earth Interactions* 2012, **16:**1–22.
15. Landsberg JJ, Waring RH: **A generalized model of forest productivity using simplified concepts of radiation use efficiency, carbon balance and partitioning.** *For Ecol Mgmt* 1997, **95:**209–228.
16. Coops NC, Waring RH, Landsberg JJ: **Assessing forest productivity in Australia and New Zealand using a physiologically-based model driven with averaged monthly weather data and satellite derived estimates of canopy photosynthetic capacity.** *For Ecol Mgmt* 1998, **104:**113–127.
17. Bachelet D, Lenihan JM, Daly C, Neilson RP, Ojima DS, Parton WJ: **MC1: a dynamic vegetation model for estimating the distribution of vegetation and associated carbon, nutrients, and water—technical documentation. Version 1.0.** Gen. Tech. Rep. PNW-GTR-508. Portland, OR: U.S. Department of Agriculture, Forest Service, Pacific Northwest Research Station; 2001:95.
18. Neilson RP: **A model for predicting continental-scale vegetation distribution and water balance.** *Ecol App* 1995, **5:**362–385.
19. Parton WJ, Schimel DS, Cole CV, Ojima DS: **Analysis of factors controlling soil organic matter levels in Great Plains grasslands.** *Soil Sci Soc Amer Jrnl* 1987, **51:**1173–1179.
20. Harmon ME, Marks B: **Effects of silvicultural treatments on carbon stores in forest stands.** *Can J For Res* 2002, **32:**863–877.
21. Mladenoff DJ: **LANDIS and forest landscape models.** *Ecol l Mod* 2004, **180:**7–19.
22. Scheller RM, Domingo JB, Sturtevant BR, Williams JS, Rudy A, Gustafson EJ, Mladenoff DJ: **Design, development, and application of LANDIS-II, a spatial landscape simulation model with flexible spatial and temporal resolution.** *Ecol Mod* 2007, **201:**409–419.
23. Hummel S, Hudak AT, Uebler EH, Falkowski MJ, Megown KA: **A comparison of accuracy and cost of LiDAR versus stand exam data for landscape management on the Malheur National Forest.** *J For* 2011, **109:**267–273.
24. Crookston NL, Dixon GE: **The forest vegetation simulator: a review of its structure, content, and applications.** *Comput Electron Agric* 2005, **49:**60–80.
25. Dixon GE: *(Comp.) Essential FVS: A user's guide to the forest vegetation simulator. Internal Report.* Fort Collins, CO: U.S. Department of Agriculture, Forest Service, Forest Management Service Center; 2008:189.
26. Rehfeldt G, Crookston N, Saenz- Romero C, Campbell E: **North American vegetation model for land-use planning in a changing climate: a solution to large classification problems.** *Ecol Appl* 2012, **22:**119–141.
27. Goetz S, Dubayah R: **Advances in remote sensing technology and implications for measuring and monitoring forest carbon stocks and change.** *Carb Mgmt* 2011, **2:**231–244.
28. Hudak A, Evans J, Smith A: **LiDAR utility for natural resource managers.** *Rem Sens* 2009, **1:**934–951.
29. Hudak A, Strand E, Vierling L, Byrne J, Eitel J, Martinuzzi S, Falkowski M: **Quantifying above forest carbon pools and fluxes from repeat LiDAR surveys.** *Rem Sens Environ* 2012, **123:**25–40.
30. Buma B, Wessman CA: **Forest resilience, climate change, and opportunities for adaptation: a specific case of a general problem.** *For Ecol Mgmt* 2013, **306:**216–225.
31. Daubenmire R: **Vegetation: identification of typal communities.** *Science* 1966, **151:**291–298.

32. Cooper SV, Neiman KE, Roberts DW: *Forest habitat types of northern Idaho: A second approximation. Gen. Tech. Rep. INT-236.* Ogden, UT: U.S. Department of Agriculture, Forest Service, Intermountain Research Station; 1991:143.

33. Kimsey MJ Jr, Garrison-Johnston MT, Johnson L: **Characterization of volcanic ash-influenced forest soils across a geoclimatic sequence.** *Soil Sci Soc Am J* 2011, **75:**267–279.

34. Evans JS, Hudak A: **A multi scale curvature algorithm for classifying discrete return lidar in forested environments.** *IEEE Trans Geosci Rem Sens* 2007, **4:**1029–1038.

35. Eskelson BNI, Temesgen H, LeMay V, Barrett TM, Crookston NL, Hudak AT: **The roles of nearest neighbor methods in imputing missing data in forest inventory and monitoring databases.** *Scand J For Res* 2009, **24:**235–246.

36. USDA Forest Service: *Climate-FVS Ready Data.* Moscow, ID: U.S. Department of Agriculture, Forest Service, Rocky Mountain Research Station, Forestry Science Laboratory; 2013. http://forest.moscowfsl.wsu.edu/climate/customData/fvs_data.php. (accessed November 2012).

37. Intergovernmental Panel on Climate Change (IPCC): *Summary for policymakers: emissions scenarios.* Special report of IPCC working group III 2000:21. ISBN 92-9169-113-5.

38. USDA Forest Service: *FVS Training Guide. NFS-Northern Region.* Fort Collins, CO: U.S. Department of Agriculture, Forest Service, Forest Management Service Center; 2011:128.

39. Rehfeldt GE, Jaquish BC: **Ecological impacts and management strategies for western larch in the face of climate-change.** *Mitig Adapt Strat Global Change* 2010, **15:**283–306.

40. Williams MI, Dumroese RK: **Preparing for climate change: forestry and assisted migration.** *J For* 2013, **111:**287–297.

41. Mátyás C: **Modeling climate change effects with provenance test data.** *Tree Physiol* 1994, **14:**797–804.

42. Halofsky JE, Hemstrom MA, Conklin DR, Halofsky JS, Kerns BK, Bachelet D: **Assessing potential climate change effects on vegetation using a linked model approach.** *Ecol Mod* 2013, **266:**131–143.

## 2

Spatially explicit analysis of field inventories for national forest carbon monitoring

David C. Marvin[*] and Gregory P. Asner

Abstract

Background: Tropical forests provide a crucial carbon sink for a sizable portion of annual global CO_2 emissions. Policies that incentivize tropical forest conservation by monetizing forest carbon ultimately depend on accurate estimates of national carbon stocks, which are often based on field inventory sampling. As an exercise to understand the limitations of field inventory sampling, we tested whether two common field-plot sampling approaches could accurately estimate carbon stocks across approximately 76 million ha of Perúvian forests. A 1-ha resolution LiDAR-based map of carbon stocks was used as a model of the country's carbon geography.

Results: Both field inventory sampling approaches worked well in estimating total national carbon stocks, almost always falling within 10 % of the model national total. However, the sampling approaches were unable to produce accurate spatially-explicit estimates of the carbon geography of Perú, with estimates falling within 10 % of the model carbon geography across no more than 44 % of the country. We did not find any associations between carbon stock errors from the field plot estimates and six different environmental variables.

Conclusions: Field inventory plot sampling does not provide accurate carbon geography for a tropical country with wide ranging environmental gradients such as Perú. The lack of association between estimated carbon errors and environmental variables suggests field inventory sampling results from other nations would not differ from those reported here. Tropical forest nations should understand the risks associated with primarily field-based sampling approaches, and consider alternatives leading to more effective forest conservation and climate change mitigation.

Keywords: Field sampling, Forest carbon stocks, Tropical forest, Carnegie Airborne Observatory

Background

More atmospheric carbon is absorbed and stored by tropical forests than any other terrestrial ecosystem on Earth [1]. This crucial ecosystem service provides a carbon sink larger than what is emitted by fossil fuel combustion across the entire European Union each year [2, 3]. Policies that monetize the amount of carbon stored annually by a hectare of tropical forest seek to incentivize forest conservation by making it more economical to leave the forest intact than to degrade or deforest the land. The resulting economic boon for landowners and countries that reduce deforestation and degradation also results

in increased carbon sequestration in the form of woody biomass, reducing global net carbon emissions. For such policies to be successful, the uncertainty in standing carbon stocks and change (flux) must be reduced.

Accurate carbon flux calculations necessitate accurate estimates of standing carbon stocks at one or more time periods, unless carbon fluxes are measured using more direct methods (e.g., eddy covariance, atmospheric inversion). Both the price of carbon and the efficacy of climate change mitigation can be negatively affected by uncertainty in our understanding of carbon stocks, requiring the deployment of methods to make highly accurate spatial and temporal estimates of forest carbon. As forest carbon stock uncertainties increase, the monetary value of that carbon is decreased through a sliding scale discount [4,

*Correspondence: dmarvin@carnegiescience.edu
Department of Global Ecology, Carnegie Institution for Science, 260 Panama St., Stanford, CA 94305, USA

5], reducing investment opportunities and the economic benefits accrued by countries and landowners. Depending on the baseline levels of deforestation and carbon storage rates, Kohl et al. [4] found in a simulation study that many countries would generate no economic benefit with total errors in carbon estimates exceeding 5 % unless baseline deforestation rates are very high. Ultimately the power of these policies to increase forest conservation and mitigate climate change may rely on our ability to accurately quantify forest carbon in a spatially explicit manner, as opposed to generalized estimates for total carbon stocks of a landscape, habitat type, or eco-region.

Spatially explicit maps of forest carbon, or carbon geographies, allow for multi-stakeholder engagement at subnational levels [6]. Landholders and agencies within tropical countries control forest assets across a variety of spatial extents. Accordingly, knowledge of carbon stocks is needed at scales commensurate with the activities of these subnational stakeholders. Providing generalized carbon stock estimates for a given area, rather than a carbon geography, removes important spatial heterogeneity in forest carbon that could lead to over- or underestimates of carbon stocks, and consequently, the potential for poor land-use decisions. Both field inventories [7] and forest fragmentation [8] may introduce bias into forest carbon stock estimates; a bias that arises when local landscape heterogeneity is disregarded.

Forest carbon inventories typically rely (or are encouraged to rely) on a network of field plots installed either on a regular systematic grid or a random stratified grid [9–11]. Importantly, the IPCC guidelines for estimating and measuring carbon direct countries to use these sampling approaches [12, 13]. Although these guidelines are primarily designed for total national or average carbon density estimates, they are sometimes used for carbon geographies as well [14, 15]. Field plots are usually 1-ha in size (the generally accepted standard for a forest plot inventory [16]) but forest inventories that use smaller plot designs run the risk of the plots being even less representative of the surrounding forest [17]. The data collected in the field can then be combined with remote sensing data in two general ways: stratify-and-multiply, and model-linked. The stratify-and-multiply approach [18] uses remote sensing data to partition a country by land cover, climate, or other environmental (or biogeochemical) strata, or uses a regular (systematic) grid in place of partitioning by strata. Each unique stratum (or grid cell for systematic sampling) is assigned the field plot-estimated carbon stock and multiplied by the total area within the stratum. Alternatively in a model-linked approach, a model is calibrated to link field plot-estimated carbon stocks with multiple environmental variables at the location of the plots (or an intermediate remote sensing product such as LiDAR tree height), and

the model applied to the entirety of the dataset for which there is no field data (or intermediate data product). The integration of remote sensing data with field data is preferable, but countries may not have the capacity to work with remote sensing data, due to a lack of either in-country expertise, funding to outsource the work, access to technology, or sufficient remotely sensed data [e.g., [19], [20]. In cases where field plot sampling cannot be supplemented by sufficient remote sensing data and/or modeling, spatial heterogeneity in forest carbon is necessarily disregarded because of the time and expense required for the massive field sampling needed to sufficiently capture such heterogeneity [7].

While field plot sampling alone is commonly used to scale ecosystem properties and processes from local-to-regional scales, the efficacy of this approach has never been assessed at the appropriate scale. As a result, we do not know whether field carbon inventories can be used to create accurate spatially-explicit maps of national carbon geography at one-ha resolution. Here, using Perú as an example country, we examine both systematic and stratified random field-based sampling designs with a LiDAR-estimated national carbon map spanning all 76 million ha of intact and recovering forest land (Fig. 1a) as a model of the country's true carbon stocks (see "Methods" section for more detail). Hereafter, we refer to this LiDAR-estimated carbon map as "the model carbon stock" or "the model aboveground carbon density (ACD)." For both sampling designs we examine the stratify-and-multiply approach (i.e., applying each sampled value to its corresponding unsampled population at the original spatial resolution of the map) and a model-linked approach using the random forest machine learning algorithm to create spatially explicit carbon geographies of the country. We assume perfect a priori knowledge of the country's stratification, and assume that each forest plot location is accessible. This is a theoretical exercise using no field inventory data; rather all values are extracted from the carbon map as if it reflected reality. We chose a 10 % interval (i.e., ±5 %) around the true carbon stock value as our threshold for an estimate to be considered accurate. We asked the following questions. (i) What sampling intensity is needed to accurately estimate the total national carbon stock? (ii) Can either sampling method accurately estimate the national carbon geography at 1-ha resolution? (iii) What topographic and climatic variables cause increased error in carbon stock estimates?

Results and discussion
Total national carbon stocks
Systematic grid
A systematic sampling grid with dimensions of ≤56 km (requiring a minimum of 236 field plots) resulted in

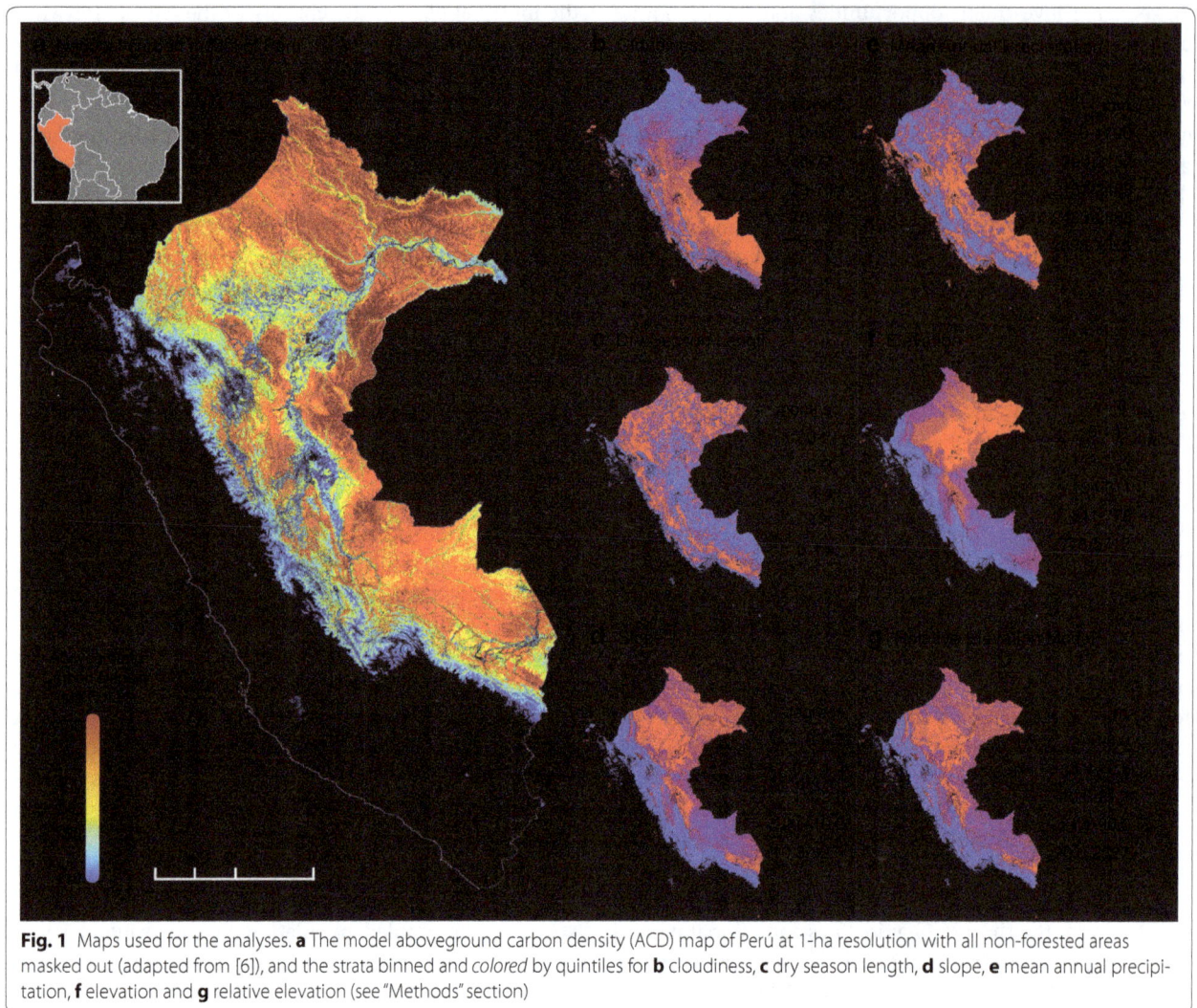

Fig. 1 Maps used for the analyses. **a** The model aboveground carbon density (ACD) map of Perú at 1-ha resolution with all non-forested areas masked out (adapted from [6]), and the strata binned and *colored* by quintiles for **b** cloudiness, **c** dry season length, **d** slope, **e** mean annual precipitation, **f** elevation and **g** relative elevation (see "Methods" section)

<2.5 % difference between the total field plot-estimated and the total model national forest carbon stock (Fig. 2). Even for systematic grids of larger spatial grains, estimates tended to stay within the 10 % threshold, with just a few of the large grid sizes falling outside that range. This demonstrates that it is possible to estimate national total forest carbon stock quite accurately with minimal sampling effort using a systematic grid.

Stratified random

Using a stratified random sampling approach, we find that only a single plot per substratum is needed to estimate the total carbon stock of Perú to within 10 % of the total model national carbon stock. A maximum of 5398 plots nationwide are needed to sample all combinations of strata representing more than 1000 ha. In fact, only the largest 15 % of the substrata (by area)

need to be sampled with a single field plot to yield an accurate estimate of the total national carbon stock, requiring a minimum of 810 plots (Additional file 1: Fig. S1).

Both sampling methods result in accurate estimates of Perú's total carbon stock, usually falling within a few percent of the model total. However, the systematic grid approach requires less sampling intensity compared to the stratified random, needing only single plots spaced as far apart as 56 km over the entire country. Systematic grid sampling is the cheapest and most efficient way to accurately sample total national carbon stocks of even a country with highly heterogeneous forest carbon distributions.

While accurate estimates of total carbon at the national and regional scale are important for understanding broad geographic patterns of carbon distributions, they are very

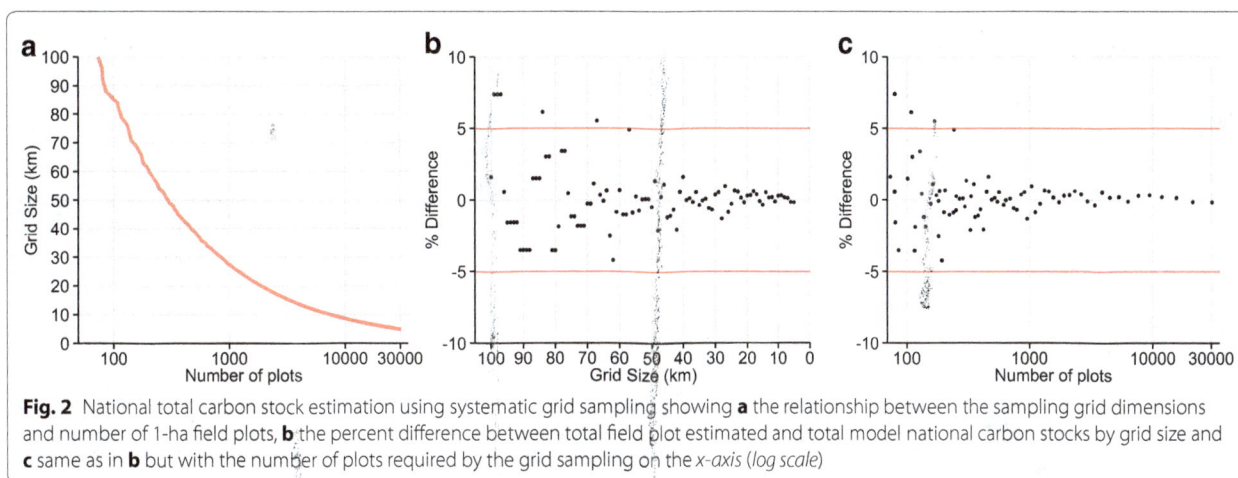

Fig. 2 National total carbon stock estimation using systematic grid sampling showing **a** the relationship between the sampling grid dimensions and number of 1-ha field plots, **b** the percent difference between total field plot estimated and total model national carbon stocks by grid size and **c** same as in **b** but with the number of plots required by the grid sampling on the x-axis (log scale)

difficult to use in applications of carbon conservation and monetization, activities that are highly spatially dependent and vary at local, landscape, and sub-regional scales. Improved knowledge of spatial variation in carbon stocks, ideally at the hectare resolution, will enhance policies that seek to incentivize conservation and climate change mitigation through carbon valuation. Maps of national local carbon geographies must be developed with particular attention to accuracy, repeatability, and cost.

Carbon geography
Systematic grid
Using a systematic sampling grid to estimate ACD and then upscaling the result at 1-ha resolution using the stratify-and-multiply approach leads to a large percentage of inaccurate estimates across the country (i.e., field plot-estimated ACD failed to be within 10 % of the model ACD) (Fig. 3a, c). Regardless of the sampling grid size, this approach produces inaccurate estimates across approximately 78 % of the country (Fig. 4a, c). The inaccurate estimates are quite consistent, 82 % using a 100 km grid (requiring just 74 1-ha plots) vs. 71 % using a 5 km grid (requiring >30,000 1-ha plots), suggesting that this method performs poorly in creating an accurate map of Perú's carbon geography. Using a model-linked random forest approach, rather than stratify-and-multiply upscaling, improves the carbon map of Perú but still results in inaccurate estimates across more than half of the country (Fig. 5a, c). Using a systematic grid of 15 km (requiring approximately 3400 1-ha plots) as the training dataset for a random forest model, 63 % of the country is still inaccurately estimated. Even using the 5 km grid results in 56 % of the country inaccurately estimated using random forest upscaling, but still requires a field plot sampling intensity of 30,459 1-ha plots to produce the training dataset (Figs. 3e, f, 5a).

Stratified random
The stratified random sampling stratify-and-multiply approach did not perform much better than the systematic grid stratify-and-multiply approach in estimating the carbon geography of Perú. The distribution of inaccurate estimates was similar to the systematic grid approach (Fig. 3b, d), and resulted in nearly the same amount of total inaccurate estimates across the country (Fig. 4b, d). Using only a single randomly placed plot per substrata (requiring around 5400 plots) leads to 79 % of the country inaccurately estimated. There is barely any improvement—with 76 % of the country inaccurately estimated—when the number of plots randomly placed inside a substrata is increased to 100 (requiring around 540,000 plots). Again the model-linked random forest upscaling approach did not substantially increase the area accurately mapped (Fig. 5b, c), with 63 % of the country inaccurately estimated regardless of the number of plots used per substratum (1, 5 or 100 plots).

Neither field plot sampling approach assessed here was able to accurately estimate the carbon geography at the 1-ha scale across more than 25–44 % of Perú. While field plot systematic sampling tended to equally under- and overestimate the model ACD across the country, the field plot stratified sampling consistently underestimated model ACD at the 1-ha scale (Fig. 4b, d). Chronically underestimating local carbon stocks will artificially deflate the per-hectare value of carbon, leading to reduced conservation incentives by landholders. Even when the total sampling is increased to an impractical number of plots (i.e., tens- to hundreds-of-thousands), the accuracy does not improve substantially. This demonstrates that neither sampling approach would be an appropriate choice for developing national maps of carbon geography, or changes in carbon geography via emissions or sequestration.

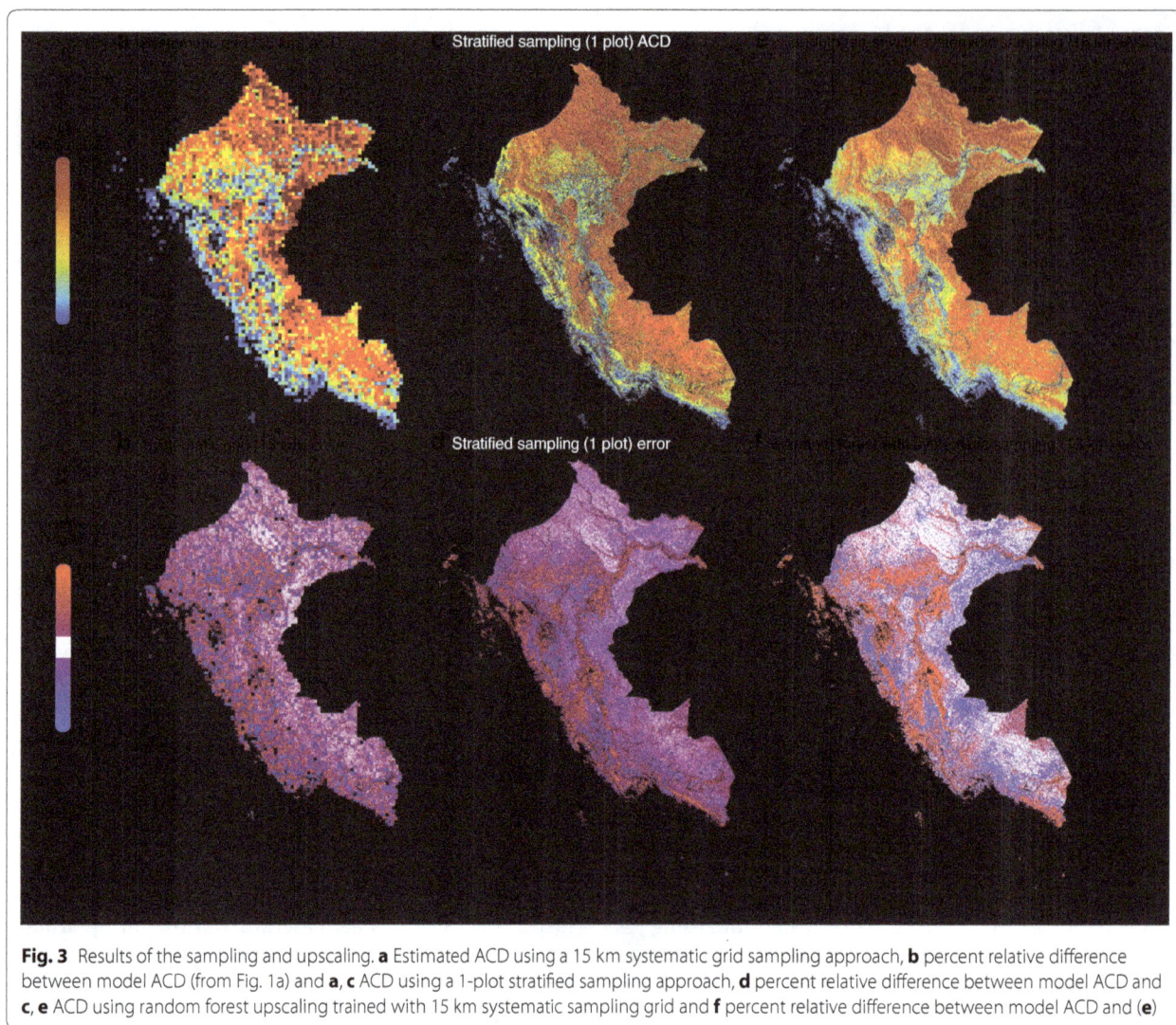

Fig. 3 Results of the sampling and upscaling. **a** Estimated ACD using a 15 km systematic grid sampling approach, **b** percent relative difference between model ACD (from Fig. 1a) and **a**, **c** ACD using a 1-plot stratified sampling approach, **d** percent relative difference between model ACD and **c**, **e** ACD using random forest upscaling trained with 15 km systematic sampling grid and **f** percent relative difference between model ACD and (**e**)

Underlying drivers of uncertainty

To understand whether there are particular strata or carbon values that lead to higher field plot estimated ACD errors, we used simulations to examine the number of field plots needed to reliably estimate (probability ≥ 0.9) the mean ACD of all substrata (unique combinations of the 6 quintile-binned strata) (Fig. 1b–g) to within 10 % of the model ACD (Fig. 6a, b). While most substrata (81 %) can be accurately estimated at a certain sampling density, some never reach the 0.9 probability threshold at any sampling density, meaning they are so heterogeneous that field plot sampling is not feasible. The median number of field plots needed to reliably sample the mean model ACD of any substratum is 43 (mean = 61; Fig. 6b).

Further examination of the simulation outcomes by strata or carbon value shows a few strong patterns. Substrata that had higher mean model ACD values require fewer randomly placed plots to accurately sample the

mean model ACD (Fig. 6c). This implies that the heteroskedasticity in the relationship used to develop the model ACD map (see "Methods" section) is not a major factor in the inability of the field plot sampling strategies to reproduce the model ACD. The strata with lower elevation and slope values require fewer plots to accurately sample the mean ACD (Fig. 6d, e), indicating that these areas are more likely to have lower heterogeneity in their carbon stocks. The other strata showed weaker or no patterns (Additional file 1: Fig. S2). Similar patterns were found when the same simulations were conducted with strata that were binned by equal range (Additional file 1: Figs. S3–S5).

We also extracted the relative percent error between the field plot sampling estimated ACD and the model ACD, and graphed this against each stratum on a hectare-by-hectare basis (Fig. 7). Both field-plot sampling methods produced similar patterns in their estimation errors

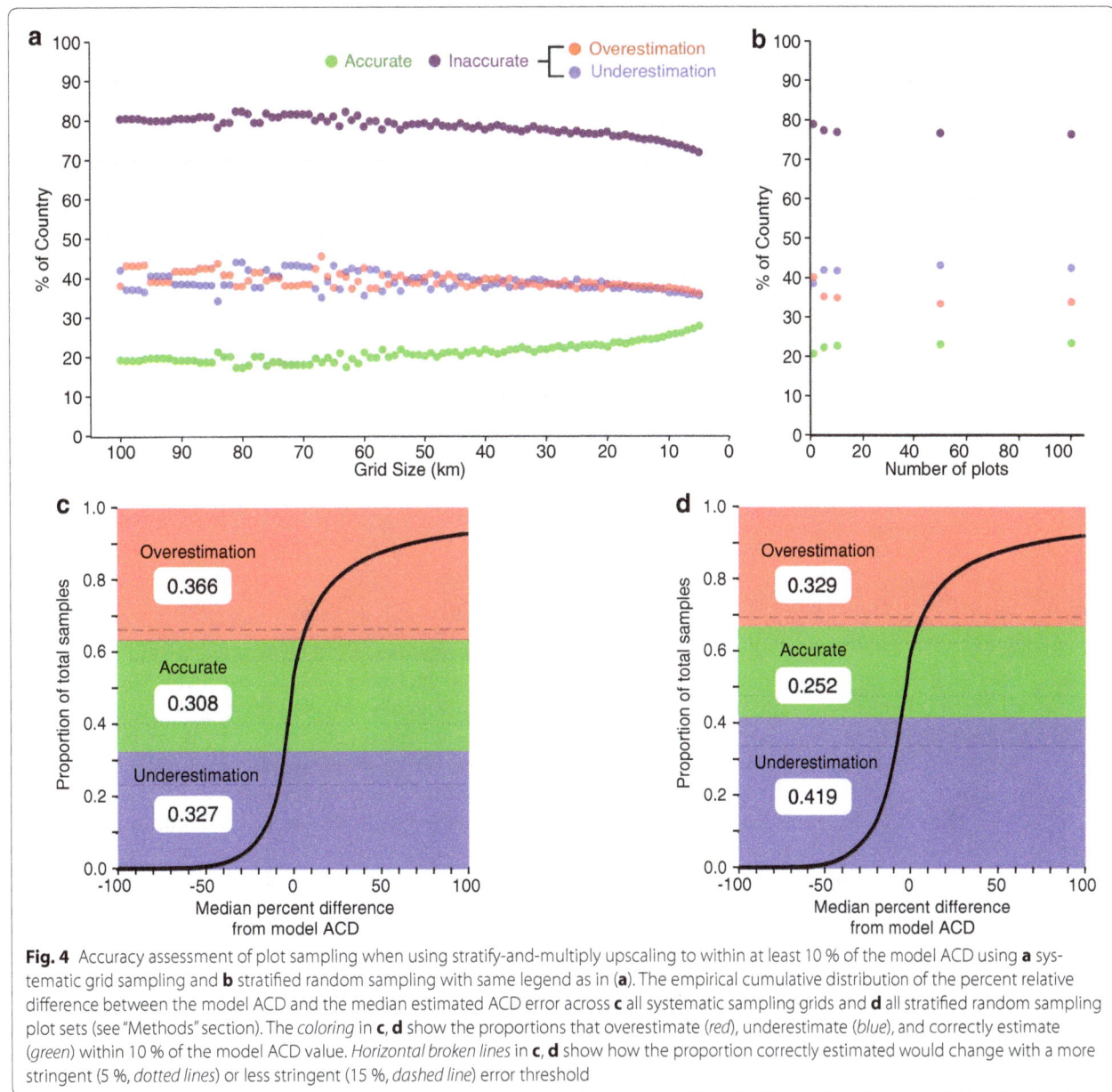

Fig. 4 Accuracy assessment of plot sampling when using stratify-and-multiply upscaling to within at least 10 % of the model ACD using **a** systematic grid sampling and **b** stratified random sampling with same legend as in (**a**). The empirical cumulative distribution of the percent relative difference between the model ACD and the median estimated ACD error across **c** all systematic sampling grids and **d** all stratified random sampling plot sets (see "Methods" section). The *coloring* in **c**, **d** show the proportions that overestimate (*red*), underestimate (*blue*), and correctly estimate (*green*) within 10 % of the model ACD value. *Horizontal broken lines* in **c**, **d** show how the proportion correctly estimated would change with a more stringent (5 %, *dotted lines*) or less stringent (15 %, *dashed line*) error threshold

with each stratum. None of the strata show a particularly clear overall association with their ACD estimation error. Instead, generally the spread in error densities seem either restricted to low strata values in the case of elevation, slope, and relative elevation, or are more widespread across the strata values in the case of cloudiness, mean annual precipitation, and dry season length. This is probably more a reflection of the underlying data distributions than patterns in the relationship between estimation error and environmental variables (i.e., where there are more data you would expect a larger spread in the errors).

The lack of strong and clear trends between the percent relative error and the strata used in this analysis further reduces the utility of plot sampling for creating accurate, spatially explicit national carbon maps. If there were particular strata or subsets of strata that were underlying the resulting errors, then these could be isolated and sampled differently from the remaining areas of the country. The large errors within and between the strata indicate that widespread ACD estimation errors will be unavoidable when using a field-plot sampling strategy.

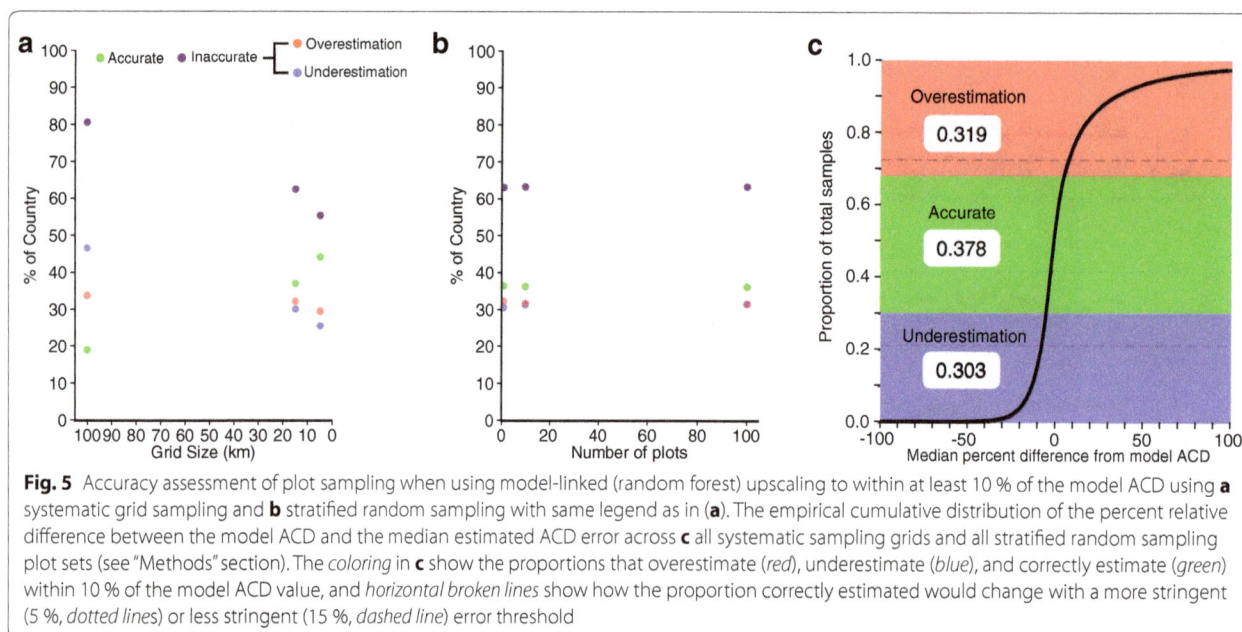

Fig. 5 Accuracy assessment of plot sampling when using model-linked (random forest) upscaling to within at least 10 % of the model ACD using **a** systematic grid sampling and **b** stratified random sampling with same legend as in (**a**). The empirical cumulative distribution of the percent relative difference between the model ACD and the median estimated ACD error across **c** all systematic sampling grids and all stratified random sampling plot sets (see "Methods" section). The *coloring* in **c** show the proportions that overestimate (*red*), underestimate (*blue*), and correctly estimate (*green*) within 10 % of the model ACD value, and *horizontal broken lines* show how the proportion correctly estimated would change with a more stringent (5 %, *dotted lines*) or less stringent (15 %, *dashed line*) error threshold

Conclusion

Creating a national carbon geography using field inventory plot sampling is unlikely to produce accurate results that can be deployed for use in spatially explicit actions to reduce carbon emissions. In this exercise we find that two common methods for large-scale field sampling fail to produce accurate (within a 10 % interval around the model ACD) carbon estimates for any more than 44 % of the total forested areas of Perú (approximately 76 million ha), based on a model of carbon geography. This holds true even when 12 layers of remote sensing imagery were used to upscale 30,459 field plot samples using a machine learning algorithm. While both sampling strategies can produce accurate estimates of the *total* carbon content of Perú using relatively few plots, the local carbon geographies of countries are far more important in a carbon valuation and carbon sequestration context. Without accurate, one-hectare resolution carbon geographies, land use decisions by subnational stakeholders (individual landowners and agencies) may be based on insufficient or biased information.

Perú is an ideal tropical country for this type of exercise because it hosts a wide range of topographic, biotic, geologic, and climatic variation resulting in highly heterogeneous landscape carbon distributions. While some countries may have very low carbon heterogeneity, most efforts to map national carbon stocks will face the issue of high sampling errors resulting from non-homogenous carbon distributions. This means our results are likely applicable to most tropical countries, or at least to substantial portions of any particular country.

Field inventories could be targeted toward more intensive sampling of areas likely to have higher carbon heterogeneity, potentially reducing the estimation uncertainties of any local carbon geography. This is the basic premise underlying any stratification approach [11], whether the basic stratification used here or those that incorporate remote sensing data. However, we did not find any variables driving the errors in local carbon estimates among the six topographic and climatic strata tested. These six strata are the environmental variables that best explain total variation in carbon across Perú [6]. Therefore, countries using field-based carbon mapping will find it difficult a priori to target sampling toward areas of potential high carbon estimation uncertainty. Moreover, even if a country were to have perfect knowledge of its strata, as in the stratification exercise presented here, field plot sampling still cannot reproduce the model ACD without incurring substantial errors across much of the country.

Countries seeking to value their forest carbon reserves for conservation and climate change mitigation must look beyond the sole use of field plot sampling. Of course, field inventory plots are critical for understanding local-scale ecological processes and for calibration/validation of remote sensing data. (We emphasize here the important distinction between the use of field plots as calibration/validation for remote sensing products, and using remote sensing products to scale field plot results to larger areas). While field plots are integral to remote sensing campaigns, they are not designed for, and do not perform well at, producing spatially explicit estimates of forest carbon stocks [21]—even when using satellite

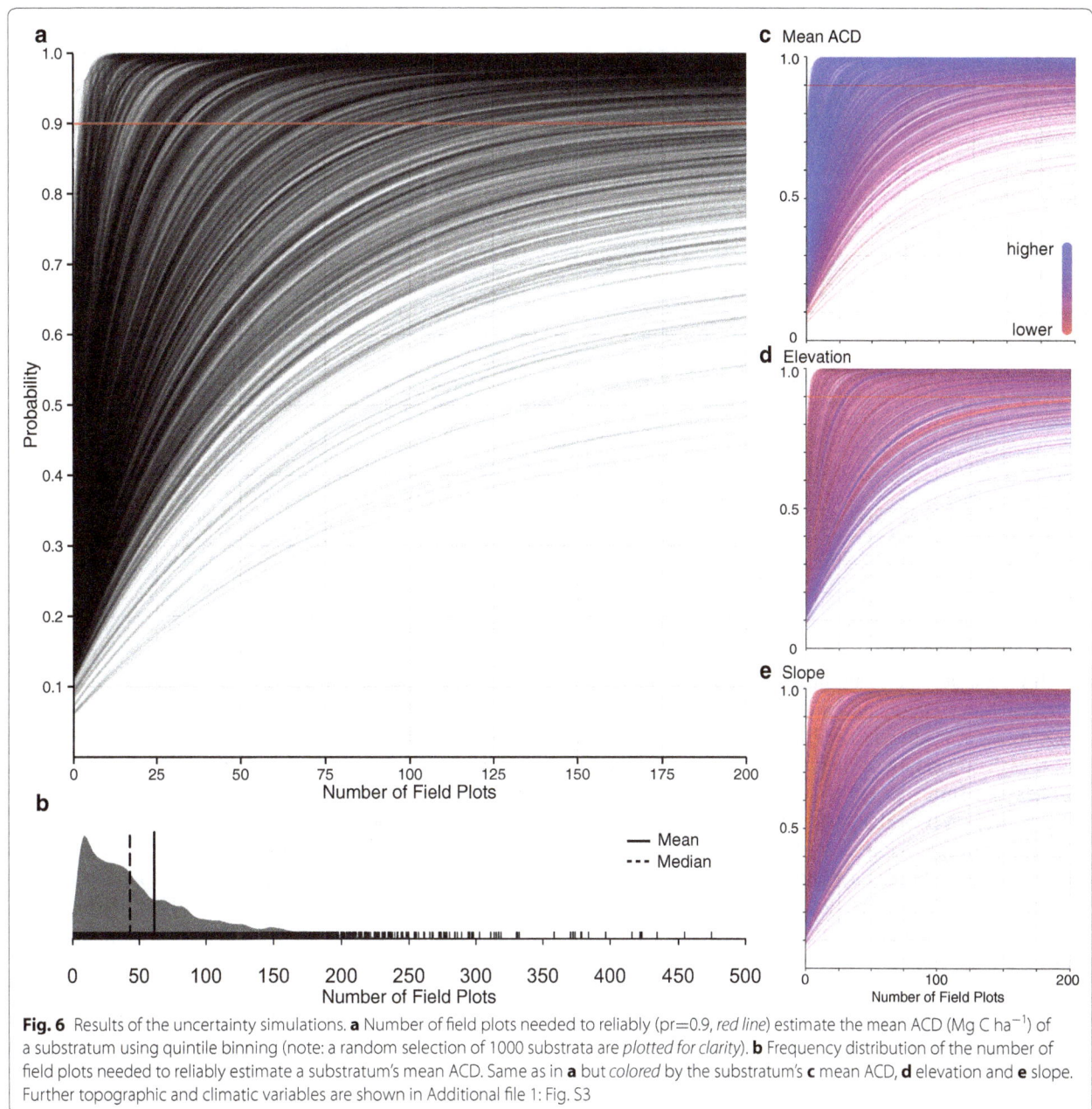

Fig. 6 Results of the uncertainty simulations. **a** Number of field plots needed to reliably (pr=0.9, *red line*) estimate the mean ACD (Mg C ha^{-1}) of a substratum using quintile binning (note: a random selection of 1000 substrata are *plotted for clarity*). **b** Frequency distribution of the number of field plots needed to reliably estimate a substratum's mean ACD. Same as in **a** but *colored* by the substratum's **c** mean ACD, **d** elevation and **e** slope. Further topographic and climatic variables are shown in Additional file 1: Fig. S3

based sensors for upscaling. Field inventories are hard pressed to adequately capture the link between remotely sensed environmental variables and estimated carbon stocks. For example, 30,000 field plots spread across Perú—an extremely ambitious sampling effort for most countries—is still only 0.5 % of the total area (6.76 million ha) sampled by airborne LiDAR and used to create the model ACD map [6].

Instead tropical countries should look toward airborne and spaceborne sensors to fulfill the need for improved wall-to-wall aboveground carbon maps at hectare scales.

The LiDAR sensor on the future ICESat-2 satellite is unlikely to accurately measure tropical forest carbon [22] and the GEDI sensor will be a short-term (1–2 year) mission on the International Space Station [23]. The lack of a long-term spaceborne sensor to measure global forest carbon suggests that we must rely on airborne platforms to carry out the bulk of national carbon sampling for at least the next decade. Airborne LiDAR can now successfully map forest carbon stocks at high-resolution at subnational to national scales at high accuracy and extremely low cost on a per-hectare basis if operated

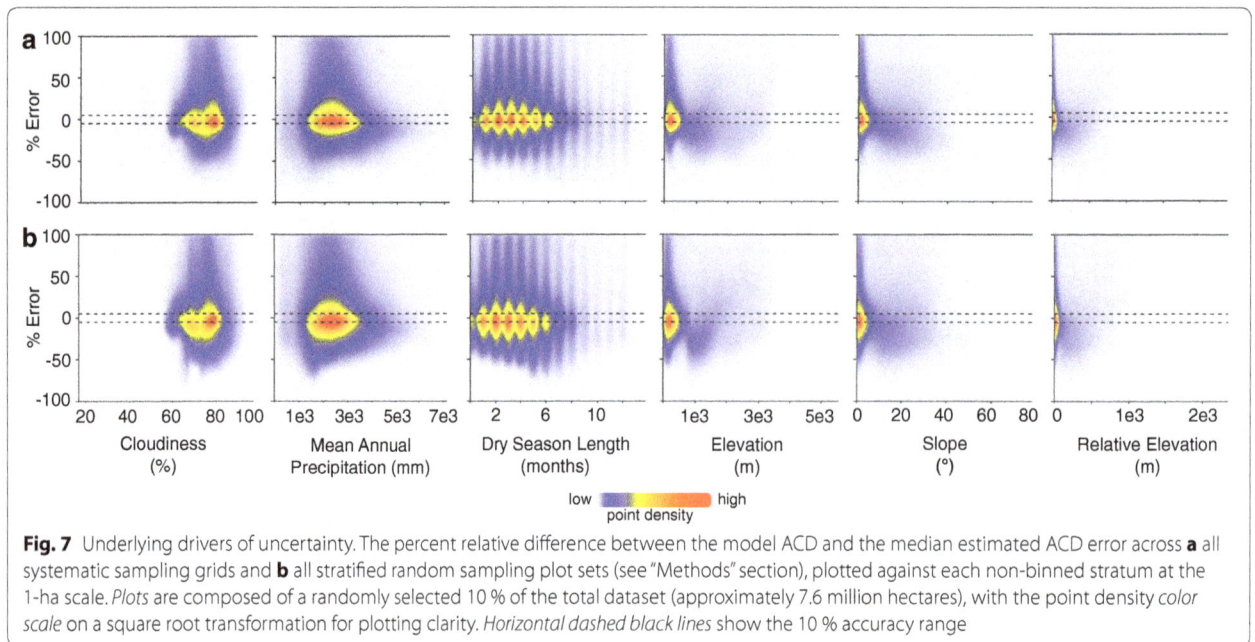

Fig. 7 Underlying drivers of uncertainty. The percent relative difference between the model ACD and the median estimated ACD error across **a** all systematic sampling grids and **b** all stratified random sampling plot sets (see "Methods" section), plotted against each non-binned stratum at the 1-ha scale. *Plots* are composed of a randomly selected 10 % of the total dataset (approximately 7.6 million hectares), with the point density *color scale* on a square root transformation for plotting clarity. *Horizontal dashed black lines* show the 10 % accuracy range

non-commercially; hundreds of thousands of hectares can be mapped per day at accuracies that approach or match field plot based sampling [24, 25]. The economies of scale achieved by a single airborne LiDAR sensor leads to far more cost-effective and accurate carbon mapping that is easily repeatable—yielding estimates of carbon net changes over time—gives this method of mapping a drastic advantage over field plot sampling.

No doubt tropical forests and their substantial carbon sink capacities will play a major role during the implementation of agreements forged during the December 2015 UNFCCC climate conference in Paris. Tropical forest nations must be ready to respond with accurate assessments of their carbon stocks and reliably monitor their changes over time. While field inventory plots have long been the standard for meeting these challenges, with tens of millions of dollars spent on plot implementation and infrastructure, this should not preclude a move toward more accurate and cost-effective forest carbon mapping.

Methods
Study area
Perú is an ideal test case for investigating whether field plot sampling can accurately map carbon because its forests span a wide range of topographic, climatic, floristic, and geologic variables. Heterogeneity in aboveground carbon density (ACD) is driven by combinations of these variables, and higher ACD heterogeneity leads to less accurate field plot estimates of ACD [7]. Results from within the diverse environmental gradients of Perú can then be applied to other countries of similar conditions.

We used the 1-ha resolution aboveground carbon density map of Perú created by Asner et al. [6] as a model of the "true" carbon stocks of the country. We refer to this LiDAR-estimated carbon map throughout the paper as "the model carbon stock" or "the model aboveground carbon density (ACD)." The model ACD map was produced from a countrywide airborne LiDAR campaign with the Carnegie Airborne Observatory (CAO; cao.carnegiescience.edu). The LiDAR data were integrated with high-resolution satellite imaging, a large field plot sampling network (to calibrate/validate the LiDAR carbon mapping), and an advanced geospatial scaling algorithm [for more details see 6]. Please refer to Fig. S5 of [6] for model validation results, which compared 536,874 ha of LiDAR-measured forest TCH that were not used to train the model-estimated TCH of those same locations. The R^2 is 0.78 and the RMSE is 3.50 m. While the underlying relationship between LiDAR measured-TCH and estimated ACD may be heteroskedastic, with a higher ACD errors found at taller TCH, the relative ACD uncertainty that results is low (i.e., <20 % for high ACD (>120 MgC ha^{-1}) lowland forests).

We masked this map to exclude areas that are not forested, using a mean per hectare threshold of >70 % photosynthetic vegetation and >2 m top-of-canopy height (TCH). The photosynthetic vegetation map was created from a national-scale mapping of Perú using the CLA-Slite algorithm with mosaicked Landsat satellite imagery [26]. The TCH map is the underlying dataset used to produce the ACD map of Perú [6]. This left a forested area of 76,457,286 ha for the analysis. While this is 59.5 % of

Perú by land area, it contains 98.5 % of the total carbon of the entire country (6.798 Pg). All analyses were performed using R statistical software [27] with geospatial manipulations and analyses performed using the 'raster' package [28].

Environmental stratification layers

We compiled a set of six co-aligned, 1-ha resolution environmental stratification layers (hereafter 'strata') comprised of topographic and climatic variables covering all of Perú. Cloudiness (%) was created from long-term (2000–2010) NASA moderate resolution imaging spectroradiometer (MODIS) data and described in further detail by [6]. Mean annual precipitation (MAP) was calculated from NASA Tropical Rainfall Measuring Mission (TRMM) 2b31 monthly data (1998–2011), and dry season length (DSL) was calculated using the same TRMM data for all months with less than 100 mm of precipitation. Elevation, slope, and a relative elevation model (REM) were created from NASA Shuttle Radar Topography Mission (SRTM) at 90 m resolution. Relative elevation is a proxy for vegetation related water resources, and is developed by calculating the height of the ground above nearest water body [29]. Each stratum was masked by the same forest mask described above.

Field plot sampling designs

We tested two commonly used field plot sampling strategies: systematic sampling and stratified random sampling. This study is not meant to be an exhaustive review of spatial sampling techniques, rather to evaluate two methods that are commonly employed or recommended to map forest carbon over large geographic areas [9, 10, 12, 13]. Therefore, we did not consider other possible spatial sampling designs (e.g., simple random sampling, cluster sampling).

Systematic sampling uses sampling units (i.e., field inventory plots) placed at regular intervals according to an ordering scheme. For the country of Perú, we chose regularly spaced square grids with dimensions ranging from 5–100 km, increasing at 1-km intervals. Only those sampling units on the grid that fall within the bounds of the forest mask were retained (hereafter referred to as systematic grid sampling).

Stratified sampling places sampling units across a region according to pre-defined subregions that are more homogenous than the region as a whole, thereby reducing inherent sampling errors. The degree to which a stratified sampling technique accurately estimates the true population depends largely upon choosing homogenous subregions from which to conduct the subsampling [30]. We used unique combinations of the six strata described above to create these subregions (hereafter

referred to as 'substrata'). First each stratum was binned by quintile, creating 5 categorical classes for each stratum (Fig. 1b–g). Then a map of all unique combinations of the six quintile-binned strata was produced, and any resulting substrata less than 1000 ha in area was excluded (hereafter referred to as 'quintile-binned substrata'). This resulted in 5398 unique substrata totaling 98.2 % of the forested area used for the analyses. We repeated the above steps, instead binning each stratum by 5 equal subsets of the total range of values of each stratum (hereafter referred to as 'range-binned substrata') (Additional file 1: Fig. S3). Again removing any substrata with less than 1000 ha, resulting in 447 substrata representing 99.9 % of the forested area used for the analyses. Sampling units were then randomly selected from within each substratum according to the simulations described below (hereafter referred to as 'stratified random sampling').

Estimating total national carbon stocks
Systematic grid

For each systematic sampling grid, the ACD value extracted from the centroid of each sampling grid cell was used as the ACD value for all of the 1-ha values in that sampling grid cell. The total number of grid cells in a systematic grid represents the number of field plots needed to create the estimated ACD map. We produced 96 different estimated ACD maps, each with a spatial resolution corresponding to the sampling grid dimensions used to create the map (5–100 km in 1 km increments). The ACD was summed across the entire country for each of the different estimated ACD maps, and compared to the total of the model ACD map.

Stratified random

We used Monte Carlo simulations to determine the probability of a randomly placed set of field inventory plots within each quintile-binned substratum to accurately estimate the total carbon stocks of Perú. For all of the 5398 substrata, a random sample of ACD values drawn from all ACD values of the substratum was selected and used to estimate the mean ACD of the substratum. The mean ACD of each substratum was multiplied by the total number of hectares in the substratum to get the estimated total ACD of that substratum. All estimates of total ACD from the substrata were then summed to get an estimated total ACD of the country, and compared to the total of the model ACD map. This was repeated 5000 times and the probability of correctly estimating the total ACD of Perú to within 10 % was determined. This simulation was run for sample sizes of 1 through 100 field plots in each substratum. We repeated this same simulation, but at each iteration progressively removing the smallest substrata (by area) in increments of 5 %. We used the mean ACD of the sampled substrata to estimate the total ACD of the substrata that were removed from the simulation. We did not test the range-binned substrata because of its poor performance in estimating the mean ACD of a substratum (see "Uncertainty simulations and extraction" section below).

Estimating national carbon geographies

Systematic grid

Each 1-ha resolution systematic grid estimated ACD map (see above) represents a stratify-and-multiply upscaling approach whereby the sampled value from the systematic grid (which can be thought of as stratification in this context) is applied to the corresponding unsampled population (the rest of the 1-ha values in the systematic grid cell). Each estimated ACD map was co-aligned with the model ACD map of Perú, subtracted from the model ACD map, and the difference divided by the model ACD map to get the relative percent difference on a per hectare basis. We also performed a model-linked upscaling approach using random forest machine learning in addition to the stratify-and-multiply upscaling approach. We used the systematic sampling grid values as the training dataset for each random forest model. For the predictors of the model, we used the same 12 contiguous remote sensing layers that were used to create the model ACD map [see 6]. For each systematic grid, a random forest model was fit with the sampled values and their co-aligned values from the remote sensing layers. Each fitted model was then applied to the entire country using the 12 remote sensing layers to create a wall-to-wall random forest estimated ACD map [for more details on the random forest approach see [31]. This map was then compared to the model ACD map in the same manner as above.

Stratified random

We chose a subset {1, 5, 10, 50, 100} of field plot sets used to randomly sample each substratum to create a stratified random estimated ACD map using a stratify-and-multiply approach. For each substratum, a field plot set was randomly drawn from all ACD values of the substratum, and the mean of the set was mapped back onto all hectares of that substratum. This produced a set of maps of estimated ACD for the entire country of Perú at 1-ha resolution based on the stratified random sampling approach. For each map created from one of the five field plot sets, the stratified random estimated ACD map was subtracted from the model ACD map, and the difference divided by the model ACD map to get the relative percent difference. Again, we did not test the range-binned strata for the reason described in the preceding section. We also applied a model-linked upscaling approach using the random forest machine learning algorithm. Here the

training datasets were composed of the stratified random sample values and their co-aligned values from the 12 remote sensing layers. Random forest models were fit and applied in the same way as described above.

Uncertainty simulations and extractions

We ran Monte Carlo simulations to find the number of field plots it would take to accurately estimate the mean ACD of a substratum for the stratified random sampling approach. For each substratum, a random sample of ACD values (representing field plots) was selected from all possible ACD values of the substratum and the mean of the sample was compared to the mean of all ACD values of the substratum. This was repeated 5000 times to find the probability that the selected number of field plots would produce an estimate accurate to within 10 % of the mean of the substratum as estimated by the model ACD map. This simulation was run for sample sizes of 1 through 100 field plots. We then mapped the carbon value (either mean or total) and the environmental strata value of each substratum onto the results of these simulations to examine potential patterns.

We co-aligned the spatially explicit maps of estimated ACD created from both sampling approaches with the original (non-binned) strata. For each sampling approach, we created a median estimated ACD map across all of the sampling grids (systematic sampling, 5–100 km) and plot sets (stratified random, {1, 5, 10, 50, 100} plots). We then extracted the percent relative difference between the median estimated ACD map and the model ACD map. We also extracted the six climatic and topographic values associated with each mapped hectare.

Additional file

Additional file 1: Figure S1. The fraction of stratified random substrata needed to estimate model total national carbon stocks. **Figure S2.** Number of field plots needed to reliably estimate the mean model substratum ACD using stratified sampling. **Figure S3.** Strata binned and colored by five equal ranges. **Figure S4.** Number of field plots needed to reliably estimate the mean model substratum ACD using equal range binning. **Figure S5.** Number of field plots needed to reliably estimate the mean model substratum ACD using equal range binning colored by environmental variables.

Abbreviations

ACD: aboveground carbon density; CAO: Carnegie Airborne Observatory; IPCC: Intergovernmental Panel on Climate Change; LiDAR: Light Detection and Ranging; MAP: mean annual precipitation; MODIS: Moderate Resolution Imaging Spectroradiometer; REM: relative elevation model; SRTM: Shuttle Radar Topography Mission; TCH: top of canopy height; TRMM: Tropical Rainfall Monitoring Mission; UNFCCC: United Nations Framework Convention on Climate Change.

Authors' contributions

DCM and GPA designed the study. GPA provided the data and DCM analyzed the data. DCM and GPA wrote the paper. All authors read and approved the final manuscript.

Acknowledgements
We thank P. Brodrick for helpful comments on the manuscript. This study was supported by the John D. and Catherine T. MacArthur Foundation and the endowment of the Carnegie Institution for Science. The Carnegie Airborne Observatory is made possible by the Gordon and Betty Moore Foundation, the John D. and Catherine T. MacArthur Foundation, Avatar Alliance Foundation, W.M. Keck Foundation, the Margaret A. Cargill Foundation, Grantham Foundation for the Protection of the Environment, Mary Anne Nyburg Baker and G. Leonard Baker Jr, and William R. Hearst III. We thank colleagues from the Perúvian Ministry of Environment (MINAM) for their long-term support and collaboration under the joint MINAM-Carnegie Institution program agreement.

Competing interests
The authors declare that they have no competing interests.

References

1. Pan Y, Birdsey RA, Fang J, Houghton R, Kauppi PE, Kurz WA, Phillips OL, Shvidenko A, Lewis SL, Canadell JG, Ciais P, Jackson RB, Pacala SW, McGuire AD, Piao S, Rautiainen A, Sitch S, Hayes D. A large and persistent carbon sink in the world's forests. Science. 2011;333:988–93.
2. LeQuere C, Moriarty R, Andrew RM, Peters GP, Ciais P, Friedlingstein P, Jones SD, Sitch S, Tans P, Arneth A, Boden TA, Bopp L, Bozec Y, Canadell JG, Chevallier F, Cosca CE, Harris I, Hoppema M, Houghton RA, House JI, Jain A, Johannessen T, Kato E, Keeling RF, Kitidis V, KleinGoldewijk K, Koven C, Landa CS, Landschützer P, Lenton A, et al. Global carbon budget 2014. Earth Syst Sci Data Discuss. 2014;7:521–610.
3. CDIAC: Global carbon project: Full global carbon budget. US. Department of Energy; 2012.
4. Köhl M, Baldauf T, Plugge D, Krug J. Reduced emissions from deforestation and forest degradation (REDD): a climate change mitigation strategy on a critical track. Carbon Balance Manag. 2009;4:10.
5. REDD Offset Working Group: California, Acre and Chiapas: partnering to reduce emissions from tropical deforestation. Green Technology Leadership Group; 2012.
6. Asner GP, Knapp DE, Martin RE, Tupayachi R, Anderson CB, Mascaro J, Sinca F, Chadwick KD, Higgins M, Farfan W, Llactayo W, Silman MR. Targeted carbon conservation at national scales with high-resolution monitoring. Proc Natl Acad Sci USA. 2014;111:E5016–22.
7. Marvin DC, Asner GP, Knapp DE, Anderson CB, Martin RE, Sinca F, Tupayachi R. Amazonian landscapes and the bias in field studies of forest structure and biomass. Proc Natl Acad Sci USA. 2014;111:E5224–32.
8. Chaplin-Kramer R, Ramler I, Sharp R, Haddad NM, Gerber JS, West PC, Mandle L, Engstrom P, Baccini A, Sim S, Mueller C, King H. Degradation in carbon stocks near tropical forest edges. Nat Commun. 2015;6:10158.
9. Tomppo E, Gschwantner T, Lawrence M, McRoberts RE. National forest inventories. Dordrecht: Springer Science & Business Media; 2009.
10. GOFC-GOLD. A sourcebook of methods and procedures for monitoring and reporting anthropogenic greenhouse gas emissions and removals associated with deforestation, gains and losses of carbon stocks in forests remaining forests, and forestation. GOFC-GOLD land cover project office, The Netherlands: Wageningen University; 2014.
11. Maniatis D, Mollicone D. Options for sampling and stratification for national forest inventories to implement REDD+ under the UNFCCC. Carbon Balance Manag. 2010;5:9.
12. IPCC: 2013 Revised Supplementary Methods and Good Practice Guidance Arising From the Kyoto Protocol. Switzerland: IPCC; 2014.
13. IPCC: Guidelines for national greenhouse gas inventories. Japan: IGES; 2006.
14. Blackard J, Finco M, Helmer E, Holden G, Hoppus M, Jacobs D, Lister A, Moisen G, Nelson M, Riemann R. Mapping U.S. forest biomass using nationwide forest inventory data and moderate resolution information. Remote Sens Environ. 2008;112:1658–77.
15. Cartus O, Kellndorfer J, Walker W, Franco C, Bishop J, Santos L, Fuentes J. A national, detailed map of forest aboveground carbon stocks in Mexico. Remote Sens. 2014;6:5559–88.
16. Phillips OL, Baker T, Feldpausch T, Brienen R: RAINFOR field manual for plot establishment and remeasurement. R Soc. 2016:1–22.
17. Réjou-Méchain M, Muller-Landau HC, DETTO M, Thomas SC, Le Toan T, Saatchi SS, Barreto-Silva JS, Bourg NA, Bunyavejchewin S, Butt N, Brockelman WY, Cao M, Cárdenas D, Chiang JM, Chuyong GB, Clay K, Condit R, Dattaraja HS, Davies SJ, Duque Á, Esufali S, Ewango C, Fernando RHS, Fletcher CD, Gunatilleke IAUN, Hao Z, Harms KE, Hart TB, Hérault B, Howe RW, et al. Local spatial structure of forest biomass and its consequences for remote sensing of carbon stocks. Biogeosci Discuss. 2014;11:5711–42.
18. Goetz SJ, Baccini A, Laporte NT, Johns T, Walker W, Kellndorfer J, Houghton RA, Sun M. Mapping and monitoring carbon stocks with satellite observations: a comparison of methods. Carbon Balance Manag. 2009;4:2.
19. Romijn E, Herold M, Kooistra L, Murdiyarso D, Verchot L. Assessing capacities of non-annex I countries for national forest monitoring in the context of REDD. Environ Sci Policy. 2012;19–20:33–48.
20. Romijn E, Lantican CB, Herold M, Lindquist E, Ochieng R, Wijaya A, Murdiyarso D, Verchot L. Assessing change in national forest monitoring capacities of 99 tropical countries. For Ecol Manage. 2015;352:109–23.
21. Saatchi S, Mascaro J, Xu L, Keller M, Yang Y, Duffy P, Espírito Santo F, Baccini A, Chambers J, Schimel D. Seeing the forest beyond the trees. Glob Ecol Biogeogr. 2015;24:606–10.
22. Luccio M. Improving global carbon estimates with LiDAR. Sensors and systems. 2014. Accessed 2 May 2016.
23. Goetz S, Dubayah R. Advances in remote sensing technology and implications for measuring and monitoring forest carbon stocks and change. Carbon Manag. 2011;2:231–44.
24. Mascaro J, Detto M, Asner GP, Muller-Landau HC. Evaluating uncertainty in mapping forest carbon with airborne LiDAR. Remote Sens Environ. 2011;115:3770–4.
25. Asner GP, Mascaro J. Mapping tropical forest carbon: calibrating plot estimates to a simple LiDAR metric. Remote Sens Environ. 2014;140:614–24.
26. Asner GP, Knapp DE, Balaji A, Paez-Acosta G. Automated mapping of tropical deforestation and forest degradation: cLASlite. J Appl Remote Sens. 2009;3:033543.
27. R Development Core Team: R: A language and environment for statistical computing. 2014.
28. Package "raster" [cran.r-project.org/web/packages/raster].
29. Asner GP, Clark JK, Mascaro J, Vaudry R, Chadwick KD, Vieilledent G, Rasamoelina M, Balaji A, Kennedy-Bowdoin T, Maatoug L, Colgan MS, Knapp DE. Human and environmental controls over aboveground carbon storage in Madagascar. Carbon Balance Manag. 2012;7:2.
30. Wang J, Haining R, Cao Z. Sample surveying to estimate the mean of a heterogeneous surface: reducing the error variance through zoning. Int J Geogr Inf Sci. 2010;24:523–43.
31. Mascaro J, Asner GP, Knapp DE, Kennedy-Bowdoin T, Martin RE, Anderson C, Higgins M, Chadwick KD. A tale of two "forests": random forest machine learning aids tropical forest carbon mapping. PLoS One. 2014;9:e85993–9.

## 3

Tree component biomass expansion factors and root-to-shoot ratio of Lebombo ironwood: measurement uncertainty

Tarquinio Mateus Magalhães[1,2]* and Thomas Seifert[2]

Abstract

Background: National and regional aboveground biomass (AGB) estimates are generally computed based on standing stem volume estimates from forest inventories and default biomass expansion factors (BEFs). AGB estimates are converted to estimates of belowground biomass (BGB) using default root-to-shoot ratios (R/S). Thus, BEFs and R/S are not estimated in ordinary forest inventories, which results in uncertainty in estimates of AGB and BGB. Here, we measured BEF and R/S values (including uncertainty) for different components of Lebombo ironwood (*Androstachys johnsonii* Prain) trees and assessed their dependence on tree size.

Results: The BEF values of tree components were unrelated or weakly related to tree size, and R/S was independent of tree size. BEF values varied from 0.02 for foliage to 1.31 Mg m^{-3} for whole tree; measurement uncertainty (SE) varied from 2.9% for stem BEF to 10.6% for whole-tree BEF. The belowground, aboveground, and whole-tree BEF-based biomass densities were 30 ± 2.3 (SE = 3.89%), 121 ± 7.84 (SE = 3.23%), and 151 ± 9.87 Mg ha^{-1}(SE = 3.27%), respectively. R/S was 0.24 with an uncertainty of 3.4%.

Conclusions: Based on the finding of independence or weak dependence of BEF on tree size, we concluded that, for *A. johnsonii*, constant component BEF values can be accurately used within the interval of harvested tree sizes.

Keywords: *Androstachys johnsonii* Prain; Mecrusse; Additivity; Belowground biomass; Forest inventory

Background

National and regional aboveground biomass (AGB) estimates are generally calculated based on estimates of standing stem volume from forest inventories and from default biomass expansion factors (BEFs). The AGB estimates are converted into belowground biomass (BGB) using default root-to-shoot ratio (R/S) values. This method is commonly used to estimate carbon stocks for national greenhouse gas (GHG) inventories [1].

However, BEF and R/S values can vary according to vegetation type, precipitation regime, mean annual temperature [2], and tree age and size [3-7]; thus, use of default values for national- or regional-scale estimates might result in unreliable assessments of biomass, carbon, and GHGs. In addition, because BEF and R/S values are not estimated during ordinary forest inventories, uncertainty in estimates of AGB and BGB is mainly attributed to these parameters [8], and it thus represents a major gap in carbon accounting at regional and national levels [9]. Few studies have provided estimates of BEF and R/S with measures of uncertainty, and although individual R/S values for specific forest and woodland types have not been widely studied, these values enable more-accurate estimates of belowground biomass [2] when compared to default ones. Therefore, estimates of BEF and R/S with uncertainty are needed for different types of woodlands.

The objective of this study was to develop tree component BEF and R/S values with known uncertainty for *A. johnsonii*.

Results

Descriptive statistics of the collected data

The number of trees recorded during the first sampling phase ranged from approximately 500 to >1000 ha^{-1}

* Correspondence: tarqmag@yahoo.com.br
[1]Departamento de Engenharia Florestal, Universidade Eduardo Mondlane, Campus Universitário Principal, Edifício no 1, Maputo, Mozambique
[2]Department of Forest and Wood Science, University of Stellenbosch, Private Bag X1 Matieland, 7602 Stellenbosch, South Africa

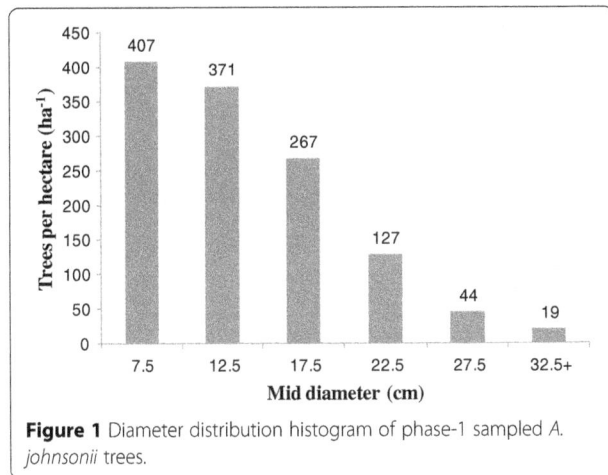

Figure 1 Diameter distribution histogram of phase-1 sampled *A. johnsonii* trees.

with an average of 1236 ha^{-1}, distributed in each diameter class as shown in the Figure 1 – diameter distribution histogram – which follows a pattern of an inverse J-shaped curve, typical of an uneven-aged forest. The size and volume of the trees varied substantially (Table 1). The average AGB per tree ($\bar{w}_{Shoot_1}$) was 97.95 kg. The dry weight of the components measured destructively during the second sampling phase, as well as Hohenadl form factor and stem volume, also varied considerably (Table 2).

Biomass expansion factors

The total tree and aboveground BEFs were approximately 131% and 105% of the stem volume, respectively (Table 3). For the major components, the stem had the highest BEF, and this value was more than two-fold higher than the BEF of crowns and roots. The standard error of all estimates was <11%; stem estimates were the most precise, and foliage estimates had the largest error (Table 3).

Using linear regression test, Pearson's correlation coefficient test of significance, and distance covariance (dcov) test of independence, the BEF of taproots, lateral roots, and foliage was found to be DBH-dependent (Tables 4 and 5) (a weak dependence). Other seven

component BEFs and total tree BEF were not found to have any kind of dependence on DBH (neither linear nor nonlinear). The strongest DBH-dependence was found for foliage BEF (adjusted R^2 = 0.2900, r = – 0.5329, dcor = 0.5874). Seven component BEFs were linearly TH-dependent (Table 6); however, using dcov test of independence, only 5 component BEFs were TH-dependent; i. e. the linear dependence of crown and shoot system BEFs on TH was not detected by dcov test of independence (Table 7). The BEF of foliage was the most strongly dependent on both DBH and TH. Component BEF values decreased with increasing TH and DBH (except for the relationship between lateral roots and DBH).

Biomass density

Total tree biomass was approximately 25% higher than AGB (Table 8). The root system, stem, and crown observed biomass densities of 29.62, 84.57, and 36.55 Mg ha^{-1}, respectively. Stem biomass density accounted for approximately 70% of AGB and 56% of the total tree biomass density. As expected, the estimates of biomass densities are as precise as the estimates of BEFs.

Root-to-shoot ratio

The average root-to-shoot ratio was 0.24 (minimum = 0.07, maximum = 0.35, SD = 0.04, CV = 16.8%). The uncertainty (SE) of the estimated R/S was 3.4% (CI = 6.78%). The root-to-shoot ratio was neither linear nor nonlinearly dependent on any of the four variables (DBH, TH, AGB, and BGB) (Tables 9 and 10). The BGB density calculated based on R/S was 29.26 Mg ha^{-1} (SE = 3.4%), which was 1.20% smaller and 13.73% more precise than the BGB density estimate based on BEF.

Discussion

Component biomass expansion factors and biomass density

A wider range of DBH was measured during the first sampling phase than during the second phase. However, the DBH of *A. johnsonii* rarely exceeds 35 cm (here, <1% of trees during the first sampling phase). Although large

Table 1 Values of variables estimated for 3574 *Androstachys johnsonii* trees in 23 plots during the first sampling phase

Statistic	Trees (ha^{-1})	Stem volume (m^3)	Stem volume (m^3/ha)	DBH (cm)	TH (m)	H (m)
Average	1236.2187	0.0933	115.3149	13.4112	10.5616	10.4272
Minimum	541.1268	0.0020	66.8999	5.0000	1.8000	1.6000
Maximum	2220.2115	1.6463	170.4468	50.0000	22.5000	22.3000
SD	476.7204	0.1153	25.4407	6.2879	2.7637	2.7901
CV	38.5628	123.6853	22.0900	46.8856	26.1678	26.7583
SE	99.4031	0.0019	5.3048	0.1052	0.0462	0.0467
SE (%)	8.0409	2.0689	4.6061	0.7843	0.4377	0.4476

DBH, diameter at breast height; TH, total height; H, stem height; SD, standard deviation; CV, coefficient of variation; SE, standard error.

Table 2 Values of variables for 93 *Androstachys johnsonii* trees (a subset of the trees from phase 1) obtained during the second phase using destructive sampling

	Second phase variables	Average	Minimum	Maximum	SD	CV
	Taproot + stump	23.651	1.474	71.926	18.926	80.019
	Lateral roots	24.083	0.746	100.815	23.945	99.428
	Root system	**47.735**	**2.545**	**162.105**	**41.210**	**86.331**
	Stem wood	124.068	4.947	357.348	99.497	80.196
	Stem bark	14.198	0.677	55.805	12.372	87.138
Component dry weight (kg)	**Stem**	**138.267**	**5.636**	**413.153**	**110.577**	**79.974**
	Branches	55.586	2.583	211.320	57.355	103.183
	Foliage	2.807	0.333	15.100	2.493	88.818
	Crown	**58.393**	**3.038**	**216.695**	**59.077**	**101.172**
	Shoot system	**196.659**	**9.823**	**590.863**	**163.713**	**83.247**
	Total tree	**244.394**	**12.484**	**752.571**	**204.330**	**83.607**
	Diameter at breast height (DBH) (cm)	17.5860	5.0000	32.0000	7.5122	42.7167
	Total height (TH) (m)	12.3230	5.0000	16.0000	2.1381	17.3508
Dendrometric variables	Stem length (H) (m)	10.7470	4.2500	14.8400	2.1381	22.7562
	Stem volume (v2) (m3)	0.1890	4.2500	0.5806	0.1512	22.7562
	Hohenadal form factor (fh)	0.4460	0.3002	0.6128	0.0592	13.2716

The major components and their values are indicated in bold font. SD, standard deviation; CV, coefficient of variation.

variation in TH was observed during both phases, <4% of the trees had TH >16 m or <5 m, which indicated that phase-2 samples were representative of the phase-1 samples, and thus the values could be extrapolated.

BEF values are generally calculated from the ratio of tree component or total tree biomass (W_h) to stem or merchantable timber volume (v) [4,6,9-12] or biomass (W_s) [7,8,13-15]. We calculated BEFs using the first option (here called BEF_1) with total stem volume. The stem BEF value was 0.7334 Mg m^{-3}, which meant that stem biomass

(in Mg) was 0.7334-fold larger than stem volume (in m^3). Therefore, BEF computed according to biomass (the second option, here called BEF_2) can be calculated as a function of BEF_1 as $BEF_2 = \frac{W_h}{0.7334 \times v} = \frac{1}{0.7334} \times BEF_1$.

Since BEF_2 is obtained by multiplying BEF_1 by a constant, the relationship between BEF_2 and tree size (DBH and TH) is the same as that between BEF_1 and tree size (both relationships will be either significant or insignificant). Therefore, trends in BEF values calculated here using the first option were compared indiscriminately to those calculated using either option.

Table 3 Component biomass expansion factors (BEF$_h$), their variances (VAR$_{BEF}$), standard errors (SE), and 95% confidence intervals (CI) for *Androstachys johnsonii* trees

#	Tree component	BEF$_h$ (Mg m^{-3})	VAR$_{BEF}$ (Mg^2 m^{-6})	SE (Mg m^{-3})	SE (%)	95% CI (Mg m^{-3})	95% CI (%)
1	Taproot + stump	0.1407	3.6E-05	0.0060	4.2382	± 0.0119	±8.4764
2	Lateral roots	0.1162	4.4E-05	0.0067	5.7232	±0.0133	±11.4465
3	**Root system (1 + 2)**	**0.2569**	**1.0E-04**	**0.0100**	**3.8930**	**±0.0200**	**±7.7860**
4	Stem wood	0.6569	3.6E-04	0.0191	2.9046	±0.0382	±5.8092
5	Stem bark	0.0765	1.3E-05	0.0036	4.7534	±0.0073	±9.5068
6	**Stem (4 + 5)**	**0.7334**	**4.4E-04**	**0.0210**	**2.8615**	**±0.0420**	**±5.7230**
7	Branches	0.2928	3.1E-04	0.0177	6.0590	±0.0355	±12.1180
8	Foliage	0.0242	6.6E-06	0.0026	10.6242	±0.0051	±21.2483
9	**Crown (7 + 8)**	**0.3170**	**3.6E-04**	**0.0190**	**5.9973**	**±0.0380**	**±11.9946**
10	**Shoot system (6 + 9)**	**1.0504**	**1.2E-03**	**0.0340**	**3.2345**	**±0.0679**	**±6.4690**
11	**Total tree (3 + 10)**	**1.3072**	**1.8E-03**	**0.0428**	**3.2736**	**±0.0856**	**±6.5472**

The major components and their values are indicated in bold font. SE, standard error; CI, confidence limit.

Table 4 Linear regression test for dependence of biomass expansion factors (BEF) on diameter at breast height (DBH) in *A. johnsonii*

$BEF = b_0 + b_1DBH$

#	Tree component	b_0 (± SE)	b_1 (± SE)	Probability	Adjusted R^2
1	Taproot + stump	0.1890 (±0.0115)	− 0.0027 (±0.0006)	0.0000	0.1768
2	Lateral roots	0.0789 (±0.0125)	0.0021 (±0.0007)	0.0017	0.0933
3	**Root system (1 + 2)**	**0.2679 (±0.0175)**	**− 0.0006 (±0.0009)**	**0.4963**	**− 0.0058**
4	Stem wood	0.6482 (±0.0176)	0.0005 (±0.0009)	0.5911	− 0.0078
5	Stem bark	0.0798 (±0.0075)	− 0.0002 (±0.0004)	0.6314	− 0.0084
6	**Stem (4 + 5)**	**0.7280 (±0.0179)**	**0.0003 (±0.0009)**	**0.7435**	**− 0.0098**
7	Branches	0.2821 (±0.0376)	0.0006 (±0.0020)	0.7583	− 0.0099
8	Foliage	0.0557 (±0.0055)	− 0.0018 (±0.0003)	0.0000	0.2870
9	**Crown (7 + 8)**	**0.3378 (±0.0411)**	**− 0.0012 (±0.0021)**	**0.5838**	**− 0.0076**
10	**Shoot system (6 + 9)**	**1.0658 (±0.0445)**	**− 0.0009 (±0.0023)**	**0.7080**	**− 0.0094**
11	**Total tree (3 + 10)**	**1.3336 (±0.0574)**	**− 0.0015 (±0.0030)**	**0.6192**	**− 0.0082**

The major components and their values are indicated in bold font. b_0 and b_1, regression parameters; SE, standard error; probability refers to the significance of the regression.

The same principle holds for BEF values calculated using merchantable timber volume or biomass; because merchantable timber volume or biomass are obtained by multiplying stem biomass or volume by the merchantable fraction of the total stem (ratio of timber volume to stem volume) [8], which is a constant. For most trees, this fraction is very close to 1 [8], which makes BEF values calculated with merchantable volume or biomass very close to those calculated with stem volume or biomass.

We preferred the use of BEF_1 to BEF_2 because stem volume is easily measured destructively than stem biomass,

and volume is the main variable of interest in most forest inventories. In addition, stem volume was preferred to merchantable volume because merchantable height is sensitive to personal judgment and thus is more subjective than stem height, especially for standing trees. Merchantable tree height measurement (e.g. to 7 cm top diameter as defined by Lehtonen et al. [4], Lehtonen et al. [9], Edwards and Christie [16], and Black et al. [17]) in standing trees is subjective and more susceptible to measurement error than total tree height, because the 7 cm top diameter on the stem is difficult to identify than the tip of the tree. Moreover, in most tropical tree species, and

Table 5 Pearson's correlation coefficient test of significance, and distance covariance test of independence of biomass expansion factors (BEF) on diameter at breast height (DBH) in *A. johnsonii*

BEF vs. DBH

#	Tree component	Pearson's correlation test		Distance covariance test of independence		
		r	Probability	dcov	dcor	Probability
1	Taproot + stump	− 0.4310	1.6E-05	0.1802	0.4668	0.0150
2	Lateral roots	0.3211	0.0017	0.1670	0.4377	0.0150
3	**Root system (1 + 2)**	**− 0.0714**	**0.4963**	**0.1154**	**0.2546**	**0.0700**
4	Stem wood	0.0564	0.5911	0.0671	0.1430	0.7650
5	Stem bark	− 0.0504	0.6314	0.0442	0.1468	0.5200
6	**Stem (4 + 5)**	**0.0344**	**0.7435**	**0.0677**	**0.1428**	**0.7450**
7	Branches	0.0323	0.7583	0.1567	0.2190	0.2950
8	Foliage	− 0.5429	1.9E-08	0.1521	0.5874	0.0150
9	**Crown (7 + 8)**	**− 0.0575**	**0.5838**	**0.1684**	**0.2274**	**0.2450**
10	**Shoot system (6 + 9)**	**− 0.0394**	**0.7080**	**0.1738**	**0.2260**	**0.3300**
11	**Total tree (3 + 10)**	**− 0.0522**	**0.6192**	**0.2132**	**0.2432**	**0.2750**

The major components and their values are indicated in bold font. r, Pearson's correlation coefficient; dcov, distance covariance; dcor, distance correlation; probability refers to the significance of the test.

Table 6 Linear regression test for dependence of biomass expansion factors (BEF) on total tree height (TH) in *Androstachys johnsonii*

$BEF = b_0 + b_1 TH$

#	Tree component	b_0 (± SE)	b_1 (± SE)	Probability	Adjusted R^2
1	Taproot + stump	0.2884 (±0.0248)	− 0.0120 (±0.0020)	0.0000	0.2787
2	Lateral roots	0.1178 (±0.0304)	− 0.0001 (±0.0024)	0.9585	− 0.0110
3	**Root system (1 + 2)**	**0.4061 (±0.0371)**	**− 0.0121 (±0.0030)**	**0.0001**	**0.1456**
4	Stem wood	0.6697 (±0.0404)	− 0.0010 (±0.0032)	0.7486	− 0.0098
5	Stem bark	0.1018 (±0.0170)	− 0.0021 (±0.0014)	0.1345	0.0137
6	**Stem (4 + 5)**	**0.7715 (±0.0409)**	**− 0.0031 (±0.0033)**	**0.3461**	**− 0.0011**
7	Branches	0.4745 (±0.0844)	− 0.0147 (±0.0068)	0.0314	0.0394
8	Foliage	0.1142 (±0.0118)	− 0.0073 (±0.0009)	0.0000	0.3911
9	**Crown (7 + 8)**	**0.5887 (±0.0900)**	**− 0.0221 (±0.0072)**	**0.0029**	**0.0835**
10	**Shoot system (6 + 9)**	**1.3603 (±0.0969)**	**− 0.0251 (±0.0077)**	**0.0016**	**0.0939**
11	**Total tree (3 + 10)**	**1.7664 (±0.1230)**	**− 0.0373 (±0.0098)**	**0.0003**	**0.1267**

The major components and their values are indicated in bold font. b_0 and b_1, regression parameters; SE, standard error; probability refers to the significance of the regression.

especially in broadleaf species (as opposed to coniferous), taking a minimum top diameter of 7 cm to define merchantable tree height is somewhat impractical because the merchantable height is limited by branching, irregular form or defects which causes inconsistence in the top diameter definition.

Because stem volume is the auxiliary variable for all tree components, estimation of biomass density based on BEF achieves the property of additivity automatically for the major components (root system, shoot system, stem, and crown) and for total tree biomass, without additional efforts, which is a great advantage.

The BEF values estimated here fall in the range of many estimates obtained worldwide e.g. [4-7,10,14,18], especially with those of whole-tree BEF. For example, Kamelarczyk [18] reported whole-tree BEF values from 0.06 to 2.90 for 17 miombo tree species in Zambia. Estimates of aboveground and total tree BEF compiled for Africa by the FAO [19] were 1.5 and 1.9, 43% and 45% larger than our estimates, respectively; FAO values of

Table 7 Pearson's correlation coefficient test of significance and distance covariance test of independence of biomass expansion factors (BEF) on total tree height (TH) in *A. johnsonii*

BEF vs. TH

#	Tree component	Pearson's correlation test		Distance covariance test of independence		
		r	Probability	dcov	dcor	Probability
1	Taproot + stump	− 0.5353	3.2E-08	0.1106	0.5672	0.0150
2	Lateral roots	− 0.0055	0.9585	0.0516	0.2679	0.1550
3	**Root system (1 + 2)**	**− 0.3936**	**9.5E-05**	**0.0918**	**0.4009**	**0.0150**
4	Stem wood	− 0.0337	0.7486	0.0370	0.1562	0.6350
5	Stem bark	− 0.1564	0.1345	0.0249	0.1636	0.5250
6	**Stem (4 + 5)**	**− 0.0988**	**0.3461**	**0.0439**	**0.1833**	**0.3750**
7	Branches	− 0.2233	0.0314	0.0844	0.2334	0.2400
8	Foliage	− 0.6306	1.2E-11	0.0835	0.6386	0.0150
9	**Crown (7 + 8)**	**− 0.3057**	**0.0029**	**0.1089**	**0.2910**	**0.0600**
10	**Shoot system (6 + 9)**	**− 0.3220**	**0.0016**	**0.1131**	**0.2911**	**0.1300**
11	**Total tree (3 + 10)**	**− 0.3691**	**0.0003**	**0.1495**	**0.3376**	**0.0400**

The major components and their values are indicated in bold font. r, Pearson's correlation coefficient; dcov, distance covariance; dcor, distance correlation; probability refers to the significance of the test.

Table 8 Biomass density (W_h), variance (VAR_{Wh}), standard error (SE), and 95% confidence intervals (CI) for each component in *Androstachys johnsonii* trees

#	Tree component	W_h (Mg ha^{-1})	VAR_{Wh} (Mg^2 ha^{-2})	SE (Mg ha^{-1})	SE (%)	95% CI (Mg ha^{-1})	95% CI (%)
1	Taproot + stump	16.2192	0.4725	0.6874	4.2382	±1.3748	±8.4764
2	Lateral roots	13.4005	0.5882	0.7669	5.7232	±1.5339	±11.4465
3	**Root sytem (1 + 2)**	**29.6197**	**1.3296**	**1.1531**	**3.8930**	**±2.3062**	**±7.7860**
4	Stem wood	75.7526	4.8413	2.2003	2.9046	±4.4006	±5.8092
5	Stem bark	8.8182	0.1757	0.4192	4.7534	±0.8383	±9.5068
6	**Stem (4 + 5)**	**84.5708**	**5.8565**	**2.4200**	**2.8615**	**±4.8400**	**±5.7230**
7	Branches	33.7612	4.1845	2.0456	6.0590	±4.0912	±12.1180
8	Foliage	2.7923	0.0880	0.2967	10.6242	±0.5933	±21.2483
9	**Crown (7 + 8)**	**36.5535**	**4.8058**	**2.1922**	**5.9973**	**±4.3844**	**±11.9946**
10	**Shoot system (6 + 9)**	**121.1243**	**15.3491**	**3.9178**	**3.2345**	**±7.8356**	**±6.4690**
11	**Total tree (3 + 10)**	**150.7440**	**24.3521**	**4.9348**	**3.2736**	**±9.8696**	**±6.5472**

The major components and their values are indicated in bold font. SE, standard error; CI, confidence limit.

eastern Africa were 2.3 for aboveground BEF and 2.9 for total BEF, which were more than two-fold higher than our estimates. However, the FAO's global-scale estimates (1.0 for aboveground BEF and 1.3 for total tree BEF) were closer to our findings [19].

Reports on the dependence of BEF values on DBH and TH vary, from strong reverse dependence [3-7] to weak reverse dependence or independence [10]. Here, we found component BEFs to be either independent or have a weak reverse dependence on DBH and TH, which indicated that small and large *A. johnsonii* trees contain approximately the same quantity of biomass per unit volume.

Ducta et al. [6] maintained that the reverse dependence of BEF on tree size is a result of an inverse relationship between wood density and tree size. We did not observe variation in stem wood and stem bark densities according to DBH and TH for *A. johnsonii* (adj. $R^2 < 0.0309$, $P > 0.05$), and a very weak relationship was found between total stem density and DBH (adj. $R^2 = 0.1342$, $P = 0.0002$) and TH (adj. $R^2 = 0.0661$, $P = 0.0072$). These results explained the independence or weak dependence of component BEF values on tree size.

Our observation of a slightly stronger relationship between BEF values and TH compared to DBH was

consistent with the findings of other researchers [4,6,8,12], but contradicted the report by Sanquetta et al. [7].

The dependence of component BEFs (taproots, lateral roots, and foliage) on DBH detected by the linear regression test and Pearson's correlation coefficient test of significance were also detected by the dcov test of independence; suggesting that, the most pronounced dependence of these component BEFs on DBH is linear, since dcov test measures all types of dependence (linear and nonlinear). On the other hand, the absence of dependence of other 7 components and total tree BEFs on DBH by either method, suggests that there is not any type of dependence (linear, nonlinear or nonmonotone) of those component BEFs on DBH.

A linear dependence of crown and shoot system BEFs on TH was detected by the linear regression test and Pearson's correlation coefficient test of significance. However, this dependence was not detected by the dcov test of independence, which may suggest that this linear dependence is casual.

The finding of independence or weak dependence of the BEF on tree size might be related to the minimum DBH measured in the phase-2 (DBH ≥ 5 cm). It has been reported by Brown et al. [3], Saquentta et al. [7], Marková

Table 9 Linear regression test of dependence of root-to-shoot ratio on diameter at breast height (DBH), total height (TH), aboveground biomass (AGB), belowground biomass (BGB), and total biomass (TB) in *Androstachys johnsonii* trees

#	Regression equation	b_0 (± SE)	b_1 (± SE)	Probability	Adjusted R^2
1	R/S = b0 + b1DBH	0.24051 (±0.01080)	0.00006 (±0.00057)	0.9154	− 0.0109
2	R/S = b0 + b1H	0.27807 (±0.02454)	− 0.00296 (±0.00196)	0.1347	0.0137
3	R/S = b0 + b1AGB	0.27015 (±0.00662)	0.00001 (±0.00003)	0.7812	− 0.0101
4	R/S = b0 + b1BGB	0.23354 (±0.00634)	0.00017 (±0.00010)	0.0997	0.0188
5	R/S = b0 + b1TB	0.23876 (±0.00659)	0.00001(±0.00002)	0.5802	− 0.0076

b_0 and b_1, regression parameters; SE, standard error; probability refers to the significance of the regression.

Table 10 Pearson's correlation coefficient test of significance and distance covariance test of independence of root-to-shoot ratio on diameter at breast height (DBH), total height (TH), aboveground biomass (AGB), belowground biomass (BGB), and total biomass (TB) in *Androstachys johnsonii* trees

#	Pair of variables	Pearson's correlation test		Distance covariance test of independence		
		r	Probability	dcov	dcor	Probability
1	R/S vs. DBH	0.0112	0.9154	0.0554	0.1573	0.8650
2	R/S vs. H	− 0.1563	0.1347	0.0473	0.2662	0.0650
3	R/S vs. AGB	0.0292	0.7812	0.2475	0.1518	0.8200
4	R/S vs. BGB	0.1718	0.0997	0.1575	0.1951	0.2900
5	R/S vs. TB	0.0581	0.5802	0.2763	0.1519	0.8100

r, Perason's correlation coefficient; dcov, distance covariance; dcor, distance correlation; probability refers to the significance of the test.

and Pokorný [10], and Soares and Tomé [20] that the decrease of the BEF with tree size reaches an asymptote at a given tree size. This is presumably due to stabilization of growth rate [7].

The finding of independence or weak dependence of the BEF on tree size suggests that, for *A. johnsonii*, constant component BEF values can be accurately used within the interval of harvested tree sizes ($5 \leq DBH \leq 32$, Table 1), in contrast to findings by Brown et al. [3] and Sanquetta et al. [7]. Here, further research would be needed to reveal the relationship between tree component BEFs and tree's $DBH \leq 5$ cm.

We defined the stem as the length from the top of the stump to the height corresponding to 2.5 cm diameter. Differences among stem definitions (e.g. different stump height or different minimum top diameter, stump considered as part of the stem) would affect the BEF estimates.

It was difficult to compare our 4 major and 6 minor component BEF and biomass density values, because few similar studies have been performed in African and Mozambican woodlands. The majority of available studies provide estimates of BEF and biomass for shoot systems and occasionally for the whole tree. Our estimated AGB density (121 Mg ha^{-1}) was within the range reported by Lewis et al. [21] for tropical African forests (114–749 Mg ha^{-1}) and by Brown [11] for hardwood forests (75–175 Mg ha^{-1}); and lower than estimates for closed tropical forests (144–513 Mg ha^{-1}) [22,23]. Our AGB density estimate was higher than Brown and Lugo's estimate for open tropical forests (50 Mg ha^{-1}) [23].

Estimates of stem-wood biomass density by Brown and Lugo [23] for undisturbed, logged, and unproductive tropical African forests were 148.6, 41.2, and 36 Mg ha^{-1}, respectively, while our estimate was 75.75 Mg ha^{-1}. Our estimated whole-tree biomass density (approximately 150 Mg ha^{-1}) was similar to those for unproductive (129 Mg ha^{-1}) and logged (179 Mg ha^{-1}) tropical African forests, and smaller than Brown and Lugo's estimate for undisturbed forests (238 Mg ha^{-1}) [23]. However, the estimates by Brown and

Lugo [22,23] were performed more than 4 decades ago, and thus, they might not reflect the current situation.

Our estimated AGB density (121 Mg ha^{-1}) are in agreement with those estimated for Mozambique by Brown [24] for dense forests in moist-dry season (120 Mg ha^{-1}) and in moist-short dry season (130 Mg ha^{-1}) but are higher compared to dense forests in dry season (70 Mg ha^{-1}). Yet, mecrusse woodlands (*A. johnsonii* stands) are typically from dry season [25-29], implying that the biomass productivity of mecrusse woodlands is, approximately, twice as larger than the average productivity of dense forests in dry season in Mozambique.

The estimated uncertainty in our BEF values (2.9%–10.6%) was lower than that of Lehtonen et al. (3%–21%) [4,9] and Jalkanen et al. (4%–13%) [30]. The component biomass and stem volume values used here to calculate BEF were obtained directly using destructive sampling, whereas Lehtonen et al. [4,9] and Jalkanen et al. [30] were based on values obtained indirectly using regression models. These different approaches might explain the differences among BEF estimates and the higher uncertainty reported by those authors, because they also incorporate uncertainty from the regression models.

The default IPCC aboveground BEF for tropical broadleaf species is 1.5 ± 0.2 (SE = 6.67%) [31]. This BEF value is 43% larger and 200% more uncertain than our estimated aboveground BEF (1.05 ± 0.07 Mg m^{-3}, SE = 3.23%). The default IPCC BEF-based AGB density was 173 ± 28 Mg ha^{-1} (SE = 8.10%), 43% larger and 107% more uncertain than our estimated AGB density (121 ± 6.47 Mg ha^{-1}, SE = 3.23%). The default IPCC BEF-based AGB density is not in agreement with the estimated AGB density for Mozambique by Brown [24].

Root-to-shoot ratios

The average root-to-shoot ratio found in this study (0.24 or 1:4) was larger than that observed by some authors, such as 1:5 (0.2) reported by Kramer [32], 1:6 (0.17) reported by Perry [33], and 0.17 reported by Sanquetta et al. [7]. The findings of these authors suggest that

AGB is 5- to 6-fold greater than BGB, but our finding that AGB is, on average, almost 4-fold higher than BGB was consistent with the default IPCC root-to-shoot ratio of 1:4 (0.25) [31]. We determined BGB by complete removal of the root system, including the root collar and fine roots. Estimates of R/S may vary greatly if the root system is partially removed, as performed by many authors e.g. [7,8,34-37], if the depths of excavation are predefined [7,37,38], if fine roots are excluded [39-41]. R/S values may also vary if root sampling procedures are applied, for example, where only a number of roots from each root system is fully excavated, and then the information from the excavated roots is used to estimate biomass for the roots not excavated [42-44]. Different estimates of R/S can also be obtained if the stump is considered as part of the stem, as in Segura and Kanninen [14].

Wang et al. [45] similarly observed little variation in the relationship between R/S and tree diameter. However, different results were obtained by Mokany et al. [2] for root-to-shoot ratios in different terrestrial biomes (forests, woodland, shrublands and grasslands), where the ratios decreased significantly with increasing shoot biomass, tree height, and DBH. Our findings were also inconsistent with those of Sanquetta et al. [7], who found that R/S decreased as DBH and TH increased. This might be presumably because A. johnsonii, as a tropical native species, has a very low and/or constant growth rate within the interval of harvested DBH, as opposed to the planted Pinus spp. studied by Sanqueta et al. [7].

As in the case of the BEF, the finding of independence of R/S on tree size might be related to the minimum DBH measured in the phase-2 (DBH ≥ 5 cm). Mokany et al. [2], Saquentta et al. [7], Jenkins et al. [46], and Zhou and Hemstrom [47] have shown that decrease of R/S with tree size reaches an asymptote at a given tree size, presumably due to stabilization of growth rate [7]. Inclusion of trees with DBH ≤ 5 cm could cause variation of R/S with tree size. Therefore, researches are also needed here to reveal the relationship between R/S and tree's DBH ≤ 5 cm.

Conclusions

The belowground, aboveground, and whole-tree BEFs were 0.26 ± 0.02 (SE = 3.89%), 1.05 ± 0.07 (SE = 3.23%), and 1.30 ± 0.09 Mg m^{-3} (SE = 3.27%), respectively; equivalent to, approximately, the following BEF-based biomass densities: 30 ± 2.31, 121 ± 7.84, and 151 ± 9.87 Mg ha^{-1}, respectively.

We observed that component BEFs in Androstachys johnsonii Prain were independent or only weakly dependent on tree size (DBH and TH), and that TH was more important that DBH in explaining BEF. Therefore, we suggested that constant component BEF values can be accurately used within the interval of harvested tree sizes. The root-to-shoot ratio (average = 0.24 ± 0.02; SE = 3.4%) was not dependent on tree height, DBH, AGB, or BGB.

Methods
Study area
Mecrusse is a forest type in which the dominant canopy species is Androstachys johnsonii, the relative cover of which varies from 80% to 100% [48]. In Mozambique (18°15′S, 35°00′E), mecrusse woodlands are mainly found in Inhambane and Gaza provinces in Massangena, Chicualacuala, Mabalane, Chigubo, Guijá, Mabote, Funhalouro, Panda, Mandlakaze, and Chibuto districts. The easternmost mecrusse forest patches, located in Mabote, Funhalouro, Panda, Mandlakaze, and Chibuto districts, were defined as the study area (Figure 2) and encompassed 4,502,828 ha [49], of which 226,013 ha (5%) were mecrusse woodlands. The climate throughout the study area is dry tropical, with the exception of humid tropical areas in western Panda and southwestern Mandlakaze districts [25-29,49]; a warm or rainy season occurs from October to March, and a cool or dry season occurs from March to September [25-29].

The mean annual temperature generally exceeds 24°C, and mean annual precipitation varies from 400 to 950 mm [25-29,49]. According to the United States Food and Agriculture Organization (FAO) classification [50], soils are mainly Ferralic Arenosols across more than 70% of the study area [49]. Arenosols, Umbric Fluvisols, and Stagnic soils are predominant in the northernmost part of the study area [49]. There is a shortage of water resources and precipitation throughout the study area; only Chibuto and Mandlakaze districts have water resources [25-29,49].

Data collection
We used a two-phase sampling design to determine stem volume and biomass. In the first phase, we measured diameter at breast height (DBH) and stem height of 3574 trees (m_1) in 23 randomly located circular plots (20-m radius) (Figure 2) for estimation of stem volume; only trees with DBH ≥ 5 cm were considered. In the second phase, 93 trees (m_2) (DBH ≥ 5 cm) were randomly selected from those analysed during the first phase for destructive measurement of biomass and stem volume. The felled trees were divided into the following components: (1) taproot; (2) lateral roots; (3) root system (1 + 2); (4) stem wood; (5) stem bark; (6) stem (4 + 5); (7) branches; (8) foliage; (9) crown (6 + 7); (10) shoot system (6 + 9); and (11) whole tree (3 + 10). Tree components were sampled and the dry weights estimated as follows.

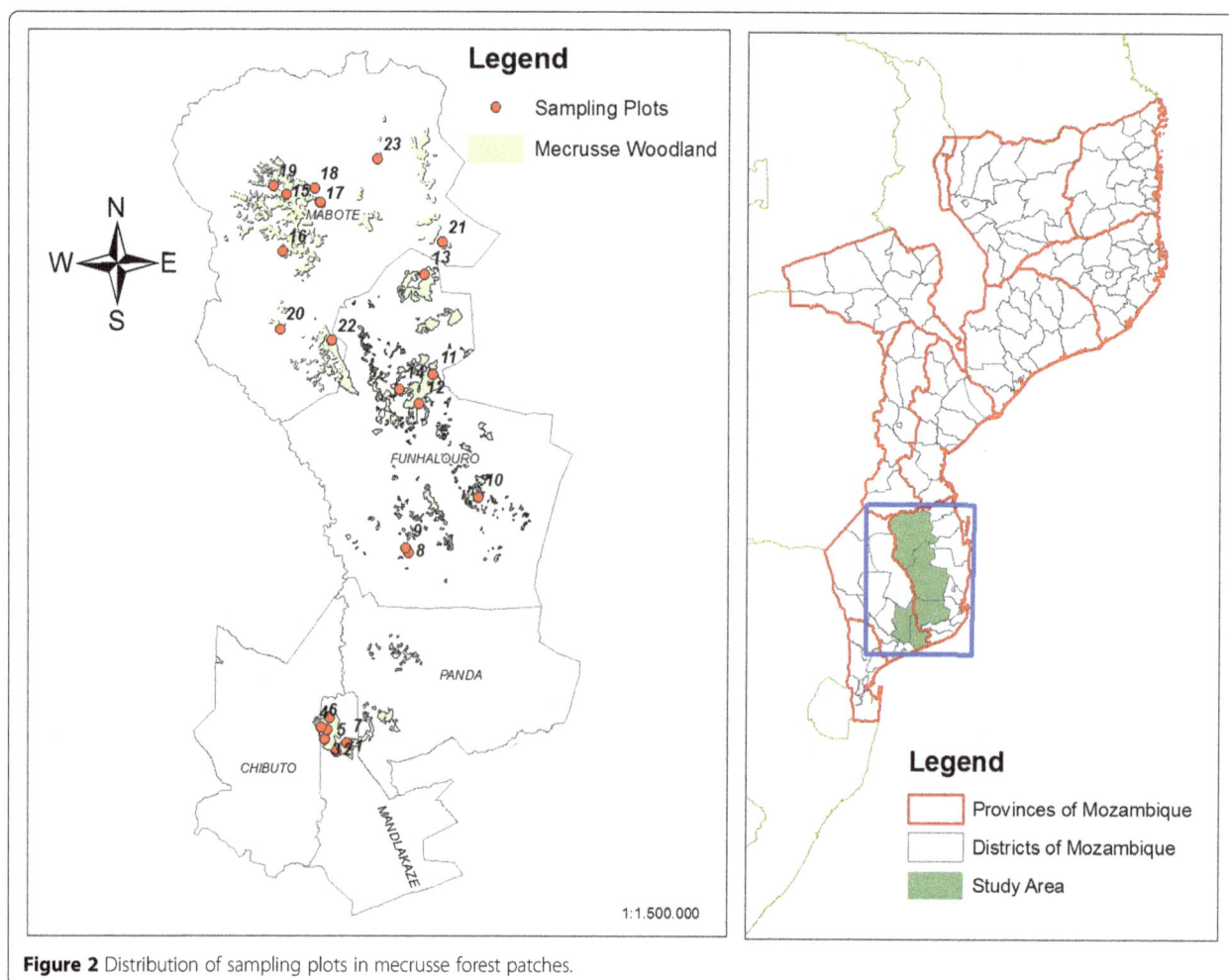

Figure 2 Distribution of sampling plots in mecrusse forest patches.

Root system

The stump height was predefined as being 20 cm from the ground level for all trees and considered as part of the taproot, as recommended by Parresol [51] and because in larger *A. johnsonii* trees this height (20 cm) is affected by root buttress; therefore, the root collar was also considered part of the taproot. The root system was divided into 3 sub-components: fine lateral roots, coarse lateral roots, and taproot. Lateral roots with diameters at insertion point on the taproot < 5 cm were considered as fine roots and those with diameters ≥ 5 cm were considered as coarse roots.

First, the root system was partially excavated to the first node, using hoes, shovels, and picks; to expose the primary lateral roots (Figure 3a, b). The primary lateral roots were numbered and separated from the taproot with a chainsaw (Figure 3a, b) and removed from the soil, one by one. This procedure was repeated in the subsequent nodes until all primary roots were removed from the taproot and the soil. Finally, the taproot was excavated and removed (Figure 3 c–f). The complete removal of the root system

was relatively easy because 90% of the lateral roots of *A. johnsonii* are located in the first node, which is located close to ground level (Figure 3 a–c); the lateral roots grow horizontally to the ground level, do not grow downwards; and because the taproots had, at most, only 4 nodes and at least 1 node (at ground level). The root system was removed completely, so the depth of excavation depended on the depth of the taproot.

Fresh weight was obtained for the taproot, each coarse lateral root and for all fine lateral roots. A sample was taken from each sub-component, fresh weighed, marked, packed in a bag, and taken to the laboratory for oven drying. For the taproot, the samples were two discs, one taken immediately below the ground level and another from the middle of the taproot. For the coarse lateral roots, two discs were also taken, one from the insertion point on the taproot and another from the middle of it. For fine roots the sample was 5 to 10% of the fresh weight of all fine lateral roots. Oven drying of all samples was done at 105°C to constant weight, hereafter, referred to as dry weight.

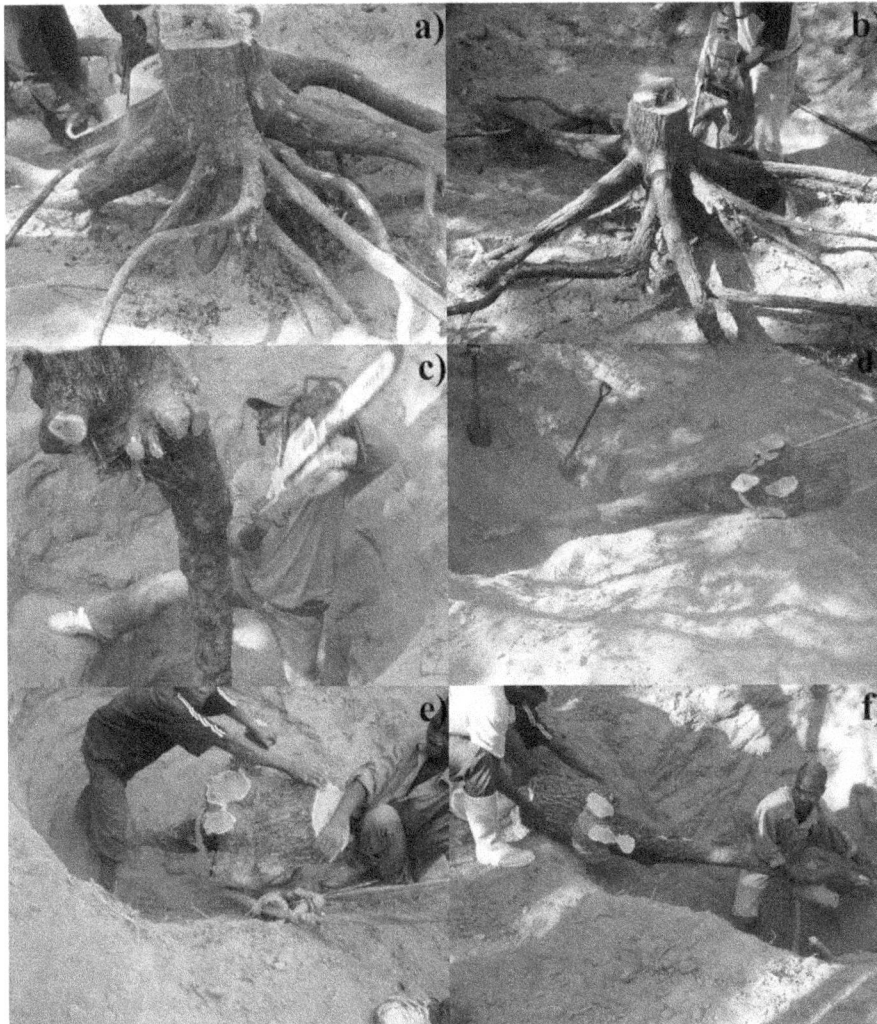

Figure 3 Separation of lateral roots from the root collar/taproot (**a, b, c**), and removal of the taproot including the root collar and the stump (**d, e, f**).

Stem wood and stem bark

Felled trees were scaled up to a 2.5 cm top diameter. The stem was defined as the length of the trunk from the stump to the height that corresponded to 2.5 cm diameter, to standardize with the definitions of fine branches. The remainder (from the height corresponding to 2.5 cm diameter to the tip of the tree) was considered a fine branch.

First, we divided the stem of each felled tree into 10 segments of equal length, and we measured the diameter of each segment at the midpoint, starting from the bottom of the stem, for volume and form factor determination using Hohenadl formula. The stem was, then, divided into sections, the first with 1.1 m length, the second with 1.7 m, and the remaining with 3 m, except the last, the remainder, which length depended on the length of the stem.

Discs were removed at the bottom and top of the first section, and on the top of the remaining sections; i.e.:

discs were removed at heights of 0.2 m (stump height), 1.3 m (breast height), 3 m, and the successive discs were removed at intervals of 3 m to the top of the stem, and their fresh weights measured using a digital scale.

Diameters over and under bark were taken from the discs in the North–south direction (previously marked on the standing tree) with the help of a ruler. The volumes over and under the bark of the stem were obtained by summing up the volumes of each section calculated using Smalian's formula [52]. Bark volume was obtained from the difference between volume over bark and volume under bark.

The discs were dipped in drums filled with water, until constant weight (3 to 4 months), for its saturation and subsequent determination of the saturated volume and basic density. The saturated volume of the discs was obtained based on the water displacement method [53] using Archimedes' principle. This procedure was done twice: before and after debarking; hence, we obtained saturated volume under and over the bark.

Wood discs and respective barks were oven dried at 105°C to constant weight. Basic density was obtained by dividing the oven dry weight of the discs (with and without bark) by the relevant saturated wood volume [54,55]. Therefore, two distinct basic densities were calculated: (1) basic density of the discs with bark and (2) basic density of the discs without bark.

We estimated the basic density at point of geometric centroid of each section using the regression function of density over height [56]. This density value was taken as representative of each section [56].

Crown

The crown was divided into two sub-components: branches and foliage. Primary branches, originating from the stem, were classified in two categories: primary branches with diameters at the insertion point on the stem ≥ 2.5 cm were classified as large branches, and those with diameter < 2.5 cm were classified as fine branches. Large branches were sampled similarly to coarse roots, and fine branches and foliage were sampled similarly to fine roots.

Tree component dry weights

We determined dry weight of the taproot, lateral roots, branches, and foliage by multiplying the ratio of fresh-to oven-dry weight of each sample by the total fresh weight of the relevant component. Dry weights of the root system and crown were obtained by summing up the relevant sub-components' dry weights. Dry weights of each stem section (with and without bark) were obtained by multiplying respective densities by relevant stem section volumes.

Stem (wood + bark) and stem wood dry weights were obtained by summing up each section's dry weight with and without bark, respectively. The dry weight of stem bark was determined from the difference between the dry weights of the stem and stem wood. We determined the dry weight of major components (root system, shoot system, and crown) and the whole tree by summing the dry weights of their constituent components.

Data processing and analysis

Stem volume was computed using Hohenadl's method (Eq. 1) [57].

$$v_{i2} = \frac{\Pi L}{40}(d_{.05}^2 + d_{.15}^2 + d_{.25}^2 + d_{.35}^2 + d_{.45}^2 + d_{.55}^2 + d_{.65}^2$$
$$+ d_{.75}^2 + d_{.85}^2 + d_{.95}^2)[m^3] \quad (1)$$

where v_{i2} is the stem volume of the ith tree from the second sampling phase, L is the stem length (in meters), and $d_{.i}$ is the diameter (in meters) measured at the proportional distance along the stem of the ith tree.

The individual stem volume of the ith tree of the jth plot from the first sampling phase (v_{ij1}) was calculated using Eq. 2 as follows:

$$v_{ij1} = \frac{\Pi}{4}DBH^2 \times H \times f_h \ [m^3] \quad (2)$$

where H is stem height and f_h is the Hohenadl form factor of the trees from the second sampling phase, obtained using Eq. 3 as:

$$f_h = 0.1\left(1 + \frac{d_{.15}^2}{d_{.05}^2} + \frac{d_{.25}^2}{d_{.05}^2} + \frac{d_{.35}^2}{d_{.05}^2} + \frac{d_{.45}^2}{d_{.05}^2}\right.$$
$$\left. + \frac{d_{.55}^2}{d_{.05}^2} + \frac{d_{.65}^2}{d_{.05}^2} + \frac{d_{.75}^2}{d_{.05}^2} + \frac{d_{.85}^2}{d_{.05}^2} + \frac{d_{.95}^2}{d_{.05}^2}\right)[\text{dimensionless}]$$
$$(3)$$

The main auxiliary variable (the first-phase variable) is the stand-level stem volume (m^3 ha^{-1}), estimated from Eq. 4 as follows:

$$V_1 = \frac{\sum_{j=1}^{n}\sum_{i=1}^{m_j} v_{ij1}}{n \times a} = \bar{v}_1 \times N_1 [m^3 ha^{-1}] \quad (4)$$

where m_j is the number of trees in the jth plot, n is the number of plots, a is the plot area (ha), $\bar{v}_1$ is the average stem volume of the trees of the first phase (m^3), and N_1 is the average number of trees per hectare estimated from the first sampling phase. Stem height of trees from the first phase was obtained by subtracting predefined stump height from the whole-tree height (TH) to standardize the definitions of stem height and stem length (for phase-1 trees).

The component biomass expansion factors for each tree (BEF_{hi}) in the second sampling phase were calculated as the ratio of tree component biomass w_{hi2} to stem volume v_{i2} [4-6,9,10] (Eq. 5) and the average was taken as the component BEF (Eq. 6) of the woodland. This process enabled us to convert stem volume to biomass. The root-to-shoot ratio (R/S) was determined as the ratio of BGB to AGB [1,2,40] for each tree (Eq.7); the average value was taken as the overall vegetation R/S (Eq. 8).

$$BEF_{hi} = \frac{w_{hi2}}{v_{i2}} \ [\text{Mg m}^{-3}] \quad (5)$$

$$BEF_h = \frac{\sum_{i=1}^{m_2} BEF_{hi}}{m_2} \ [\text{Mgm}^{-3}] \quad (6)$$

$$(R/S)_i = \frac{w_{Root_{i2}}}{w_{Shoot_{i2}}} \text{ [dimensionless]} \qquad (7)$$

$$R/S = \frac{\sum_{i=1}^{m_2} (R/S)_i}{m_2} \text{ [dimensionless]} \qquad (8)$$

where BEF_{hi} is the BEF of the hth component of the ith tree; w_{hi2} is biomass of the hth component of the ith tree measured during the second phase; BEF_h is the average BEF of the hth component; R/S_i is the root-to-shoot ratio of the ith tree; $w_{Root_{i2}}$ and $w_{Shoot_{i2}}$ represent BGB and AGB, respectively, of the ith tree of the second phase; and m_2 is the total number of trees in the second sampling phase.

The average tree component biomass density W_h (Mg ha^{-1}) was estimated as the product of the respective component BEF_h values and V_1 (Eq. 9):

$$W_h = BEF_h \times V_1 = W_{hi} \times N_1 \text{ [Mg ha}^{-1}\text{]} \qquad (9)$$

where

$$W_{hi} = BEF_h \times \bar{v}_1 \text{ [Mg]} \qquad (10)$$

is the estimated average component biomass per tree, which yields W_h when multiplied by the number of trees per hectare.

BEF_{hi} (Eq. 5) is the ratio of biomass of a tree component to stem volume; therefore, BEF_h (Eq. 6) is a mean ratio (not a ratio of means). These variables represent double sampling with mean-of-ratios estimators and dependent phases, and the uncertainty (variance and standard error) of the estimated BEF_h and W_h must be computed accordingly (as for R/S).

We calculated the variance of the estimated W_{hi} (Eq. 10) according to Freese [58,59]:

$$VAR_{W_{hi}} = \bar{v}_1^2 \left(\frac{S_{BEF_h}^2}{m_2}\right)\left(1 - \frac{m_2}{m_1}\right)$$
$$+ \frac{S_{w_{h2}}^2}{m_1}\left(1 - \frac{m_1}{M}\right) \text{ [Mg}^2\text{]} \qquad (11)$$

Rearranging Eq. 10 as $BEF_h = \frac{W_{hi}}{\bar{v}_1}$, the variance of the estimated BEF_h becomes [60]:

$$VAR_{BEF_h} = \frac{VAR_{W_{hi}}}{\bar{v}_1^2}$$
$$= \left(\frac{S_{BEF_h}^2}{m_2}\right)\left(1 - \frac{m_2}{m_1}\right) + \left(\frac{1}{\bar{v}_1^2}\right)\left(\frac{S_{w_{h2}}^2}{m_1}\right)$$
$$\times \left(1 - \frac{m_1}{M}\right) \text{ [Mg}^2\text{m}^{-6}\text{]} \qquad (12)$$

Similarly, the variance of the estimated W_h is [60]:

$$VAR_{W_h} = N_1^2$$
$$\times \left[\bar{v}_1^2\left(\frac{S_{BEF_h}^2}{m_2}\right)\left(1 - \frac{m_2}{m_1}\right) + \frac{S_{w_{h2}}^2}{m_1}\left(1 - \frac{m_1}{M}\right)\right]$$
$$\text{[Mg}^2\text{ha}^{-2}\text{]} \qquad (13)$$

and the variance of the estimated R/S is

$$VAR_{R/S} = \left(\frac{S_{R/S}^2}{m_2}\right)\left(1 - \frac{m_2}{m_1}\right) + \left(\frac{1}{\bar{w}_{Shoot_1}^2}\right)$$
$$\times \left(\frac{S_{w_{Root2}}^2}{m_1}\right)\left(1 - \frac{m_1}{M}\right) \text{[dimensionless]} \qquad (14)$$

where

$$S_{BEF_h}^2 = \frac{\sum BEF_{hi}^2 - \frac{\left(\sum BEF_{hi}\right)^2}{m_2}}{m_2 - 1} \text{ [Mg}^2\text{m}^{-6}\text{]} \qquad (15)$$

is the variance of BEF_h for the second phase;

$$S_{w_{h2}}^2 = \frac{\sum w_{hi2}^2 - \frac{\left(\sum w_{hi2}\right)^2}{m_2}}{m_2 - 1} \text{ [Mg}^2\text{]} \qquad (16)$$

is the variance of w_{h2}; w_{h2} is the component biomass for the second phase;

$$S_R^2 = \frac{\sum (R/S)_i^2 - \frac{\left(\sum (R/S)_i\right)^2}{m_2}}{m_2 - 1} \text{[dimensionless]} \qquad (17)$$

is the variance of R/S for the second phase;

$$S_{w_{Root2}}^2 = \frac{\sum w_{Root_{i2}}^2 - \frac{\left(\sum w_{Root_{i2}}\right)^2}{m_2}}{m_2 - 1} \text{ [Mg}^2\text{]} \qquad (18)$$

is the variance of w_{Root2}; w_{Root2} is the BGB of trees of the second phase; $\bar{w}_{Shoot_1}$ is the average AGB per tree for the first sampling phase; and m_1, m_2, and M are the number of trees in the first sampling phase, the second sampling phase, and the entire population, respectively. The finite population correction factor $\left(1 - \frac{m_1}{M}\right)$ was eliminated because m_1 was very small relative to M, which was unknown.

The square root of Eqs. 12 and 13 is the absolute standard error of the estimated BEF_h and W_h, respectively; dividing these values by BEF_h and W_h and then multiplying them by 100 provides the respective percent standard error. The absolute and percent 95% confidence limits (CI) are computed by multiplying the absolute and percent standard error by the Student's t-value

(t). The absolute and percent standard errors of R/S are computed analogously.

The percent 95% confidence limit (Eq. 19) is also referred as percent sampling error [61].

$$95\%CI_\% = E_\% = \pm \frac{t \times SE}{\bar{X}} \times 100 \ [\%] \tag{19}$$

where SE is the standard error and $\bar{X}$ is the average BEF_h, W_h or R/S.

In this study, uncertainty is expressed as the percent SE and as the percent 95% CI to facilitate comparison with existing studies as, for our knowledge, the existing studies reporting BEFs and R/S with known uncertainty use either percent SE [4,9,30] or percent 95% CI [31,39,62] to express the uncertainty.

The dependence of the component BEF values on DBH and TH was analysed by linear regression of BEF_h against DBH and TH and testing the significance of the regression against the null hypothesis of slope = 0 using Student's t-tests; and by testing the significance of the Pearson's correlation coefficient. However, the linear regression and the Pearson's correlation coefficient detect only linear dependence; do not detect nonlinear or nonmonotone dependencies [63]. Therefore, we used distance correlation, distance covariance, and distance covariance test of independence [63,64] to address possible nonlinear dependencies between the variables under study. Distance correlation is a new dependence coefficient that measures all types of dependence between random vectors X and Y in arbitrary dimension [63]. Therefore, the distance covariance test of independence detects any nonlinear and nonmonotone dependence between two random variables [63]. We examined the relationship of R/S to DBH, TH, AGB, BGB, and total biomass by the same procedures. All analyses were performed at the 5% significance level using Microsoft Excel Data Analysis Tools and using the "Energy" package [65] in R [66].

Further, the default IPPC aboveground BEF ($BEF_{h (IPCC)}$) for tropical braodleaf species and the respective BEF-based biomass density ($W_{h(IPCC)}$) (computed using the default BEF and our estimated volume) were compared with the aboveground BEF from this study and the respective BEF-based biomass density. As for the respective uncertainties.

The default BEF-based biomass density is computed as follows (Eq. 20):

$$W_{h(IPCC)} = BEF_{h(IPCC)} \times V_1 \ [Mg^2 \ ha^{-1}] \tag{20}$$

The $BEF_{h(IPCC)}$ and V_1 are obtained from independent samples (separate surveys), therefore, the uncertainty (percent SE) of $W_{h(IPCC)}$ can be computed as in Eq. 21 [30,31,39,58,59,62,67]:

$$SE\% = \sqrt{SE\%^2_{BEF_{h(IPCC)}} + SE\%^2_{V_1}} \ [\%] \tag{21}$$

where $SE^2_{BEF_{h(IPCC)}}$, $SE^2_{V_1}$ are percent standard errors associated with $BEF_{h(IPCC)}$ and V_1, respectively.

Abbreviations
AGB: Aboveground biomass; BEF: Biomass expansion factor; BGB: Belowground biomass; DBH: Diameter at breast height; R/S: Root-to-shoot ratio.

Competing interests
All authors declare that they have no competing interests.

Authors' contributions
TM and TS jointly designed the methodology. TM collected and analysed the data and wrote the manuscript. TS reviewed the manuscript. Both authors read and approved the final manuscript.

Acknowledgments
This study was funded by the Swedish International Development Cooperation Agency (SIDA). Thanks are extended to Professor Almeida Sitoe for his advices during the preparation of the field work; and to the field and laboratory team involved in felling, excavation, fresh- and dry-weighing of trees and tree samples: Albino Américo Mabjaia, Salomão dos Anjos Baptista, José Alfredo Amanze, Amélia Saraiva Monguela, João Paulino, Alzido Macamo, Jaime, Bule, Adolfo Zunguze, Viriato Chiconele, Murrombe, Paulo Goba, Dinísio Júlio, Gerente Guarinare, Sá Nogueira Lisboa, Francisco Ussivane, Nkassa Amade, and Cândida Zita. We would also like to thank the members of local communities, community leaders, and Madeiraarte Forest Concession for the unconditional help. Acknowledges are also due to two anonymous reviewers whose insightful comments and suggestions have improved considerably this paper.

References
1. IPCC. Intergovernmental Panel on Climate Change. Guidelines for National Greenhouse Gas Inventories 2006 [http://www.ipcc.ch]
2. Mokany K, Raison RJ, Prokushkin AS. Critical analysis of root: shoot ratios in terrestrial biomes. Global Change Biol. 2006;12:84–96.
3. Brown S, Gillespie AJR, Lugo AE. Biomass estimation methods for tropical forests with application to forest inventory data. Forest Sci. 1989;35(4):881–902.
4. Lehtonen A, Mäkipää R, Heikkinen J, Sievänen R, Lisk J. Biomass expansion factors (BEFs) for Scots pine, Norway spruce and birch according to stand age for boreal forests. Forest Ecol Manage. 2004;188:211–24.
5. Cháidez JJN. Allometric equations and expansion factors for tropical dry forest trees of Eastern Sinaloa, Mexico. Trop Subtrop Agroecosys. 2009;10:45–52.
6. Dutca I, Abrudan IV, Stancioiu PT, Blujdea V. Biomass conversion and expansion factors for young Norway spruce (Picea abies (L.) Karst.) trees planted on non-forest lands in Eastern Carpathians. Not Bot Hort Agrobot Cluj. 2010;38(3):286–92.
7. Sanquetta CR, Corte APD, Silva F. Biomass expansion factors and root-to-shoot ratio for Pinus in Brazil. Carbon Bal Manage. 2011;6:1–8.
8. Levy PE, Hale SE, Nicoll BC. Biomass expansion factors and root: shoot ratios for coniferous tree species in Great Britain. Forestry. 2004;77(5):421–30.
9. Lehtonen A, Cienciala E, Tatarinov F, Mäkipää R. Uncertainty estimation of biomass expansion factors for Norway spruce in the Czech Republic. Ann For Sci. 2007;64:133–40.
10. Marková I, Pokorný R. Allometric relationships for the estimation of dry mass of aboveground organs in young highland Norway spruce stand. Acta Univ Agric Silvic Mendel Brun. 2011;59(6):217–24.
11. Brown S. Measuring carbon in forests: current and future challenges. Environ Pollut. 2002;116:363–72.
12. Silva-Arredondo FM, Návar-Cháidez JJN. Factores de expansión de biomasa en comunidades florestales templadas del Norte de Durango. México Rev Mex Cien For. 2010;1(1):55–62.
13. FAO. State of the World's forests 1997. FAO, Rome: Food and Agriculture Organisation of the United Nations; 2007.

14. Segura M, Kanninen M. Allometric models for tree volume and total aboveground biomass in a tropical humid forest in Costa Rica. Biotropica. 2005;37(1):2–8.

15. Somogyi Z, Cienciala E, Mäkipää R, Muukkonen P, Lehtonen A, Weiss P. Indirect methods of large-scale forest biomass estimation. Eur J Forest Res. 2007;126:197–207.

16. Edwards PN, Christie JM. Yield models for forest management. London: HMSO; 1981.

17. Black K, Tobin B, Siaz G, Byrne KA, Osborne B. Allometric regressions for an improved estimate of biomass expansion factors for Ireland based on a Sitka spruce chronosequence. Irish Forestry. 2004;61(1):50–65.

18. Kamelarczyk KBF. Carbon stock assessment and modelling in Zambia: a UN-REDD programme study. Zambia: United Nations–Reducing Emissions from Deforestation and forest Degradation; 2009.

19. FAO. State of the World's forests 2007. FAO, Rome: Food and Agriculture Organisation of the United Nations; 2007.

20. Soares P, Tome M. Biomass expansion factores for *Eucalyptus globulus* stands in Portugal. Forest system. 2012;21(1):141–52.

21. Lewis SL et al. Aboveground biomass and structure of 260 African tropical forests. Phil Trans R Soc B. 2013;368:1–14.

22. Brown S, Lugo AE. The storage and production of organic matter in tropical forests and their role in the global carbon cycle. Biotropica. 1982;14:161–87.

23. Brown S, Lugo AE. Biomass of tropical forests: a new estimate based on forest volumes. Science. 1984;223:1290–3.

24. Brown S. Estimating biomass and biomass change of tropical forests: a primer. FAO Forest Paper 134, 1997.

25. MAE. Perfil do distrito de Chibuto, província de Gaza. Maputo: Ministério da Administração Estatal; 2005a. p. 44.

26. MAE. Perfil do distrito de Funhalouro, província de Inhambane. Maputo: Ministério da Administração Estatal; 2005b. p. 44.

27. MAE. Perfil do distrito de Mabote, província de Inhambane. Maputo: Ministério da Administração Estatal; 2005c. p. 43.

28. MAE. Perfil do distrito de Mandhlakaze, província de Gaza. Maputo: Ministério da Administração Estatal; 2005d. p. 45.

29. MAE. Perfil do distrito de Panda, província de Inhambane. Maputo: Ministério da Administração Estatal; 2005e. p. 44.

30. Jalkanen A, Mäkipää R, Stahl G, Lehtonen A, Petersson H. Estimation of the biomass stock of trees in Sweden: comparison of biomass equations and age-dependent biomass expansion factors. Ann For Sci. 2005;62:845–51.

31. IPCC. Intergovernmental Panel on Climate Change. Good Practice Guidance for Land Use, Land-Use Change and Forestry 2003. [http://www.ipcc.ch]

32. Kramer PJ, Kozlowski TT. Physiology of Woody Plants. 2nd ed. San Diego: Academic; 1979.

33. Perry TO. The ecology of tree roots and the practical significance thereof. J Arboricult. 1982;8(8):197–211.

34. Soethe N, Lehmann J, Engels C. Root tapering between branching points should be included in fractal root system analysis. Ecol Model. 2007;207:363–6.

35. Kalliokoski T, Nygren P, Sievänen R. Coarse root architecture of three boreal tree species growing in mixed stands. Silva Fennica. 2008;42(2):189–210.

36. Kalliokoski T. Root system traits of Norway spruce, Scots pine, and silver birch in mixed boreal forests: an analysis of root architecture, morphology, and anatomy. In: PhD thesis. Finland: Department of Forest Sciences, University of Helsinki; 2011. p. 67.

37. Ruiz-Peinado R, del Rio M, Montero G. New models for estimating the carbon sink of Spanish softwood species. Forest Sys. 2011;20(1):176–88.

38. Paul KI, Roxburgh SH, England JR, Brooksbank K, Larmour JS, Ritson P, et al. Root biomass of carbon plantings in agricultural landscapes of southern Australia: development and testing of allometrics. For Ecol Manag. 2014;318:216–27.

39. Green C, Tobin B, O'Shea M, Farrel EP, Byrne KA. Above- and belowground biomass measurements in an unthinned stand of Sitka spruce (*Picea sitchensis* (Bong) Carr.). Eur J Forest Res. 2007;126:179–88.

40. Ryan CM, Williams M, Grace J. Above- and belowground carbon stocks in a Miombo woodland landscape in Mozambique. Biotropica. 2010;11(11):1–10.

41. Bolte A, Rahmann T, Kuhr M, Pogoda P, Murach D, Gadow K. Relationships between tree dimension and coarse root biomass in mixed stands of European beech (*Fagus sylvatica* L.) and Norway spruce (*Picea abies* [L.] Karst.). Plant Soil. 2004;264:1–11.

42. Mugasha WA, Eid T, Bollandsås OM, Malimbwi RE, Chamshama SAO, Zahabu E, et al. Allometric models for prediction of above- and belowground

biomass of trees in the miombo woodlands of Tanzania. For Ecol Manag. 2013;310:87–101.

43. Kuyah S, Dietz J, Muthuri C, Jamnadass R, Mwangi P, Coe R, et al. Allometric equations for estimating biomass in agricultural landscapes: II. Belowground biomass. Agr Ecosyst Environ. 2012;158:225–34.

44. Niiyama K, Kajimoto T, Matsuura Y, Yamashita T, Matsuo N, Yashiro Y, et al. Estimation of root biomass based on excavation of individual root systems in a primary dipterocarp forest in Pasoh Forest Reserve, Peninsular Malaysia. J Trop Ecol. 2010;26:271–84.

45. Wang J, Zhang C, Xia F, Zhao X, Wu L, Gadow K. Biomass structure and allometry of *Abies nephrolepis* (Maxim) in Northeast China. Silva Fennica. 2011;45(2):211–26.

46. Jenkins JC, Chojnacky DC, Heath LS, Birdsey RA. National-scale biomass estimators for United States tree species. For Sci. 2003;49(1):12–35.

47. Zhou X, Hemstrom MA. Estimating aboveground tree biomass on forest land in the Pacific Northwest: a comparison of approaches. Res. Rap. PNW-RP-584. Portland, OR: U. S Department of Agriculture, Forest Service, Pacific Northwest Research Station; 2009.

48. Mantilla J, Timane R. Orientação para maneio de mecrusse. Maputo: SymfoDesign; 2005.

49. DINAGECA. Mapa digital de uso e cobertura de terra. Maputo: Cenacarta; 1997.

50. FAO. FAO Map of World Soil Resources. Rome: Food and Agriculture Organisation of the United Nations; 2003.

51. Parresol BR. Additivity of nonlinear biomass equations. Can J For Res. 2001;31(1):865–78.

52. Husch B, Beers TW, Kershaw Jr JA. Forest mensuration. 4th ed. Hoboken, New Jersey: John Wiley & Sons, Inc; 2003. p. 443.

53. Brasil MAM, Veiga RAA, Timoni JL. Erros na determinação da densidade básica da madeira. CERNE. 1994;1(1):55–7.

54. de Gier IA. Forest mensuration (fundamentals). The Netherlands: International Institute for Aerospace Survey and Earth Sciences (ITC); 1992. p. 67.

55. Bunster J. Commercial timbers of Mozambique. Mozambique: Technological catalogue. Traforest Lda, Maputo; 2006. p. 63.

56. Seifert T, Seifert S. Modelling and Simulation of Tree Biomass. In: Seifert T, editor. Bioenergy from Wood: Sustainable Production in the Tropics, vol. 26. Dordrecht: Springer, Managing Forest Ecosystems; 2014. p. 42–65.

57. Machado SA, Figueiredo Filho A. Dendrometria. Paraná: Unicentro; 2005.

58. Freese F. Elementary forest sampling. Washington DC: United States Department of Agriculture; 1962.

59. Freese F. Statistics for land managers. Edinburgh: Paeony Press; 1984.

60. de Vries PG. Sampling theory for forest inventory. New York: Springer; 1986.

61. Sanquetta CR, Watzlawick LF, Cortê APD, Fernandes LAV. Inventários florestais: planejamento e execução. Curitiba: Multi-Graphic Gráfica e Editora; 2006.

62. Tobin B, Nieuwenhuis M. Biomass expansion factors for Sitka spruce (*Picea sitchensis* (Bong.) Carr.) in Ireland. Eur J Forest Res. 2007;126:189–96.

63. Székelly GJ, Rizzo ML. Brownian distance covariance. Ann Appl Stat. 2009;3(2):1236–65.

64. Székelly GJ, Rizzo ML, Bakirov NK. Measuring and testing independence by correlation of distances. Ann Stat. 2007;35(6):2769–94.

65. Rizzo ML, Székelly GJ. Energy. Vienna, Austria: R package version 1.6.2. R Foundation for Statistical Computing; 2015.

66. R Core Team. A language and environment for statistical computing. Vienna, Austria: R Foundation for Statistical Computing; 2015.

67. Chave J, Condit R, Aguilar S, Hernandez A, Lao S, Perez R. Error propagation and scaling for tropical forest biomass estimates. Philos Trans R Soc Lond B. 2004;359:409–20.

## 4

Carbon accretion in unthinned and thinned young-growth forest stands of the Alaskan perhumid coastal temperate rainforest

David V. D'Amore[1]*⊙, Kiva L. Oken[2], Paul A. Herendeen[3], E. Ashley Steel[4] and Paul E. Hennon[1]

Abstract

Background: Accounting for carbon gains and losses in young-growth forests is a key part of carbon assessments. A common silvicultural practice in young forests is thinning to increase the growth rate of residual trees. However, the effect of thinning on total stand carbon stock in these stands is uncertain. In this study we used data from 284 long-term growth and yield plots to quantify the carbon stock in unthinned and thinned young growth conifer stands in the Alaskan coastal temperate rainforest. We estimated carbon stocks and carbon accretion rates for three thinning treatments (basal area removal of 47, 60, and 73 %) and a no-thin treatment across a range of productivity classes and ages. We also accounted for the carbon content in dead trees to quantify the influence of both thinning and natural mortality in unthinned stands.

Results: The total tree carbon stock in naturally-regenerating unthinned young-growth forests estimated as the asymptote of the accretion curve was 484 (±26) Mg C ha^{-1} for live and dead trees and 398 (±20) Mg C ha^{-1} for live trees only. The total tree carbon stock was reduced by 16, 26, and 39 % at stand age 40 y across the increasing range of basal area removal. Modeled linear carbon accretion rates of stands 40 years after treatment were not markedly different with increasing intensity of basal area removal from reference stand values of 4.45 Mg C ha^{-1} year^{-1} to treatment stand values of 5.01, 4.83, and 4.68 Mg C ha^{-1} year^{-1} respectively. However, the carbon stock reduction in thinned stands compared to the stock of carbon in the unthinned plots was maintained over the entire 100 year period of observation.

Conclusions: Thinning treatments in regenerating forest stands reduce forest carbon stocks, while carbon accretion rates recovered and were similar to unthinned stands. However, that the reduction of carbon stocks in thinned stands persisted for a century indicate that the unthinned treatment option is the optimal choice for short-term carbon sequestration. Other ecologically beneficial results of thinning may override the loss of carbon due to treatment. Our model estimates can be used to calculate regional carbon losses, alleviating uncertainty in calculating the carbon cost of the treatments.

Keywords: Carbon sequestration, Young-growth, Stand management, Ecosystem productivity, Natural resource management

Background

Forests play a key role in the global carbon cycle, containing an estimated 861 Pg C and providing a sink of 1.1 Pg C year^{-1} [1]. Forests are critical sinks for atmospheric greenhouse gases [2], and carbon fluxes occur across many carbon pools in forests, including live biomass, soils, and woody debris [3, 4]. The terrestrial carbon stock is generally stable over time scales of decades and can only slowly alter the total terrestrial carbon balance through gains or losses [4]. Disturbances that alter forest stands can provide dramatic departures from this characteristic pattern. An example is removal of carbon due to clearcut harvesting of forests, leading

---

*Correspondence: ddamore@fs.fed.us
[1] U.S. Department of Agriculture, Forest Service, Pacific Northwest Research Station, Juneau Forestry Sciences Laboratory, 11175 Auke Lake Way, Juneau, AK 99801, USA
Full list of author information is available at the end of the article

to a large loss of terrestrial carbon. The increase in biomass, or carbon accretion, as stands regenerate and grow after harvest is unknown in many forests. Thinning is a common silvicultural practice for increasing growth of individual trees and maintaining or increasing wildlife habitat. However, the influence of thinning on the carbon balance in young forests is uncertain in southeast Alaska. Carbon fluxes need to be evaluated across a range of management options to understand and estimate the short and long-term impacts of silvicultural treatments on carbon pools.

Estimates of carbon flux in young-growth stands are needed to address land management planning goals and regional, national [5] and international carbon accounting protocols [6]. Mandates to understand the potential for forests to mitigate increasing concentrations of atmospheric CO_2 require accurate accounting of forest carbon fluxes. The USDA Forest Service, for example, has prioritized understanding carbon dynamics in forests

as part of an overall strategy to protect the long-term health of forests [7]. Necessary information about carbon cycling is particularly lacking in the perhumid coastal temperate rainforests (PCTR) of the northeast pacific coastal margin [8] (Fig. 1).

Widespread commercial forest harvest has occurred across southeast Alaska for over 50 years. However, there is no estimate of the potential carbon sequestration across the ~452,000 ha [9] of young-growth forests in the region. Natural regeneration in PCTR forests is generally vigorous and leads to rapid and nearly complete occupation of space by conifer seedlings and saplings [10]. Densely-stocked stands can produce wood products similar to thinned stands [11], but the loss of light and density of overstory trees degrades the wildlife habitat [12, 13]. A common management intervention to alleviate the high stand density is thinning [14]. Felling of a portion of the stand basal area across a specific or variable [15] spacing can be applied to achieve maximum

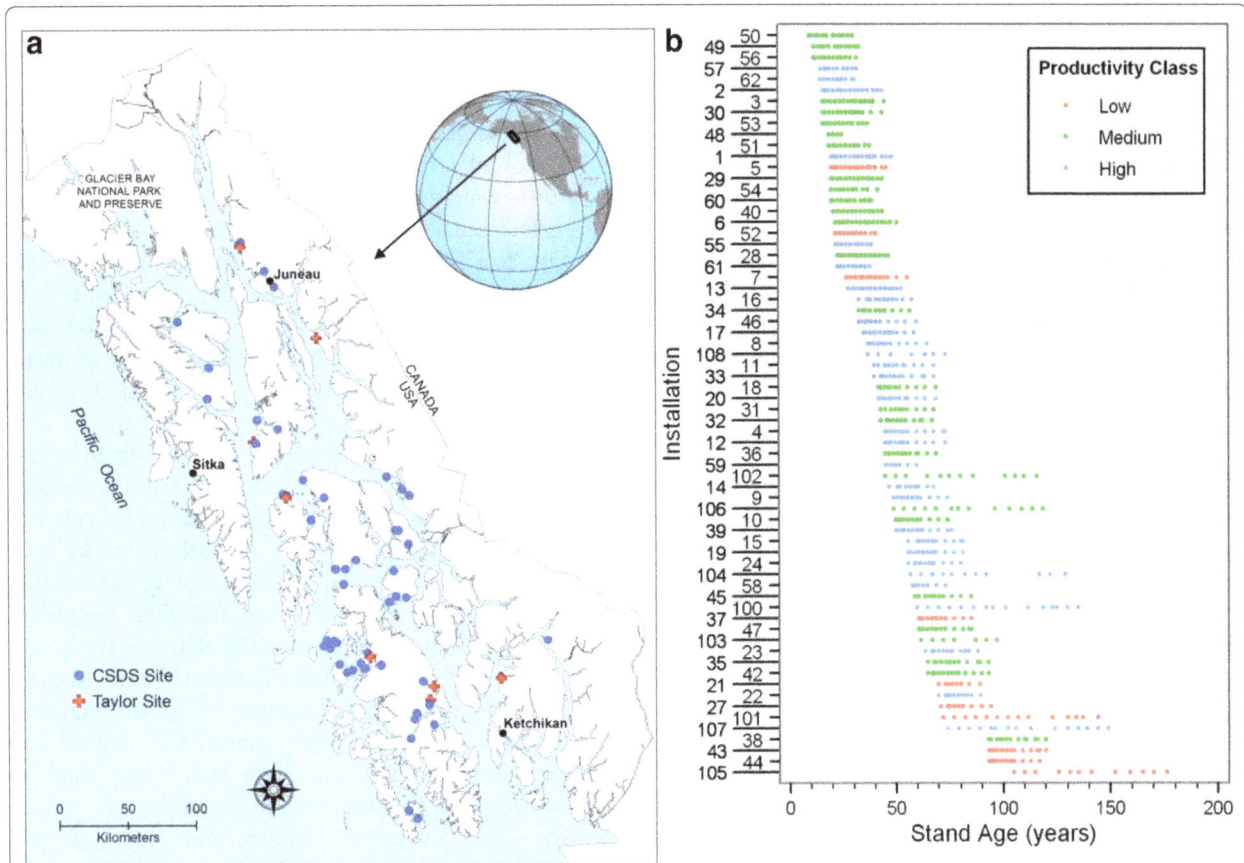

Fig. 1 Locations of the 68 Farr and 12 Taylor installations in southeast Alaska (**a**). Each CSDS ("Farr") installations consists of four plots: a control plot, a low-intensity thinned plot, a medium-intensity thinning plot, and a heavily-thinned plot. The Taylor installations consist of unthinned plots only and are generally in older stands. For a full description of the plots and thinning treatments see [19]. Data from the CSDS study arranged by age of stand at time of plot establishment (**b**). *Numbers* on Y axis refer to installation number with Farr plots <100 and Taylor plots ≥100. Productivity classes are the tertiles of the observed range for these sites as reported in [19]. Each *symbol* in an installation represents a measurement

individual tree growth. However, thinning also alters the carbon accretion trajectory of the stand [4]. When left on site, the carbon content of thinned trees, and any trees that die naturally, can be accounted for by estimating decomposition rates. The impact of stand thinning and subsequent loss of biomass via decomposition are key components in calculationg a carbon sequestration rate for use in land management planning.

Quantifying the effects of young-growth forest management on carbon storage is challenging. Allometric equations linked to direct tree measurements can be used to estimate aboveground biomass production [16–18] across stand age, and this can be converted to carbon accretion. Estimation of the long-term differences between forests with varying management treatments requires remeasurement of the same plots over decades. Long-term plots provide an excellent source of information on biomass accretion over time where plots have been maintained and re-measured.

Experimental plots maintained by the USDA Forest Service Pacific Northwest Research Station [19] offer an opportunity to estimate carbon change over time with varying levels of thinning. This dataset includes 284 plots across 68 installation sites, remeasured over several decades and spanning stand ages up to 161 years. The temporal and geographic breadth of these experimental plots provides an excellent foundation for investigating carbon standing stocks and carbon accretion rates in young-growth forests of the PCTR. In addition, the plot system allows analysis of the effects of forest thinning on carbon storage through the combination of allometric equations and repeated tree measurements over decades. We designed this study to address the critical need for an improved understanding of carbon storage in young-growth forest of the PCTR and to quantify the effects of thinning on carbon gain or loss. We hypothesized that while thinning may increase carbon accretion in individual trees, across whole stands thinning will have a neutral to negative impact on the sequestration of carbon, depending on the intensity of thinning.

Methods overview

We utilized data from two long-term silvicultural datasets young-growth forests of southeast Alaska to estimate total tree carbon stock and accretion rate. One set of plots was started in the 1920's and were not thinned ("Taylor plots", 12 of 284). The other plot system included unthinned controls and thinning treatments applied at three intensities in a randomized block design ("Farr plots", 272 of 284). Plot measurements included both live and dead trees, so estimates for both pools were calculated to account for the loss of dead tree carbon

decomposing over time in both unthinned and treated forest stands. A new allometric model for small diameter trees was developed to fill a needed information gap in determination of carbon in small trees.

Results

Live and dead tree carbon pools in naturally-regenerating young-growth stands

Live-tree carbon increased in unthinned young-growth stands across the stand age gradient and reached an asymptote of 398 ($\pm$20) Mg C ha^{-1} based on a best fit, non-linear mixed effect model (NLME) (Fig. 2a). The estimated asymptotic maximum carbon stock in the stands increased to 484 ($\pm$26) Mg C ha^{-1} with the inclusion of dead-tree carbon (Table 1) Dead trees in unthinned plots typically represent suppression mortality as tree density decreases through time. However, these mean carbon stock estimates for the measured plots have a great deal of uncertainty. A prediction interval was derived by considering observed variability within- and among-plots, in addition to the parameter uncertainty around the asymptote described above. The 90 % prediction intervals for the asymptotic carbon stock ranged from 145 to 653 Mg C ha^{-1} for the live-tree carbon model and 161–808 Mg C ha^{-1} for the model including both live- and dead-tree carbon.

We plotted carbon accretion as the change in the carbon pool over time in plots with only live tree carbon and calculated a peak at age 34.7 years ($\pm$0.5, bootstrap SE; Fig. 2b). The carbon accretion peaked at 39.3 years ($\pm$0.5, bootstrap SE; (Fig. 2c) for the model with both live and dead tree carbon. These carbon accretion rates varied dramatically across the chronosequence of measurements in the sampled stands (Fig. 2b, c). The high variability makes it difficult to estimate quantities with any reasonable level of precision directly from accretion data. While carbon accretion was more variable than carbon stock estimates, carbon accretion can also be estimated as the derivative of carbon stocks over time. The general shape of the data cloud suggests that accretion rates peak at 39 years and then decreases, tapering off at about 100 years. The shape of the accretion curves (Fig. 2d, e) derived directly from the fitted model for the total carbon stock (Fig. 2a) indicates that accretion peaks in young stands between 35 and 40 years and then tapers off as the stands age. The estimated weighted average carbon accretion rate based on the fitted model to total carbon [45] was 3.53 ($\pm$0.17) Mg C ha^{-1} year^{-1} for the live-tree carbon model and 3.81 ($\pm$0.20) Mg C ha^{-1} year^{-1} for the model that included live- and dead-tree carbon over the 150 years age span of measured trees.

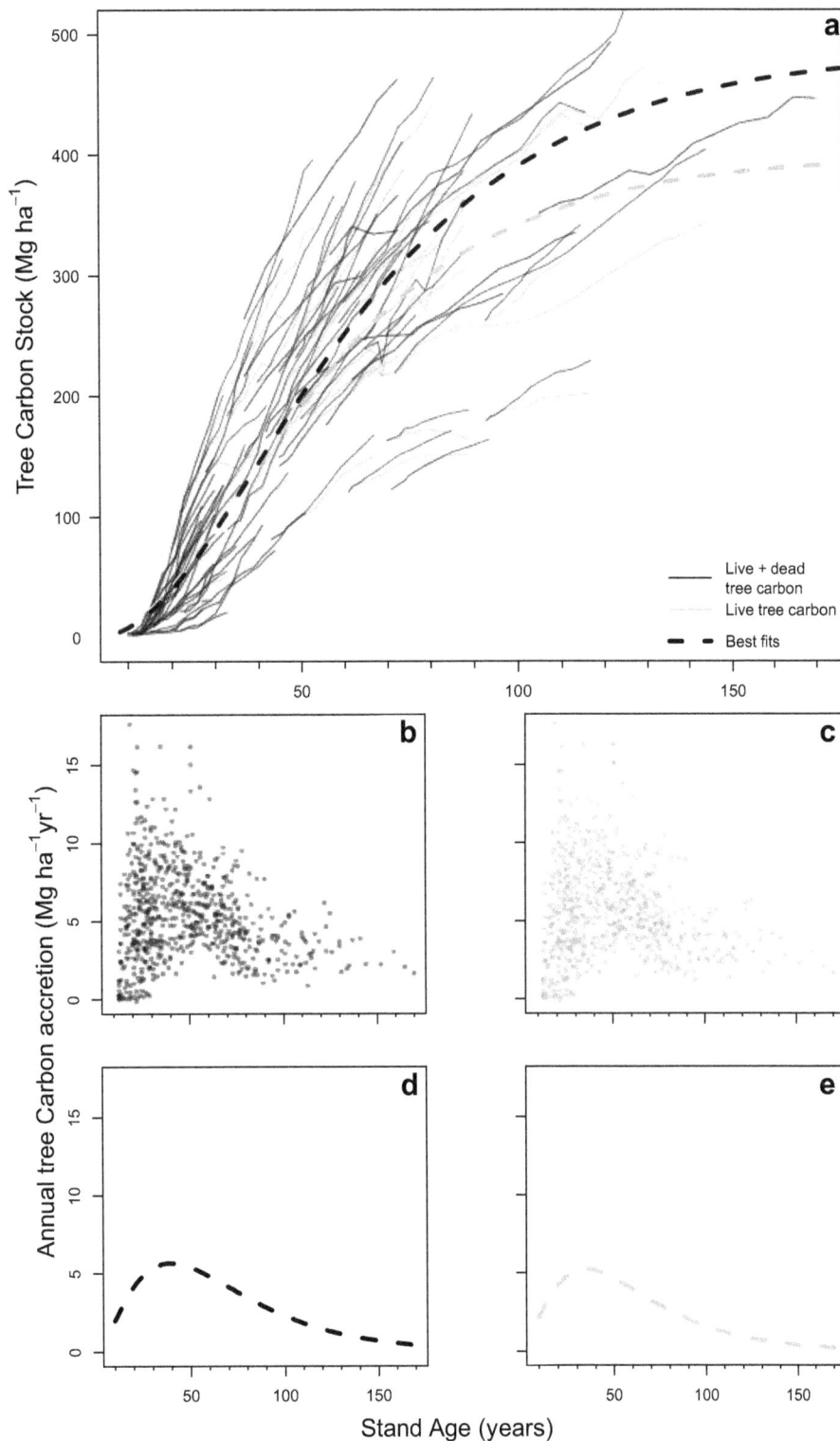

Fig. 2 Carbon stock and carbon accretion in naturally-regenerating plots (Farr study control plots and Taylor plots). **a** Measured and modeled carbon stock through time. *Solid lines* describe carbon accretion measurements within individual plots across a range of ages. *Dashed lines* are NMLE best fit models for all tree carbon (live and dead trees) and for live tree carbon. Note that the plots with low carbon stock values are all located in the same sites which occurred on the lowest productivity areas that were sampled. Observed carbon accretion rates across stand age for **b** for all carbon (live and dead trees) and for **c** for live trees only. Implied carbon accretion (derivative of the NMLE model) as a function of stand age for **d** for all carbon (live and dead trees) and for **e** for live trees only

Table 1 Parameter estimates (±SE) for best fit of carbon accretion in unthinned control stands for live tree only and for live + dead tree components of carbon using NMLE model

Model components	B_0	B_1	B_2
Control (live + dead trees)	484.58 (25.90)	0.026 (0.0008)	0.636 (0.011)
Control (live trees only)	398.24 (20.03)	0.029 (0.0009)	0.634 (0.0119)

B_0 is an estimate of the asymptote based on Eq. 3 where DBH is replaced by age and Mg C is calculated rather than height. B_1 and B_2 describe growth rate and the inflection point of the modeled relationship, respectively

Influence of thinning on carbon accretion in young-growth stands

There was a systematic decrease in the total stand carbon correlated with increasing intensity of thinning (Fig. 3). The portion of the carbon stock data in untreated young-growth stands that is nearly linear (20–100 years) was used as a basis for comparison between treated stands. The estimated average carbon pool in the unthinned control plots (Farr plots only, see methods) was greater than the estimated average carbon pool in any of the three thinning treatments at 40 years (Table 2); estimated

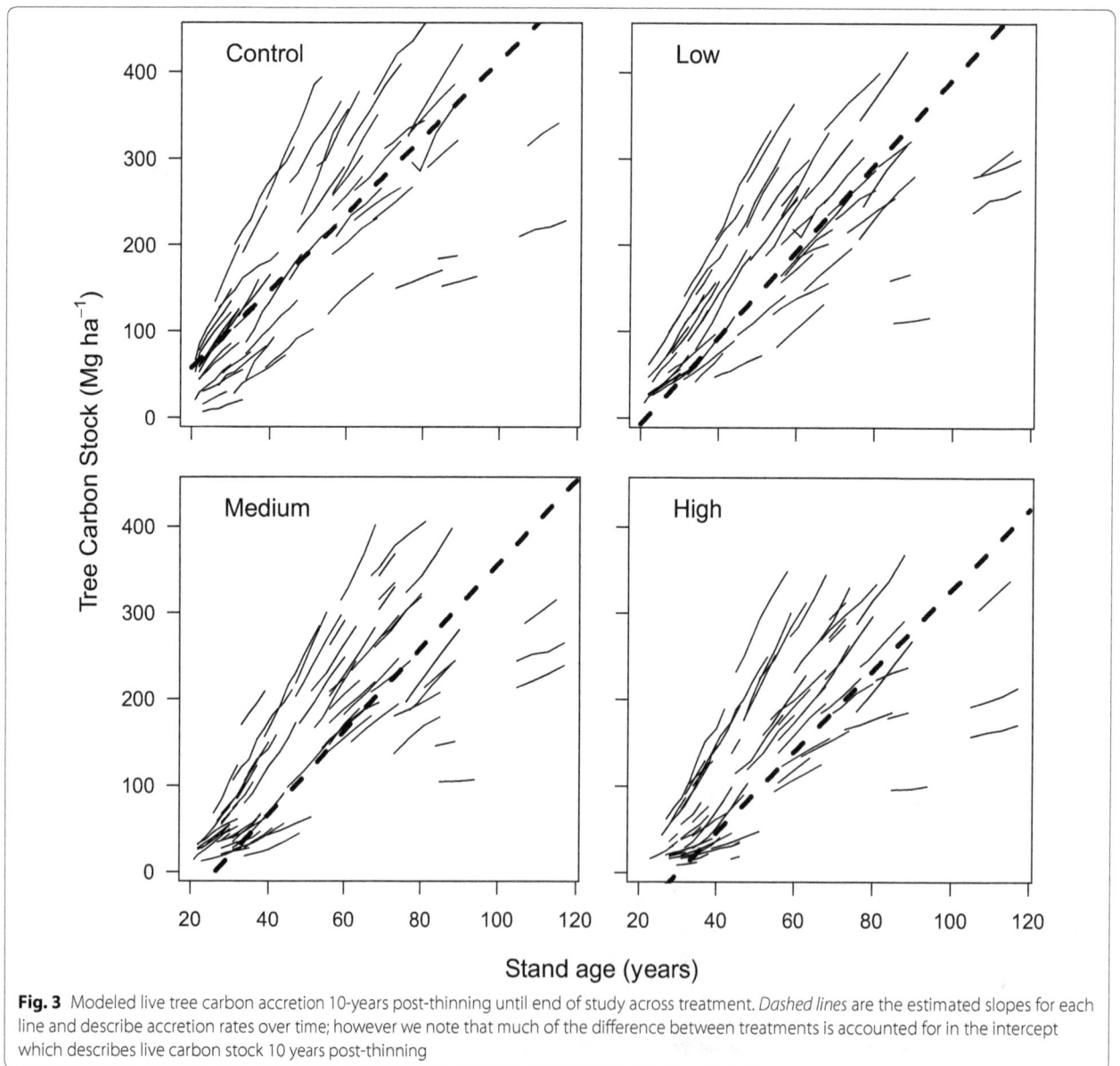

Fig. 3 Modeled live tree carbon accretion 10-years post-thinning until end of study across treatment. *Dashed lines* are the estimated slopes for each line and describe accretion rates over time; however we note that much of the difference between treatments is accounted for in the intercept which describes live carbon stock 10 years post-thinning

Table 2 Estimates of accretion rate and mean carbon density 40 y after thinning based on live tree carbon only

Treatment	Carbon accretion rate Mg C ha^{-1} y^{-1}	Among plot standard error of accretion rate	Residual standard error	Carbon density at 40 y Mg C ha^{-1}
Control	4.45 (0.31)	2.16	1.09	146.9 (7.4)
Low	5.01 (0.25)	1.87	1.28	93.4 (7.0)
Medium	4.83 (0.27)	2.20	1.04	66.3 (6.8)
Heavy	4.68 (0.31)	2.49	1.10	46.0 (6.2)

Low, medium, and heavy treatments refer to thinning intensity. Values in parentheses are parameter standard errors

average carbon pools at a given age consistently decreased with thinning intensity of treatments from low to high (Fig. 3; Table 2). The slope of the linear model fit to the data describes the stand-scale accretion rate. This accretion rate systematically decreased with thinning when decomposition of cut trees and any trees that died naturally is included and the total carbon stock was reduced by 16, 26, and 39 % across the low to high intensity thinning treatments at 40 years (Tables 2, 3). However, no major pattern between accretion rate and thinning intensity was noted with only live trees (Table 2; Fig. 4). We note that several plots, both control and treatment, displayed particularly low accretion rates and these plots were generally all located on one set of sites (Figs. 3, 4).

The residual model error in the linear models fit was similar across treatments for live trees using all plots (Tables 2, 3). This was also the case in models for live trees, cut trees, and natural mortality using plots for which cut tree data were available (221 of 284). This residual model error describes variability in carbon stocks within a plot over time after accounting for the effects of stand age and treatment. This standard error among plots for the accretion rate increased somewhat predictably across the three treatments suggesting that at more intensive levels of management it might be more difficult to predict accretion rates for an individual plot. Control plots were intermediate in their across-plot variability. We also note that residuals for both the live-tree and live-tree plus cut and natural dead tree models showed no trends over stand age, indicating that the linear model accurately described the underlying effect

of stand age on carbon stocks, but residuals did show a somewhat increasing trend over chronological time indicating a potential increase in variability of carbon stocks in recent years.

Simulation of stand carbon dynamics immediately after thinning

We simulated a hypothetical carbon accretion scenario under different thinning intensities, all of which occur when stands reach 20 years of age, based on our fitted statistical models (Fig. 5). The simulated carbon stock at the plot scale accumulates at a rapidly accelerating pace in all plots until the stands are subjected to a simulated thinning at age 20. This thinning leads to the rapid drop in the carbon stock of thinned stands, as we only accounted for the carbon in the remaining live trees in the stands for the simulation. Stands in all four treatments begin accumulating carbon again after the thinning treatment is applied according to the linear models. We expect that increased growth rates of individual trees lead to a more rapid rate of carbon accretion after thinning on a per-tree basis. Note, however, that while individual trees may accrete carbon at a more rapid rate after thinning due to increased growth rates, there are many fewer trees accreting carbon in a thinned plot. Overall, at the plot scale, there is an initial loss of carbon in the thinned stands and a similar accretion rate to the control stands (Table 2). The simulated carbon stock in all thinning treatments remains lower than that of unthinned plots up to 100 years (Fig. 5).

Table 3 Estimates of accretion rate and mean carbon density 40 y after thinning based on live and dead tree carbon

Treatment	Carbon accretion rate (Mg C ha^{-1} y^{-1})	Among plot standard error of accretion rate	Residual standard error	Carbon density at 40 years (Mg C ha^{-1})
Control	5.27 (0.32)	2.26	0.93	144.3 (8.1)
Low	5.16 (0.28)	1.97	1.09	120.7 (6.9)
Medium	5.00 (0.31)	2.27	0.94	107.0 (6.8)
Heavy	4.78 (0.32)	2.39	1.15	88.1 (5.9)

These estimates are based on the 164 of 215 study plots for which cut tree data were recorded. Low, medium, and heavy treatments refer to thinning intensity. Values in parentheses are parameter standard errors

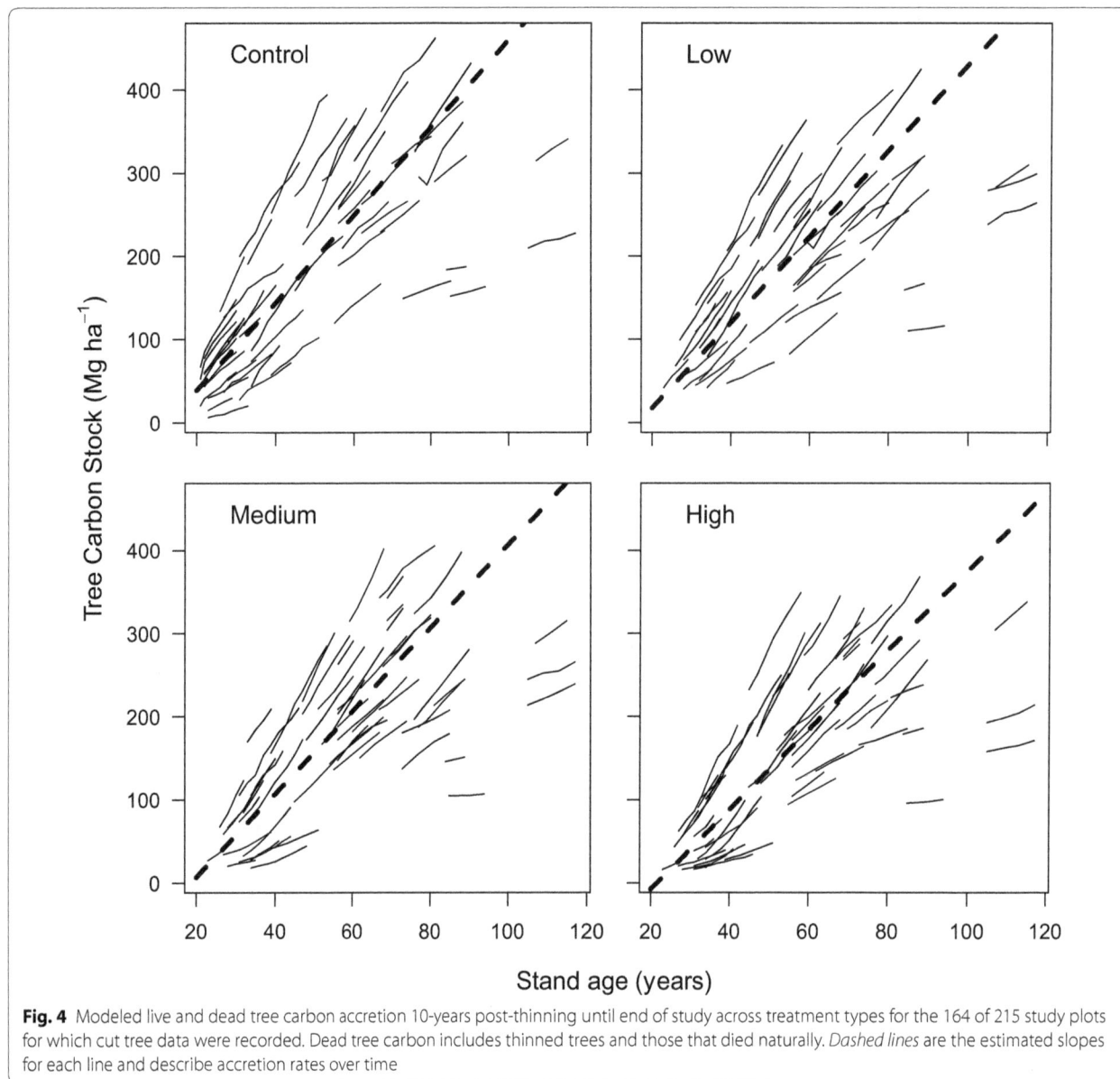

Fig. 4 Modeled live and dead tree carbon accretion 10-years post-thinning until end of study across treatment types for the 164 of 215 study plots for which cut tree data were recorded. Dead tree carbon includes thinned trees and those that died naturally. *Dashed lines* are the estimated slopes for each line and describe accretion rates over time

Discussion

Carbon balance in unthinned forest stands

The rate and location of terrestrial carbon sinks is critical to understanding the global carbon balance. Young-growth forests sequester carbon in biomass, but at widely varying rates and over different timeframes. The calculation of total carbon stock and estimated accretion rates across the age gradient of the naturally regenerating young-growth forests of southeast Alaska fills a critical information gap for this region. The loss of live carbon after thinning in naturally-regenerating stands must be considered in calculating carbon sequestration estimates for young-growth forests. Thinning treatments are

applied to achieve many ecosystem services in addition to carbon sequestration goals; therefore, our quantitative estimates of the loss of carbon after thinning enable evaluation of the carbon cost of a range of management actions for young-growth stand improvement.

Model calibration is essential for obtaining accurate carbon balance estimates across large regions [20]. Forest carbon models need to consider the entire range of stand types and ages to accurately portray the balance of carbon stock across the landscape [21]. Mature forest stands (>200 years) can accumulate carbon at an estimated $2.4\,\mathrm{Mg\,C\,ha^{-1}\,year^{-1}}$ [22]. The carbon stock in young-growth stands is particularly critical in these estimates as these stands are generally the

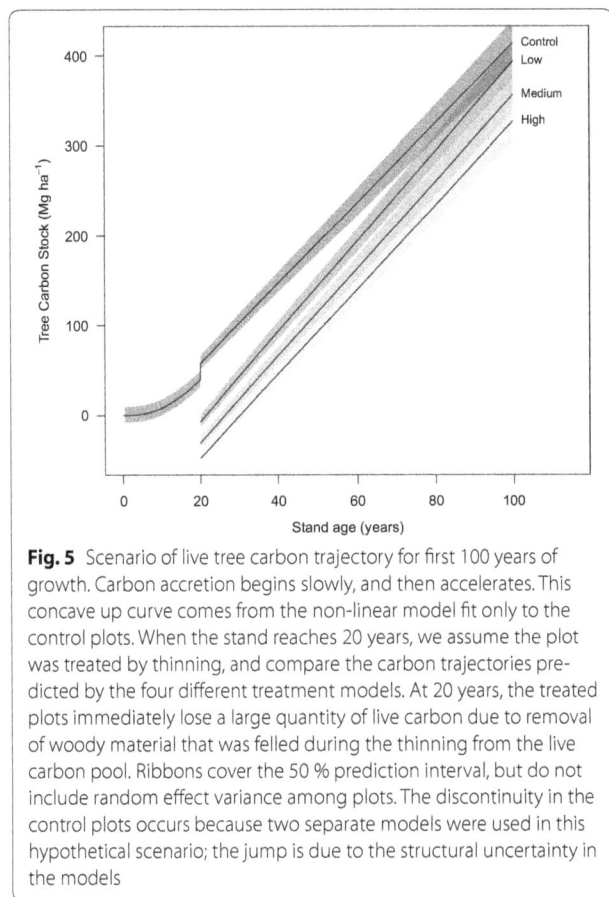

Fig. 5 Scenario of live tree carbon trajectory for first 100 years of growth. Carbon accretion begins slowly, and then accelerates. This concave up curve comes from the non-linear model fit only to the control plots. When the stand reaches 20 years, we assume the plot was treated by thinning, and compare the carbon trajectories predicted by the four different treatment models. At 20 years, the treated plots immediately lose a large quantity of live carbon due to removal of woody material that was felled during the thinning from the live carbon pool. Ribbons cover the 50 % prediction interval, but do not include random effect variance among plots. The discontinuity in the control plots occurs because two separate models were used in this hypothetical scenario; the jump is due to the structural uncertainty in the models

most active zones of carbon change on the landscape due to rapid biomass accumulation and carbon storage in trees [23]. The estimated mean accretion rate of 3.53 Mg C ha^{-1} year^{-1} over 150 year in our study area confirms the strong net gain in carbon in young-growth stands in the Alaskan PCTR. This rate is higher than the 40 years mean of 2.71 Mg C ha^{-1} year^{-1} estimated in young-growth stands in the PCTR of British Columbia [24]. Frustratingly, the uncertainty in determining the response of an individual stand is high, which limits the usefulness of model predictions for site specific estimates of carbon stock, often needed to evaluate specific management scenarios. Our models are most appropriately applied across an entire population of stands for regional and national carbon assessments. Site-specific descriptors (e.g., site productivity) that might help stratify the data and provide more accurate predictions of carbon pools will need to be applied in order to help refine our predictions of carbon accretion rates in particular locations.

Carbon balance in thinned young-growth stands

Maximizing the carbon stored in forests is a key goal of climate change mitigation programs [25]. The majority of the young-growth forest in the Alaskan PCTR result

from harvest that occurred from 1960 to 1990 [9]. Thinning young-growth stands in the PCTR is a common management strategy to improve stand structure and wood production [14] and to improve wildlife forage production [13, 26]. Renewable energy recommendations for the Alaskan PCTR highlight the potential for wood energy projects using this stock of young-growth forest [27]. However, the usual management scenario for these young-growth stands is thinning at 15–20 years [13] and nearly half of the 25–50 year old stands have been precommercially thinned [9, 14]. Therefore, recognizing the tradeoff between thinning for stand improvement, biomass energy, and carbon sequestration in young-growth forest stands is important for making land management decisions. A key finding in our study is that thinning persistently reduces the carbon stock in young growth stands. The rate of carbon accretion in thinned stands is higher than control plots after the initial carbon loss; but, the gap created by the initial carbon loss is maintained and the total stock of carbon in thinned stands does not equal the stock of carbon in the control plots over the entire 100 year period of observation. This is consistent with the observation that the reduction in total stand carbon stock may not change the net ecosystem exchange between pre- and post-thinning [28]. The maintenance of tree growth would explain the similarity in the trajectory of carbon accretion among the treatments after the initial period of disturbance.

Reduced carbon stocks due to thinning have been recognized in other forests [4, 29, 30], but is not often included in forest carbon accounting or management actions due to the lack of adequate stand response data. Our quantification of the reduction in carbon stock across a range of thinning treatments allows estimates of the effects of thinning on regional carbon stocks. The systematic variation in the carbon stock related to thinning intensity may offer a mitigation measure for achieving benefits for wildlife, wood quality, or understory abundance and diversity in managed stands. The enhanced growth of understory plants after thinning represents a tradeoff of energy from trees to forest floor and a reduction in overstory carbon compared to unthinned stands. Benefits of thinning young growth need to be balanced with the desire to maximize carbon storage in forests. For example, the less intensive thinning treatments maintain more carbon, but still provide a benefit for other desired conditions in a stand. As demonstrated by our comparison, the unthinned option provides the greatest carbon accretion of all of the thinning prescription options.

Limitations of analysis and information gaps

The carbon values provided in our study will be critical for estimating the carbon stock in the pool of young-growth forest in southeast Alaska, but, there is still considerable uncertainty in the range of carbon accretion values among the stands in our analysis. Therefore, site specific projects will need an improved model that is able to better reflect local conditions to carbon flux values. Factors that influence the variability in forest productivity among the sites or the response to thinning were included as random effects, but not specifically as predictive variables. Possible interactions with temperature [28], geology [31], soil saturation [32], nutrients [33] or other site-specific factors may play a role in site productivity. This uncertainty might be addressed by obtaining further information on the site factors that may influence the productivity of the plots such as soil, hydrology, or climate variables.

Potential alternate trajectories in the carbon accretion of thinned stands may arise that lead to different conclusions related to unthinned stands. We applied the same allometric equations to both unthinned and thinned stands in our analysis. It is possible that tree growth forms differ by thinning treatment and so biomass allocation would change in thinned stands. We are not aware of any existing allometric models for thinned stands of the PCTR. Therefore, we rely on the literature from other regions to support our conclusions and highlight that thinning has been found to primarily impact the biomass of the bole [34] and crown [35] of the thinned trees. Thinned stands can shift biomass accumulation from branch to leaf, but measured changes in bole biomass have been demonstrated to be small [36] unless very heavily thinned [37]. These observations provide some confidence that the total biomass calculated by our approach will not substantially change, but may be redistributed within the tree after thinning.

The residual trees left after thinning grow at an accelerated rate, but these trees are generally left in a condition where they do not maximize the growing space for many years. Thinning goals such as increased individual tree growth and allocation of energy to the forest floor for plant diversity lead to lower overstory biomass accumulation in thinned plots. While the growth rate for individual trees is greater in these plots, the amount of biomass accumulation that would be required by the individual residual trees to match the loss in biomass of similar unthinned stands would be physiologically difficult to attain. The difference is illustrated in our evaluation of the stands at 40 years in Tables 2 and 3. There could be cases where a light thinning leaves a higher density than other thinning treatments, in which case, the thinned stand may accumulate biomass similar to unthinned stands due to the additional growth of the residual trees. However, this scenario is unlikely to be applied under most operational applications.

Conclusions

Knowledge of the stock and rate of carbon accretion greatly enhances the understanding of carbon dynamics in the coastal forests of Alaska. The loss of carbon due to thinning can be used in the evaluation of management scenarios that address young-growth stand improvement. Regional carbon budgets will also be improved with estimates that include the carbon pool in young-growth stands of the PCTR.

Methods
Source of data

This study used data from the Cooperative Stand Density Study (CSDS; Fig. 1), comprised of two long-term silvilcultural field studies, previously compiled and published [19, 38, 39] and an earlier study implemented by Ray Taylor ("Taylor plots"; Fig. 1). Most data (272 of 284 plots) were from a study of thinning treatments on even-aged young-growth (<100 years) stands begun in 1974 and with remeasurements continuing until 2003 ("Farr plots"). The remaining 12 plots ("Taylor plots") were located in older even-aged stands initiated by windthrow or early timber harvest in the late 19th century. The original intent of the studies was to measure sites that represented commercially harvested forests. Both the harvested landscapes and the plots in this study are weighted towards higher productivity classes. The Taylor plots were first measured in the late 1920s, with remeasurement occurring periodically through 2003. The Farr plots were established to examine growth and yield and how regenerating forest stands were impacted by light (mean 47.7 % BA removal); medium (mean 60.9 % BA removal); and heavy (mean 73.5 % BA removal), thinning at varying stand ages across varying productivity classes (Fig. 1). A complete description of thinning prescriptions is available in [19]. Most of these stands initiated following clear-cut harvest, with a smaller number of the older stands initiated by windthrow. All plots were dominated by western hemlock (*Tsuga heterophylla (Raf.) Sarg*) and Sitka spruce (*Picea sitchensis (Bong.) Carr*), with small amounts of western redcedar (*Thuja plicata*) and red alder (*Alnus rubra*). Stand age at thinning treatment ranged from 10 to 93 years (Fig. 2b). In general, the four treatments (control, light, medium, and heavy thinning) were applied in a randomized block design across 62 installations. Plot age, productivity class, and remeasurement dates are shown in Fig. 1b.

Estimating biomass of live trees

Tree species and diameter at breast height (DBH) were recorded for each tree in the original study and at each remeasurement interval (roughly 2–5 years). A subset of trees (7308 of the 27562) was measured for height during each remeasurement using a clinometer and tape or laser. Tree heights were estimated from diameter and height relationships for the remaining trees (Additional file 1: Appendix A). DBH and height were used to estimate carbon using allometric equations of the form:

$$B = b_0 + b_1 d^2 h \qquad (1)$$

where d is the diameter at breast height (DBH, in meters), h is the height above breast height (m), and B is the dry biomass (kg) for all the aboveground and below-ground components of the tree [17]. The constant b_0 is the biomass of a tree at breast height and b_1 is related to the tree's density. The constants b_0 and b_1 are species-specific. We separately accounted for red alder (*Alnus rubra* Bong.), Shore pine (*Pinus contorta* var. *contorta* Douglas ex. Loudon), western redcedar (*Thuja plicata* Donn ex D. Don), Sitka spruce (*Picea sitchensis*) and western hemlock (*Tsuga heterophylla*). Calculations for any other species were done with the western hemlock equations from [17]. Note that Sitka spruce and western hemlock account for more than 98 % of all tree measurements.

The equations developed by Standish et al., [19] had a minimum tree diameter of 3.1 to 5.3 cm, and due to the large intercept terms, did not accurately estimate the biomass of small trees. The presence of many small diameter trees in our database required the development of a new equations We developed allometirc biomass equations for small trees by sampling 60 small diameter Sitka spruce and western hemlock and calculating the total biomass based on whole tree harvest and weighing (Additional file 2: Appendix B). These empirical biomass relationships for small diameter trees were based on Sitka spruce and western hemlock trees (<7.5 cm dbh) sampled in three locations arrayed across the geographic region of the database (Additional file 2: Appendix B). The dbh threshold for using our empirical biomass estimates for small trees versus the constants from Standish et al. [19], suitable for larger trees, was defined by the intersection of our local parameterization curve and the Standish parameterization under the assumed height-diameter relationship (Additional file 2: Appendix B). Because the height-diameter relationship and allometric parameterizations were species-specific, the diameter threshold that determined which biomass equation to apply was also unique to each species.

Estimating biomass of dead trees

Dead trees, both those cut during thinning and left on site and those that died from natural mortality, are often ignored in estimates of forest carbon pools and fluxes. In our analysis all cut trees were considered to be left on site to decompose. Cut trees were recorded in 164 of 215 treatment plots. Most plots missing cut tree data were reported in [19] as lacking pre-thinning data. The

exceptions are the 16 treatment plots of installation 62 ("Staney Creek"), for which no explanation of the missing cut tree data is given. In all cases, analysis that considered the effect of management on dead trees was based on the 164 plots for which cut tree data were available.

We estimated carbon content of dead trees using the following deterministic relationship previously parameterized for the region in a study of the decomposition rate of thinning slash [40]:

$$B = B_0[0.3870 \exp(-0.1429\, t) + 0.6198 \exp(-0.00223\, t)] \qquad (2)$$

where B_0 is the estimated biomass at the time of death in kilograms, and t is time since tree death in years. This equation was used for both trees that were cut at the beginning of the study during initial thinning and left on site as well as for trees that died of natural causes, typically from suppression, at some point during the study's duration. For the latter case, we assumed the tree died and began decomposition at the midpoint between the date on which the last live measurement was taken and the date on which it was marked as dead.

Estimating carbon at the plot level from individual tree biomass

We assumed that carbon made up 48 % of the dry biomass [41] of an individual tree for both live and dead trees and that the root to aboveground biomass ratio was 0.2 [17]. Carbon estimates over all trees within a plot were aggregated into a single estimate of megagrams of carbon per hectare.

Ingrowth

Due to irregular inclusion of ingrowth measurements, our analysis of carbon estimates did not account for biomass additions due to ingrowth of new trees. We evaluated the potential impact of excluding ingrowth in our carbon estimates for plots with available ingrowth measurements. In 95 % of the measurements, the contribution of ingrowth was <5 % of total plot carbon. However, the error from excluding ingrowth likely increases with stand age as these forests begin to reach the understory re-initiation phase [42].

How does carbon accretion change with stand age in naturally-regenerating forests?

To understand basic underlying carbon dynamics of young growth stands in the PCTR, we first evaluated naturally-regenerating plots. By combining data from the Farr control plots and the Taylor plots, none of which were thinned, we had a very long chronosequence of naturally-regenerating plots (Fig. 1b), measured between 1926 and 2000 that were 10 to 170 years of age. We fit an asymptotic nonlinear equation to relate carbon content to stand age [43]:

$$TC = A\left[1 - \exp\left(b_1\ age\right)\right]^{\frac{1}{1-b_2}}, \qquad (3)$$

where TC is total carbon in a stand ($Mg\ ha^{-1}$) and stand age (*age*) is measured in years. We used non-linear mixed effects models to account for correlation among repeated measures within plots, thereby allowing the stand index to implicitly enter the model as a random feature of each plot. The random effect was placed on the asymptotic amount of carbon in the plot, consistent with the idea that the random effect reflects differences in site productivity index. Models were fit using the nlme package in R [44]. The model was first fit using estimates of carbon from live trees only. We then fit the model again to estimate carbon based on both live and dead trees.

We estimated the weighted average rate of carbon accretion as:

$$\frac{Ab_1}{2b_2 + 2}$$

Using this equation [45], we weight the instantaneous rate of accretion, so that the steeper portion the curve, is most influential when accounting for overall carbon. Estimates of parameter uncertainty were derived using parametric bootstrapping.

How does carbon accretion change with thinning?

We examined the impact of the three thinning treatments on carbon accretion using the 272 Farr plots. We did not include data from the Taylor plots in this analysis as there were no equivalent examples of older thinned plots. Carbon dynamics in the first 10 years after thinning were nonlinear due to the rapidly deccelerating pace of decomposition of cut trees. These early data describe a different ecological process than data from >10 years post-thinning and were therefore excluded from our model. We excluded the first 10 years of measurements from control plots in the same blocks to balance the design. Within this age range of approximately 20–100 year-old stands, the carbon stock increased linearly among all four treatments. Therefore, we fit a linear mixed effects model to this data set. A random effect was placed on both the intercept and the slope, which was supported by likelihood ratio test, $P < 0.001$. These slopes describe the estimated average carbon accretion rate for stands within each treatment.

Additional files

Additional file 1: Appendix A. Estimating tree height.

Additional file 2: Appendix B. Estimating biomass for small trees.

Authors' contributions

DVD and PEH designed and implemented the study. KLO and PAH organized the database, prepared files for analysis, and created figures. KLO analyzed the data. EAS contributed to the design of the data analysis and interpretation of the model results. DVD wrote the initial draft of the manuscript with methods provides by KLO and PAH. All contributed to writing and editing drafts and preparing the final manuscript. All authors read and approved the final manuscript.

Author details

[1] U.S. Department of Agriculture, Forest Service, Pacific Northwest Research Station, Juneau Forestry Sciences Laboratory, 11175 Auke Lake Way, Juneau, AK 99801, USA. [2] Quantitative Ecology and Resource Management, University of Washington, Box 355020, Seattle, WA 98195, USA. [3] Graduate Degree Program in Ecology, Colorado State University, Fort Collins, CO 80523, USA. [4] U.S. Department of Agriculture, Forest Service, Pacific Northwest Research Station, 400 N 34th Street, Suite 201, Seattle, WA 98103, USA.

Acknowledgements

We would like to acknowledge the work of Ray Taylor, Bill Farr, and Mike McClellan for providing stewardship of the CSDS plots and data over 80 years. We would like to thank Dave Bassett and other workers for their dedication to re-measurement of the plots, and Mark Nay for comments on an earlier version of this manuscript and two anonymous reviewers for their review of the manuscript. We also thank Frances Biles for assistance with Fig. 1.

Competing interests
The authors declare that they have no competing interests in this manuscript.

References

1. Pan Y, Birdsey RH, Fang J, Houghton R, Kauppi PE, Kurz WA, Phillips OL, Shvidenko A, Lewis SL, Canadell JG, Ciais P, Jackson RB, Pacala SW, McGuire AD, Piao S, Rautiainen A, Sitch S, Hayes D. A large and persistent carbon sink in the world's forests. Science. 2011;333:988–93.
2. IPCC. Climate Change 2013: The physical science basis. Contribution of working group I to the fifth assessment report of the intergovernmental panel on climate change. In: Stocker, TF, Qin D, Plattner G-K, Tignor M, Allen SK, Boschung J, Nauels A, Xia Y, Bex V, Midgley PM (eds) Cambridge University Press, Cambridge. 2013. p 1535.
3. Randerson JT, Chapin FS, Harden JW, Neff JC, Harmon ME. Net ecosystem production: a comprehensive measure of net carbon accumulation by ecosystems. Ecol Appl. 2002;12:937–47.
4. Ryan MG, Harmon ME, Birdsey RA, Giardina CP, Heath LS, Houghton RA, Jackson RB, McKinley DC, Morrison JG, Murray BC, Pataki DE, Skog KE. A synthesis of the science on forests and carbon for US Forests. Issues in Ecology, Ecological Society of America, Report Number 13, 2010.
5. EISA, Energy Independence and Security Act. Public Law 110-140, United States Congress. 2007. http://www.gpo.gov/fdsys/pkg/PLAW-110publ140/pdf/PLAW-110publ140.pdf.
6. NACP, North American Carbon Program. http://nacarbon.org.
7. Federal Register. National Forest System Land Management Planning. Department of Agriculture, Forest Service. 36 CFR Part 219. 2012. http://www.fs.usda.gov/internet/fse_documents/stelprdb5362536.pdf.
8. Alaback PB. Comparative ecology of temperate rainforests of the Americas along analogous climatic gradients. Revista Chilena Historia Naturel. 1991;64:399–412.
9. USDA Forest Service. Tongass Young-Growth Management Strategy. Tongass National Forest, Region 10. 2014.
10. Harris AS, Farr WA. The forest ecosystem of southeast Alaska. 7: Forest ecology and timber management. 1974;Gen. Tech. Rep. PNW-25. Portland: USDA Forest Service, Pacific Northwest Forest and Range Experiment Station.
11. Lowell E, Dykstra C, Monserud R. Evaluating effects of thinning on wood quality in southeast Alaska. West J Appl For. 2012;27:72–83.
12. Hanley TA, Robbins CT, Spalinger DE. Forest habitats and the nutritional ecology of Sitka black-tailed deer: a research synthesis with implications

for forest management. 1989;Gen. Tech. Rep. PNW-GTR-230. Portland: US Department of Agriculture, Forest Service, Pacific Northwest Research Station. p 52.

13. Deal RL, Farr WA. Composition and development of conifer regeneration in thinned and unthinned natural stands of western hemlock and Sitka spruce in southeast Alaska. Can J For Res. 1994;24:976–84.

14. McClellan MH. Recent research on the management of hemlock-spruce forest in southeast Alaska for multiple values. Landscape Urban Planning. 2005;72:65–78.

15. Carey AB. Biocomplexity and restoration of biodiversity in temperate coniferous forest: inducing spatial heterogeneity with variable-density thinning. Forestry. 2003;76:127–36.

16. Jenkins JC, Chojnacky DC, Heath LS, Birdsey RA. National scale biomass estimators for United States tree species. For Sci. 2003;49:12–35.

17. Standish JT, GH Manning, JP Demaerschalk. Development of biomass equations for British Columbia tree species. Info. Rep. BC-X-264. Victoria: Canadian Forest Service, Pacific Forest Resource Center. 1985, p 47.

18. Woodall CW, Heath LS, Domke GM, Nichols MC. Methods and equations for estimating aboveground volume, biomass, and carbon for trees in the US forest inventory. Gen. Tech. Rep. NRS-88. Newtown Square: US Department of Agriculture, Forest Service, Northern Research Station. 2011, p 30.

19. DeMars DJ. Stand-density study of spruce-hemlock stands in southeastern Alaska. 2000; General Technical Report PNW-GTR-496. USDA Forest Service, Pacific Northwest Research Station, Portland, Oregon.

20. McGuire AD, Melillo JM, Kicklighter DW, Joyce LA. Equilibrium responses of soil carbon to climate change: empirical and process-based estimates. J Biogeograph. 1995;22:785–96.

21. Harmon ME, Krankina ON, Yatskov M, Matthews E. Predicting broadscale carbon stores of woody detritus from plot-level data. In: Lai R, Kimble J, Stweart BA, editors. Assessment methods for soil carbon. New York: CRC Press; 2001. p. 533–52.

22. Lyssaert S, Schulze ED, Borner A, Knohl A, Hessenmooler D, Law BE, Ciais P, Grace J. Old-growth forests as global carbon sinks. Nature. 2008;455:213–5. doi:10.1038/nature07276.

23. Hudiburg T, Law B, Turner DP, Campbell J, Donato D, Duane M. Carbon dynamics of Oregon and Northern California forests and potential land-based carbon storage. Ecol Appl. 2009;19:163–80.

24. Hember RA, Kurz WA, Metsaranta JM, Black TA, Guy RD, Coops NC. Accelerating regrowth of temperate-maritime forests due to environmental change. Glob Change Biol. 2012;18:2026–40. doi:10.1111/j.1365-2486.2012.02669.x.

25. Malmsheimer RW, Bowyer JL, Fried JS, Gee E, Izlar RL, Miner RA, Munn IA, Oneil E, Stewart WC. Managing forests because carbon matters: integrating energy, products, and land management policy. J For. 2011;109: Number 7S.

26. Hanley TA, McClellan MH, Barnard JC, Friberg MA. Precommercial thinning: Implications of early results from the Tongass-Wide Young-Growth Studies experiments for deer habitat in southeast Alaska. 2013; Res. Pap. PNW-RP-593. Portland: U.S. Department of Agriculture, Forest Service, Pacific Northwest Research Station. p 64.

27. Southeast Alaska Integrated Resource Plan (SEIRP). Alaska Energy Authority project, Black and Veatch report no. 172744. 2011.

28. Saunders M, Tobin B, Black K, Gioria M, Nieuwenhuis M, Osborne BA. Thinning effects on the net ecosystem exchange of a Sitka spruce forest are temperature dependent. Agric For Meteor. 2012;157:1–10. doi:10.1016/j.agrformet.2012.01.008.

29. Eriksson E. Thinning operations and their impact on biomass production in stands of Norway spruce and Scots pine. Biomass Bioenergy. 2006;30:848–54.

30. Clark J, Sessions J, Krankina O, Maness T. Impacts of thinning on carbon stores in the PNW: a plot level analysis. Corvallis: Oregon State University, College of Forestry. 2011, p 61.

31. Hahm WJ, Riebe CS, Lukens CE, Araki S. Bedrock composition regulates mountain ecosystems and landscape evolution. PNAS. 2014;111:3338–43.

32. Neiland BJ. The forest-bog complex of southeast Alaska. Vegetatio. 1971;22:1–64.

33. Sidle RC, Shaw CG. III. Evaluation of planting sites common to a southeast Alaska clear-cut. I. Nutrient status. Can J For Res. 1983;13:1–8.

34. Wittwer RF, Lynch TB, Huebschmann MM. Thinning improves growth of crop tree in natural shortleaf pine stands. South J Appl For. 1996;4:182–7.

35. Peterson JA, Seiler JR, Nowak J, Ginn SE, Kreh RE. Growth and physiological responses of young loblolly pine stands to thinning. For Sci. 1997;43:529–34.

36. Ritchie MW, Zhang J, Hamilton TA. Aboveground tree biomass for *Pinus ponderosa* in Northeastern California. Forests. 2013;4:179–96.

37. Gyawali N. Aboveground biomass partitioning due to thinning in naturally regenerated even-aged shortleaf pine (Pinus echinata Mill.) stands in southeast Oklahoma. 2003; M.S. thesis, Oklahoma State University, p 76.

38. Poage NJ, Marshall DD, McClellan MH. Maximum stand-density index of 40 western hemlock-sitka spruce stands in southeast Alaska. West J Appl For. 2007;22:99–104.

39. Poage NJ. Long-term basal area and diameter growth responses of western hemlock-sitka spruce stands in southeast Alaska to a range of thinning intensities. In: Deal, R.L, tech. editors. Integrated restoration of forested ecosystems to achieve multiresource benefits: proceedings of the 2007 national silviculture workshop. 2008; Gen.Tech. Rep. PNW-GTR-733. Portland: US Department of Agriculture, Forest Service, Pacific Northwest Research Station: 271–280.

40. McClellan MH, Hennon PE, Heuer PG, Coffin KW. Conditions and deterioration rate of precommercial thinning slash at False Island, Alaska. 2013; Res. Pap. PNW-RP-594. Portland: U.S. Department of Agriculture, Forest Service, Pacific Northwest Research Station. p 29.

41. Lamlon S, Savidge R. A reassessment of carbon content in wood: variation within and between 41 North American species. Biomass Bioenergy. 2003;25:381–8.

42. Oliver CD, Larson BC. Forest stand dynamics. John Wiley and Sons Inc., New York. 1996 (ISSN:0471138339).

43. Leighty WW, Hamburg SP, Caouette J. Effects of management on carbon sequestration in forest biomass in southeast Alaska. Ecosystems. 2006;9:1051–65.

44. Pinheiro J, Bates D, DebRoy S, Sarkar D, R Core Team. nlme: linear and Nonlinear Mixed Effects Models. 2012; R package version 3.1-105.

45. Richards FJ. A flexible growth function for empirical use. J Exp Bot. 1959;10:290–301.

## 5

Evaluating revised biomass equations: are some forest types more equivalent than others?

Coeli M. Hoover[*] and James E. Smith

Abstract

Background: In 2014, Chojnacky et al. published a revised set of biomass equations for trees of temperate US forests, expanding on an existing equation set (published in 2003 by Jenkins et al.), both of which were developed from published equations using a meta-analytical approach. Given the similarities in the approach to developing the equations, an examination of similarities or differences in carbon stock estimates generated with both sets of equations benefits investigators using the Jenkins et al. (For Sci 49:12–34, 2003) equations or the software tools into which they are incorporated. We provide a roadmap for applying the newer set to the tree species of the US, present results of equivalence testing for carbon stock estimates, and provide some general guidance on circumstances when equation choice is likely to have an effect on the carbon stock estimate.

Results: Total carbon stocks in live trees, as predicted by the two sets, differed by less than one percent at a national level. Greater differences, sometimes exceeding 10–15 %, were found for individual regions or forest type groups. Differences varied in magnitude and direction; one equation set did not consistently produce a higher or lower estimate than the other.

Conclusions: Biomass estimates for a few forest type groups are clearly not equivalent between the two equation sets—southern pines, northern spruce-fir, and lower productivity arid western forests—while estimates for the majority of forest type groups are generally equivalent at the scales presented. Overall, the possibility of very different results between the Chojnacky and Jenkins sets decreases with aggregate summaries of those 'equivalent' type groups.

Keywords: Biomass estimation, Allometry, Forest carbon stocks, Tests of equivalence, Individual-tree estimates by species group

Background

Nationally consistent biomass equations can be important to forest carbon research and reporting activities. In general, the consistency is based on an assumption that allometric relationships within forest species do not vary by region. Essentially, nearly identical trees even in distant locations should have nearly identical carbon mass. In 2003, Jenkins et al. published a set of 10 equations for estimating live tree biomass, developed from existing equations using a meta-analytical approach, which were intended to be applicable over temperate forests of the United States [1]. These equations were developed to support US forest carbon inventory and reporting, and had several key elements: (1) a national scale, so that regional variations in biomass estimates due to the use of local biomass equations was eliminated, (2) the exclusion of height as a predictor variable, and (3) in addition to equations to estimate aboveground biomass, a set of component equations allowing the separate estimation of biomass in coarse roots, stem bark, stem wood, and foliage. Since their introduction, these equations have been incorporated into the Fire and Fuels Extension of the Forest Vegetation Simulator as a calculation option [2],

*Correspondence: choover@fs.fed.us
USDA Forest Service, Northern Research Station, Durham, NH, USA

utilized in NED-2 [3], and have provided the basis for calculating the forest carbon contribution to the US annual greenhouse gas inventories for submission years 2004–2011 (e.g., see [4]). Researchers in Canada [5, 6] and the US (e.g. [7–9]) have also employed the equations while other investigators have adopted the component ratios to estimate biomass in coarse roots or other components (e.g. [10, 11]).

In 2014, Chojnacky et al. [12] introduced a revised set of generalized biomass equations for estimating aboveground biomass. These equations were developed using the same underlying data compilations and general approaches to developing the individual tree biomass estimates as for Jenkins et al. [1], but with greater differentiation among species groups, resulting in a set of 35 generalized equations: 13 for conifers, 18 for hardwoods, and 4 for woodland species. Important distinctions are: the database used to generate the revised equations was updated to include an additional 838 equations that appeared in the literature since the publication of the 2003 work or were not included at that time, taxonomic groupings were employed to account for differences in allometry, and taxa were further subdivided in cases where wood density varied considerably within a taxon. The only component equation revised by Chojnacky et al. [12] was for roots; equations were fitted for fine and coarse roots, in contrast to Jenkins et al. [1] where fine roots were not considered separately.

Based on the similarity of the equation development approach, it is likely that applications using the Jenkins et al. [1] set would have essentially the same basis for employing the revised equations. Since the primary objective of Chojnacky et al. [12] was to present the updated equations and describe the nature of the changes, only a brief discussion of the behavior of the updated equations vs. the Jenkins et al. [1] equation set was included. The authors noted that at a national level results were similar, while differences occurred in some species groups, for example, western pines, spruce/fir types, and woodland species. Given the limited information provided in Chojnacky et al. [12] we felt that a more thorough investigation of the differences in carbon stock estimates as generated with both sets of equations was needed.

One potentially practical result from a comparison of the two approaches is to identify where one set effectively substitutes for the other, which then suggests that revising or updating estimates would change little from previous analyses. For this reason we applied equivalence tests to determine the effective difference of the Chojnacky-based estimates relative to the Jenkins values. Note that hereafter we label the respective equations and species groups as Chojnacky and Jenkins (i.e., in reference to their products not the publications, per se).

In this paper, we: (1) provide a roadmap for applying the Chojnacky equations to the tree species of the US Forest Service's forest inventory [13], (2) present results of equivalence testing for carbon stock estimates computed using both sets of equations, and (3) provide general guidance on the circumstances when the choice of equation is likely to have an important effect on the carbon stock estimate. Note that we do not attempt any evaluation of relative accuracy or the relative merit of one approach relative to the other.

Results and discussion

We conducted multiple equivalence tests on data aggregated at various levels of resolution. As noted by Chojnacky et al. [12], at a national level the carbon density predicted by both equations was the same when grouped by just hardwoods and softwoods, while some type groups showed differences (though no statistical comparisons were conducted). Relative differences emerged as four regions (Fig. 1) relative to the entire United States were used to summarize total carbon stocks in the aboveground portion of live trees as shown in Fig. 2. Totals for the US as well as separate summaries according to either softwood or hardwood forest type groups (not shown) are about 1 % different. This similarity in aggregate values between the two approaches holds for the Rocky Mountain and North regions, where there is less than a 1 % difference between the two. There are more sizeable differences in the Pacific Coast and South regions, notably differing in direction and magnitude. The largest difference is in the South. Note that our results are presented in terms of carbon mass rather than biomass.

To examine the drivers of those differences, we carried out equivalence tests by forest type group at both the national and regional levels on the mean density of carbon in aboveground live trees; a summary of the results

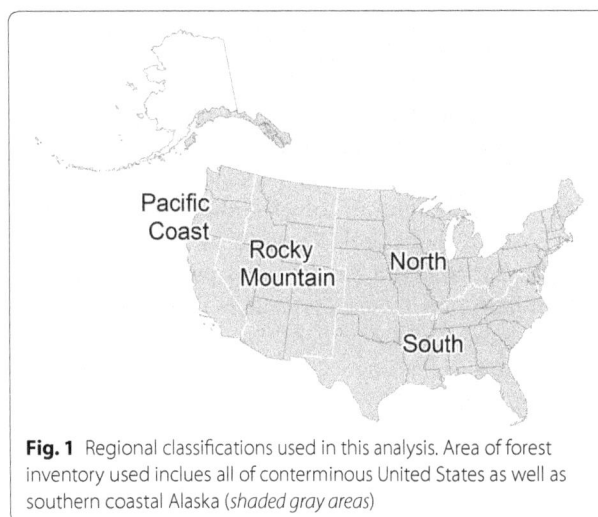

Fig. 1 Regional classifications used in this analysis. Area of forest inventory used inclues all of conterminous United States as well as southern coastal Alaska (*shaded gray areas*)

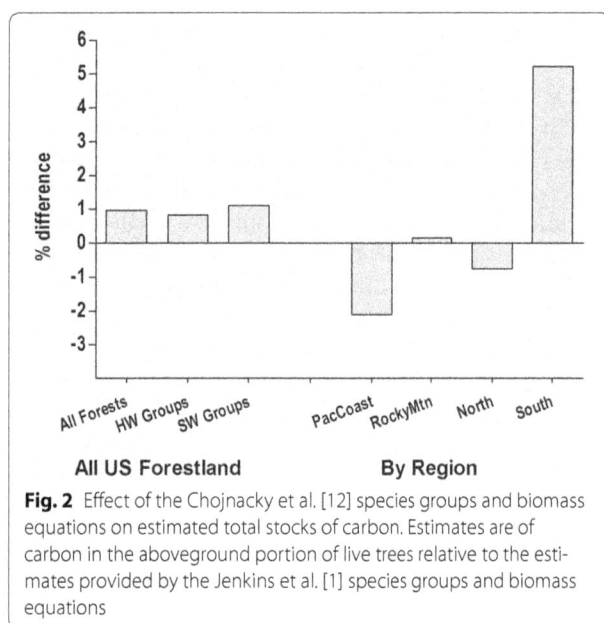

Fig. 2 Effect of the Chojnacky et al. [12] species groups and biomass equations on estimated total stocks of carbon. Estimates are of carbon in the aboveground portion of live trees relative to the estimates provided by the Jenkins et al. [1] species groups and biomass equations

is given in Table 1. The quantity tested is mean difference (Chojnacky − Jenkins) in plot level tonnes carbon per hectare; the test for equivalence was based on the percentage difference relative to the Jenkins based estimate (i.e. 100 × ((Chojnacky − Jenkins)/Jenkins)). The 5 (or 10) % of Jenkins, which was set as the equivalence interval, was put in units of tonnes per hectare for comparison with the 95 % confidence interval for the α = 0.05 (or α = 0.1) two one-sided tests (TOST) of equivalence. Of the 26 forest type groups included in the analysis, 20 are equivalent (at 5 or 10 %) at the national level, with most equivalent at 5 %. The exceptions are: spruce/fir, longleaf/slash pine, loblolly/shortleaf pine, pinyon/juniper, other western softwoods, and woodland hardwoods. At a regional level, differences emerge; in the North, only spruce/fir and loblolly/shortleaf pine are not equivalent (too few plots were available in pinyon/juniper for a reliable test statistic) while in the South, the pine types lacked equivalence, as did pinyon/juniper. This is very likely a reflection of the fact that the Chojnacky equations divide some taxa by specific gravity, while the Jenkins equations do not; softwoods generally display a larger range of specific gravity values within a species group than do hardwoods [14]. Researchers have noted considerable variability in the estimates produced by different southern pine biomass equations [15], even between different sets of local equations. Specific gravity, as mentioned above, is a factor, (southern pines exhibit considerable variability in specific gravity), as well as stand origin, and the mathematical form of the equation itself. Melson et al. [16], in their investigation of the effects of model selection on carbon stock estimates in northwest

Oregon, noted that the national level Jenkins [1] equations produced biomass estimates for *Picea* that were consistently lower than from approaches developed by the investigators, and hypothesized that differences in form between *Picea* species introduced bias into the generalized equation.

Pinyon/juniper was not equivalent in any region in which it was tested. While fir/spruce/mountain hemlock was not equivalent in the Rocky Mountains, the stock estimates were equivalent to 5 % in the Pacific Coast region, likely a function of the species and size classes that dominate the groups in each of these regions. The elm/ash/cottonwood category is represented in each region, and was equivalent to 5 % in all areas except the Pacific Coast. The woodland class has been less well studied than the others, and so less data and fewer equations are available to construct generalized equations like those in Jenkins et al. [1] and Chojnacky et al. [12]. Consequently, the woodland equations are not equivalent at the national level or in any region.

We also explored the effect of size class on equation performance, testing each combination of forest type group and stand size class and found notable differences among size classes, though no evidence of a systematic pattern. A summary of the results is given in Fig. 3a and 3b; the error bars represent the 95 % confidence interval transformed to percentage. Not every combination is shown; groups with results similar to another or comprising a very small proportion of plots are not included. While some groups such as ponderosa pine, oak/hickory, lodgepole pine, and white/red/jack pine show small differences between size classes and are equivalent (or nearly so), others such as loblolly/shortleaf pine, longleaf/slash pine (data not shown), woodland hardwoods, and spruce/fir show a strong pattern of increasing differences with increasing stand size, with a lack of equivalence between the small and large sawtimber classes. Note that both the direction and magnitude of the differences were variable across the forest type groups. Hemlock/Sitka spruce displayed a strong trend in the opposite direction, with large differences between the two approaches for the small and medium size classes, and a very small difference in the large sawtimber class. The difference between the two sets of estimates for the woodland group that is shown in Table 1 is readily apparent in Fig. 3a, with a large increase in the percent difference as the stand size class increases. This may be due to the lack of woodland biomass equations based on diameter at root collar (drc) and the difficulty of obtaining accurate drc measurements. Bragg [17] and Bragg and McElligott [15] have discussed the importance of diameter at breast height (dbh) in some detail, comparing the performance of local, regional, and national equations for southern

Table 1 Mean stock of carbon in aboveground live tree biomass as computed using the equations from Jenkins et al. [1] and Chojnacky et al. [12]

Forest type group	All US[a]		North		South		Rocky Mountain		Pacific Coast	
	Jenkins	Chojnacky	Jenkins	Chojnacky	Jenkins	Chojnacky	Jenkins	Chojnacky	Jenkins	Chojnacky
White/red/jack pine	68.7**	67.2**	67.7**	66.2**	92.4**	93.5**				
Spruce/fir	45.8	40.1	47.5	41.6					20.5*	18.9*
Longleaf/slash pine	35.4	40.6			35.4	40.6				
Loblolly/shortleaf pine	47.0	54	59.0	67.1	47.2	54.1				
Pinyon/juniper	18.4	22.5	◊15.5	◊17.2	11.5	13.3	19.6	24.1	21.4	23.4
Douglas-fir	114.5*	108.0*					71.4*	66.5*	148.6*	140.9*
Ponderosa pine	50.0**	50.7**	37.3**	37.9**			46.3**	47.1**	53.5**	54.2**
Western white pine	66.2**	67.6**							◊74.6	◊76.7
Fir/spruce/mtn hemlock	92.2*	87.1*					71.8	64.4	119.4**	117.4**
Lodgepole pine	48.6**	48.2**					48.2**	47.2**	49.5**	49.7**
Hemlock/sitka spruce	155.1**	151.0**					108.8*	101.4*	159.7**	155.9**
Western larch	62.6**	65.2**					55.4**	57.5**	69.6	72.6
Redwood	236.2**	235.3**							236.2**	235.3**
Other western softwoods	27.0	35.3					43.2*	45.8*	19.5	30.4
California mixed conifer	134.7**	132.8**							134.7**	132.8**
Oak/pine	54.1**	56.6**	64.4**	65.5**	50.9*	53.9*				
Oak/hickory	72.7**	72.8**	78.7**	78.8**	65.2**	65.3**				
Oak/gum/cypress	78.1**	79.7**	86.9**	85.2**	78.5**	80.3**				
Elm/ash/cottonwood	56.6**	56.6**	60.6**	59.8**	50.4**	52.2**	48.8**	48.2**	82.3	71.8
Maple/beech/birch	80.7**	80.3**	80.1**	79.7**	82.1**	83.3**				
Aspen/birch	45.3**	43.2**	43.9**	41.8**			52.8**	50.4**	38.0**	36.5**
Alder/maple	98.5**	100.1**							99.4**	101.0**
Western oak	64.7*	61.1*							64.7**	61.1**
Tanoak/laurel	131.2**	134.6**							131.2**	134.6**
Other hardwoods	49.6**	51.2**	43.0*	45.8*	43.2*	45.9*			67.5**	66.3**
Woodland hardwoods	8.6	11.1			5.0	7.0	12.7	15.7	22.1	29.5

Values followed by a double asterisk (**) are equivalent at 5 %; values followed by a single asterisk (*) are equivalent at 10 %. Regions are as shown in Fig. 1. A diamond preceding a value indicates that the sample size was too small for a reliable test of equivalence. Data not shown for categories represented by fewer than 10 plots

[a] As shown in Fig. 1

pines across a range of diameters. While most equations returned fairly similar estimates for trees up to 50 cm dbh, equation behavior diverged at larger diameters, in some cases returning estimates that were considerably different. In these examples, the national level Jenkins equations [1] did not produce extreme estimates, they were intermediate to those returned by local and regional equations. Melson et al. [16] also noted that considerable error could be introduced when applying equations to trees with a dbh value outside the range on which the equations were developed.

Equivalence was not tested at the level of the individual tree, though a random subset of individual tree estimates were plotted for each species group to compare tree-level biomass estimates. These plots reflect the patterns demonstrated above, with one method producing values consistently higher or lower than the other, the differences becoming more apparent at larger diameters. Tree data were also classified by east and west to further explore equation behavior within species groups where there are considerable differences in the range of tree diameters, east versus west. In many cases, no trends were revealed, but there are some key differences; a notable example is shown in Fig. 4a, b, which show the results of tree-level carbon estimates by each set of equations, categorized as east and west. In Fig. 4a, the eastern US, the Jenkins estimates are larger than those produced from the Chojnacky equations, while in Fig. 4b, the western US, the Jenkins estimates are generally somewhat lower, with the exception of the "Abies; LoSG" group. Figure 5 shows similar data for the woodland taxa; again, there is a considerable difference between the estimates

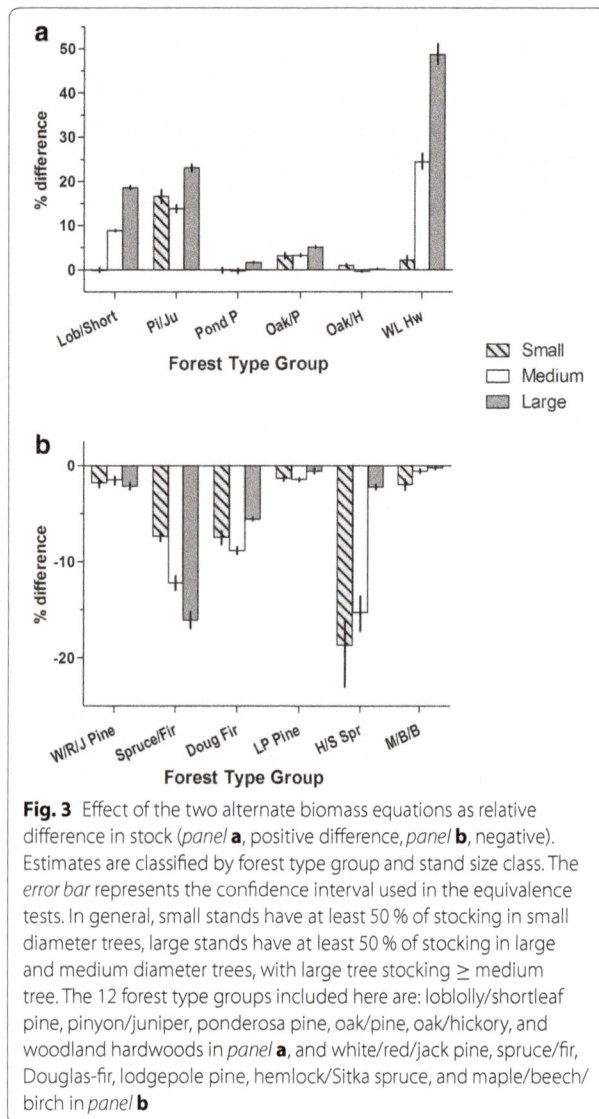

Fig. 3 Effect of the two alternate biomass equations as relative difference in stock (*panel* **a**, positive difference, *panel* **b**, negative). Estimates are classified by forest type group and stand size class. The *error bar* represents the confidence interval used in the equivalence tests. In general, small stands have at least 50 % of stocking in small diameter trees, large stands have at least 50 % of stocking in large and medium diameter trees, with large tree stocking ≥ medium tree. The 12 forest type groups included here are: loblolly/shortleaf pine, pinyon/juniper, ponderosa pine, oak/pine, oak/hickory, and woodland hardwoods in *panel* **a**, and white/red/jack pine, spruce/fir, Douglas-fir, lodgepole pine, hemlock/Sitka spruce, and maple/beech/birch in *panel* **b**

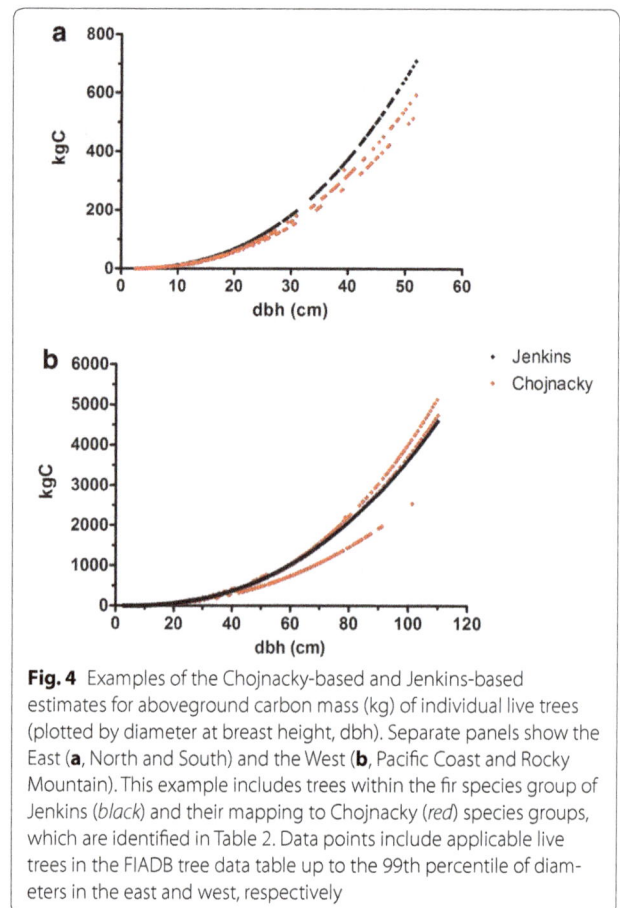

Fig. 4 Examples of the Chojnacky-based and Jenkins-based estimates for aboveground carbon mass (kg) of individual live trees (plotted by diameter at breast height, dbh). Separate panels show the East (**a**, North and South) and the West (**b**, Pacific Coast and Rocky Mountain). This example includes trees within the fir species group of Jenkins (*black*) and their mapping to Chojnacky (*red*) species groups, which are identified in Table 2. Data points include applicable live trees in the FIADB tree data table up to the 99th percentile of diameters in the east and west, respectively

computed with the two methods, with the Jenkins equations producing consistently lower estimates than the Chojnacky equations. In this case, we see no obvious differences between the predictions in the East or West.

As mentioned above, the belowground component equations were also revised in the 2014 publication, and while not divided according to hardwood and softwood, the revised root component equations are subdivided by coarse and fine roots. There are important differences in the shape of the root component curve between the two approaches (Fig. 6), and the Jenkins hardwood equation yields a consistently lower proportion than the Chojnacky equation. This suggests that adopting the Chojnacky estimates for full above- and belowground tree would add up to an additional 2–3 % of biomass for hardwoods but would also affect some softwood estimates.

A preliminary analysis did show an effect on the test for the 5 % equivalence for some categories. However, our emphases here are the various species groups/equations and not the components.

Conclusions

The revised approach to developing these biomass equations has the effect of providing better regional differentiation/representation at the plot/stand level summaries by allowing for separation within the taxonomic classes according to wood properties or growth habit. The emergence of Southern pines as distinctly different under the Chojnacky groups is one example. It is challenging to provide specific criteria for choosing one set of equations over the other, since validating any biomass equation requires the destructive sampling of multiple stems across a range of diameters. The Chojnacky groups appear to provide greater resolution across forest types and regions. From this, investigators working in southern pine, northern spruce-fir, pinyon-juniper, and woodland types may be advised to use the updated equations [12], which provide more taxonomic resolution. It should also be noted that estimates of change over time

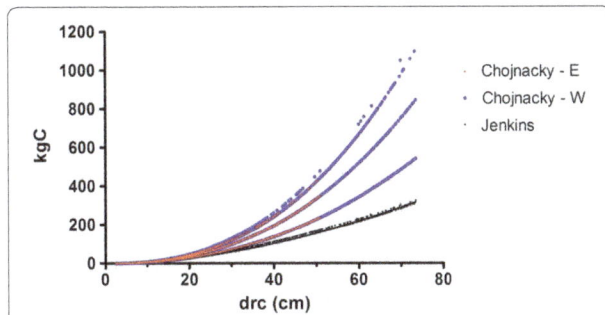

Fig. 5 Examples of the Chojnacky-based and Jenkins-based estimates for aboveground carbon mass (kg) of individual live trees by dbh. This example includes all trees within the woodland species group of Jenkins (*black*) and their mapping to Chojnacky species groups (not identified) in the East (*red*, North and South) and the West (*blue*, Pacific Coast and Rocky Mountain). Data points include all applicable live trees in the FIADB tree data table up to the 99th percentile of diameters in the East and West, respectively

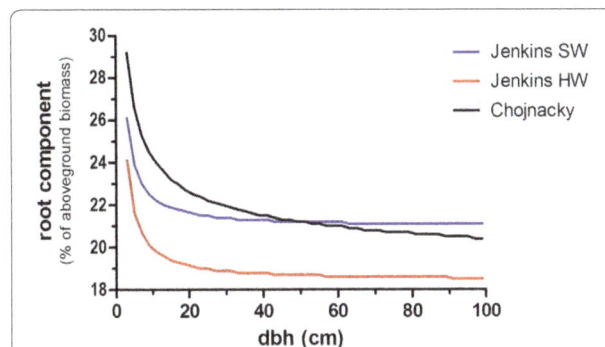

Fig. 6 Root component by diameter of the Chojnacky-based estimates (*black*) relative to the softwood (*blue*) and hardwood (*red*) root components of the Jenkins-based estimates. Root biomass is calculated as equal to a proportion of aboveground biomass

are somewhat less sensitive to equation choice than stock estimates, so if change is the primary variable of interest, the user can select either equation set, based on personal preference.

Individual large diameter trees can be very different—Chojnacky relative to Jenkins—given the general trends of the tree-level estimates (Figs. 4 and 5 in this manuscript as well as Figs. 2, 3, and 4 in Chojnacky et al. [12]). This effect of one or a very few larger trees can result in very different estimates even in an "equivalent" forest type group, and this potential for larger differences is reflected in plot-level data. For example, in some eastern hardwood type groups, which were consistently identified as equivalent, up to one-third of the plots were individually more than 5 % different. The oak/gum/cypress type group in the South had 8 % of the plots with greater carbon density by over 5 % with the Jenkins estimates, while 27 % of plots had over 5 % greater carbon.

The remaining 65 % of the individual plots are within the 5 % bounds (data not shown here). This is consistent with our observation about similarities between the two sets and scale (Fig. 2)—the sometimes obvious and large differences for some forest type groups (all scales) become obscured when summed to total live tree carbon for the US. Singling out the correct or most accurate equations is beyond the scope here; however, caution is always warranted when applying equations to trees that are considerably outside the range of diameters used to construct the equations [16].

Our results point to a few forest type groups that are clearly not equivalent—southern pines, northern spruce-fir, and lower productivity arid western forests—while the majority of forest type groups are generally equivalent at the scales presented. Overall, the possibility of very different results between the Chojnacky and Jenkins sets decreases with aggregate summaries of those 'equivalent' type groups.

Methods
Tree data source
In order to implement the revised biomass equations and identify applications where they are effectively interchangeable, or equivalent, we used the Forest Inventory and Analysis Data Base (FIADB) compiled by the Forest Inventory and Analysis (FIA) Program of the US Forest Service [13]. The data are based on continuous systematic annualized sampling of US forest lands, which are then compiled and made available by the FIA program of the US Forest Service [18]; the specific data in use here were downloaded from http://apps.fs.fed.us/fiadb-downloads/datamart.html on 02 June 2015. Surveys are organized and conducted on a large system of permanent plots over all land within individual states so that a portion of the survey data is collected each year on a continuous cycle, with remeasurement at 5 or 10 years depending on the state. The portion of the data used here include the conterminous United States (i.e., 48 states), and the portion of southern coastal Alaska that has the established permanent annual survey plots (the gray areas in Fig. 1).

Our focus here is on the tree data of the FIADB, and for this analysis we present the Chojnacky and Jenkins estimates in terms of carbon mass (i.e., kg carbon per tree or tonnes per hectare per plot). We use the entire tree data table to assure that all applicable species (the gray areas in Fig. 1) are represented. All other summaries are based on the most recent (most up-to-date) set of tree and plot data available per state, with the Chojnacky and Jenkins estimates expressed as tonnes of carbon per hectare in live trees on forest inventory plots. These plot-level values are expanded to population totals, that is, total carbon stock per state, as provided within the FIADB as the basis for the result presented in Fig. 2. A subset of the current forest plot level summaries where the entire plot is identified as forested (i.e., single condition forest plots) is the basis for the results provided in Table 1 and Fig. 3.

Application of Chojnacky et al. [12] to the FIADB

Chojnacky et al. [12] provided a revised and expanded set of biomass equations following the approach of Jenkins et al. [1]. The revised equations are based on an approach similar to that of Jenkins et al. [1] and with an expanded database of published biomass equations; see Chojnacky et al. [12] for details. The new set of 35 Chojnacky species groups are based on taxon (family or genera), growth habit, or average wood density. See Table 2 for the links between species in the FIADB and the Jenkins and Chojnacky classifications. This allocation to the newer categories is not a simple mapping of the 10 Jenkins groups to Chojnacky groups. That is, while Jenkins groups are split among Chojnacky groups, so also the Chojnacky groups are in some cases composed of species from different Jenkins groups. While Chojnacky et al. [12] developed the set of new groups based on the FIADB, similar to Jenkins et al. [1], a very small percentage of hardwood species were not explicitly named (i.e., families were not listed [12]). We assigned these to the "Cor/Eri/Lau/Etc" group (Table 2).

In order to systematically assign all the biomass estimates presented in Chojnacky et al. [12] to trees in the FIADB (as in this analysis), we present a short set of steps to make this link. Note that these include our interpretation of some of the assignments of species to groups that are not explicit such as some assignments to the woodland groups or allocation to deciduous versus evergreen. These seven steps, which also include application of the revised root component, are the basis for the biomass equation group assignments in Table 2. Note that tables and figures referenced in this list refer to those in Chojnacky et al. [12]:

1. Overall, follow the placement of taxa as suggested within the manuscript (i.e., as in Tables 2, 3, 4, and Figs. 2, 3, and 4).
2. If a tree record is one of the five families (of Table 4) and the tree diameter is measured as diameter at root collar then one of the Table 4 woodland equations applies. Otherwise, if one of the five (Table 4) families and diameter is dbh then use the appropriate equation from Tables 2 or 3. If not one of the five Table 4 families but tree diameter is provided as a root collar measurement, then convert drc to dbh following information provided in Fig. 1 before applying a Table 2 or 3 equation.

3. The calculations for the woodland (Table 4) Cupressaceae ("Cupre; WL") uses the "2nd juniper" equation from footnote #2 in Table 5.
4. The Fabaceae/Juglandaceae split into the two groups—"Fab/Jug/Carya" and "Fab/Jug"—is according to the genus *Carya* versus all others (i.e., not-*Carya*).
5. Fagaceae's deciduous/evergreen split—"Faga; Decid" and "Faga; Evergrn"—sets deciduous as the default. The Fagaceae allocated to evergreen are those five species explicitly listed as evergreen in Table 3 and those identified as evergreen from the USDA PLANTS database [19], which currently includes the addition of three live oak species.
6. The 6-family general equation at the middle of page 136 (in Table 3 of Chojnacky et al. [12])—"Cor/Eri/Lau/Etc"—is assigned trees by family from 3 sources: (a) the six families listed in Table 3; (b) the five additional families noted in the Fig. 3 caption, and (c) any additional formerly unassigned hardwood species.
7. Roots—the Chojnacky estimates use both of the belowground root equations of Table 6 (the sum of the two is generally equivalent to the original Jenkins root component). Note these are dbh-based, so a drc tree should first convert drc-to-dbh according to Fig. 1. Also note, all other (other than root) components of the original Jenkins et al. [1] are applicable here.

Identifying equivalence between the alternate biomass estimates

Tests of equivalence of the plot level (tonnes carbon per hectare) representation of the Jenkins and Chojnacky groups are included principally as guidance as to where the choice of biomass equations may matter. The analysis does not address relative accuracy of the two alternatives. Specifically, we focused on equivalence tests of the mean difference between the two estimates at the plot, or stand, level according to region and forest type groups. While these are species (group) level equations, any practical effect (of interest) is at plot to landscape to national (carbon reporting) levels. Equivalence tests are appropriate where the questions are more directly "are the groups similar, or effectively the same?" and not so much "are they different?" [20, 21]. This distinction follows from the idea that failure to reject a null hypothesis of no difference between populations does not necessarily indicate that the null hypothesis is true. The essential characteristic of an equivalence test is that the null hypothesis is stated such that the two populations are different [22, 23] which can be viewed as the reverse of the more common approach to hypothesis testing. The specific measure, or threshold, of where two populations can be considered

Table 2 Guide to applying Chojnacky species groups (as shown in Table 5, Chojnacky et al. [12]) to US species

Scientific name	Common name	Jenkins group	Chojnacky et al. parameters when diameter is measured at	
			Breast height	Root collar
Abies spp.	Fir spp.	T Fir/Hem	Abies; HiSG	Pinac; WL
A. amabilis	Pacific silver fir	T Fir/Hem	Abies; HiSG	Pinac; WL
A. balsamea	Balsam fir	T Fir/Hem	Abies; LoSG	Pinac; WL
A. bracteata	Bristlecone fir	T Fir/Hem	Abies; HiSG	Pinac; WL
A. concolor	White fir	T Fir/Hem	Abies; HiSG	Pinac; WL
A. fraseri	Fraser fir	T Fir/Hem	Abies; HiSG	Pinac; WL
A. grandis	Grand fir	T Fir/Hem	Abies; HiSG	Pinac; WL
A. lasiocarpa var. arizonica	Corkbark fir	T Fir/Hem	Abies; HiSG	Pinac; WL
A. lasiocarpa	Subalpine fir	T Fir/Hem	Abies; LoSG	Pinac; WL
A. magnifica	California red fir	T Fir/Hem	Abies; HiSG	Pinac; WL
A. shastensis	Shasta red fir	T Fir/Hem	Abies; HiSG	Pinac; WL
A. procera	Noble fir	T Fir/Hem	Abies; HiSG	Pinac; WL
Chamaecyparis spp.	White-cedar spp.	Cedar/Larch	Cupr; MedSG	Cupre; WL
C. lawsoniana	Port Orford cedar	Cedar/Larch	Cupr; MedSG	Cupre; WL
C. nootkatensi	Alaska yellow cedar	Cedar/Larch	Cupr; HiSG	Cupre; WL
C. thyoides	Atlantic white cedar	Cedar/Larch	Cupr; MedSG	Cupre; WL
Cupressus spp.	Cypress	Woodland	Cupr; HiSG	Cupre; WL
C. arizonica	Arizona cypress	Woodland	Cupr; HiSG	Cupre; WL
C. bakeri	Baker/Modoc cypress	Woodland	Cupr; HiSG	Cupre; WL
C. forbesii	Tecate cypress	Woodland	Cupr; HiSG	Cupre; WL
C. macrocarpa	Monterey cypress	Woodland	Cupr; HiSG	Cupre; WL
C. sargentii	Sargent's cypress	Woodland	Cupr; HiSG	Cupre; WL
C. macnabiana	MacNab's cypress	Woodland	Cupr; HiSG	Cupre; WL
Juniperus spp.	Redcedar/juniper spp.	Cedar/Larch	Cupr; HiSG	Cupre; WL
J. pinchotii	Pinchot juniper	Woodland	Cupr; HiSG	Cupre; WL
J. coahuilensis	Redberry juniper	Woodland	Cupr; HiSG	Cupre; WL
J. flaccida	Drooping juniper	Woodland	Cupr; HiSG	Cupre; WL
J. ashei	Ashe juniper	Woodland	Cupr; HiSG	Cupre; WL
J. californica	California juniper	Woodland	Cupr; HiSG	Cupre; WL
J. deppeana	Alligator juniper	Woodland	Cupr; HiSG	Cupre; WL
J. occidentalis	Western juniper	Woodland	Cupr; HiSG	Cupre; WL
J. osteosperma	Utah juniper	Woodland	Cupr; HiSG	Cupre; WL
J. scopulorum	Rocky Mtn. juniper	Woodland	Cupr; HiSG	Cupre; WL
J. virginiana var. silcicola	Southern redcedar	Cedar/Larch	Cupr; HiSG	Cupre; WL
J. virginiana	Easterm redcedar	Cedar/Larch	Cupr; HiSG	Cupre; WL
J. monosperma	Oneseed juniper	Woodland	Cupr; HiSG	Cupre; WL
Larix spp.	Larch spp.	Cedar/Larch	Larix	Pinac; WL
L. laricina	Tamarack	Cedar/Larch	Larix	Pinac; WL
L. lyallii	Subalpine larch	Cedar/Larch	Larix	Pinac; WL
L. occidentalis	Western larch	Cedar/Larch	Larix	Pinac; WL
Calocedrus decurrens	Incense-cedar	Cedar/Larch	Cupr; MedSG	Cupre; WL
Picea spp.	Spruce spp.	Spruce	Pice; HiSG	Pinac; WL
P. abies	Norway spruce	Spruce	Pice; HiSG	Pinac; WL
P. breweriana	Brewer spruce	Spruce	Pice; HiSG	Pinac; WL
Picea engelmannii	Englemann spruce	Spruce	Pice; LoSG	Pinac; WL
P. glauca	White spruce	Spruce	Pice; HiSG	Pinac; WL
P. mariana	Black spruce	Spruce	Pice; HiSG	Pinac; WL

Table 2 continued

Scientific name	Common name	Jenkins group	Chojnacky et al. parameters when diameter is measured at	
			Breast height	Root collar
P. pungens	Blue spruce	Spruce	Pice; HiSG	Pinac; WL
P. rubens	Red spruce	Spruce	Pice; HiSG	Pinac; WL
P. sitchensis	Sitka spruce	Spruce	Pice; LoSG	Pinac; WL
Pinus spp.	Pine spp.	Pine	Pinu; LoSG	Pinac; WL
P. albicaulis	Whitebark pine	Pine	Pinu; LoSG	Pinac; WL
P. aristata	Rocky Mtn. bristlecone pine	Pine	Pinu; LoSG	Pinac; WL
P. attenuata	Knobcone pine	Pine	Pinu; LoSG	Pinac; WL
P. balfouriana	Foxtail pine	Pine	Pinu; LoSG	Pinac; WL
P. banksiana	Jack pine	Pine	Pinu; LoSG	Pinac; WL
P. edulis	Common/two-needle pinyon	Pine	Pinu; HiSG	Pinac; WL
P. clausa	Sand pine	Pine	Pinu; HiSG	Pinac; WL
P. contorta	Lodgepole pine	Pine	Pinu; LoSG	Pinac; WL
P. coulteri	Coulter pine	Pine	Pinu; LoSG	Pinac; WL
P. echinata	Shortleaf pine	Pine	Pinu; HiSG	Pinac; WL
P. elliottii	Slash pine	Pine	Pinu; HiSG	Pinac; WL
P. engelmannii	Apache pine	Pine	Pinu; LoSG	Pinac; WL
P. flexilis	Limber pine	Pine	Pinu; LoSG	Pinac; WL
P. strobiformis	Southwestern white pine	Pine	Pinu; LoSG	Pinac; WL
P. glabra	Spruce pine	Pine	Pinu; LoSG	Pinac; WL
P. jeffreyi	Jeffrey pine	Pine	Pinu; LoSG	Pinac; WL
P. lambertiana	Sugar pine	Pine	Pinu; LoSG	Pinac; WL
P. leiophylla	Chihauhua pine	Pine	Pinu; LoSG	Pinac; WL
P. monticola	Western white pine	Pine	Pinu; LoSG	Pinac; WL
P. muricata	Bishop pine	Pine	Pinu; HiSG	Pinac; WL
P. palustris	Longleaf pine	Pine	Pinu; HiSG	Pinac; WL
P. ponderosa	Ponderosa pine	Pine	Pinu; LoSG	Pinac; WL
P. pungens	Table Mountain pine	Pine	Pinu; HiSG	Pinac; WL
P. radiata	Monterey pine	Pine	Pinu; LoSG	Pinac; WL
P. resinosa	Red pine	Pine	Pinu; LoSG	Pinac; WL
P. rigida	Pitch pine	Pine	Pinu; HiSG	Pinac; WL
P. sabiniana	Gray pine	Pine	Pinu; LoSG	Pinac; WL
P. serotina	Pond pine	Pine	Pinu; HiSG	Pinac; WL
P. strobus	Eastern white pine	Pine	Pinu; LoSG	Pinac; WL
P. sylvestris	Scotch pine	Pine	Pinu; LoSG	Pinac; WL
P. taeda	Loblolly pine	Pine	Pinu; HiSG	Pinac; WL
P. virginiana	Viginia pine	Pine	Pinu; HiSG	Pinac; WL
P. monophylla	Singleleaf pinyon	Pine	Pinu; LoSG	Pinac; WL
P. discolor	Border pinyon	Pine	Pinu; LoSG	Pinac; WL
P. arizonica	Arizona pine	Pine	Pinu; LoSG	Pinac; WL
P. nigra	Austrian pine	Pine	Pinu; LoSG	Pinac; WL
P. washoensis	Washoe pine	Pine	Pinu; LoSG	Pinac; WL
P. quadrifolia	Four leaf pine	Pine	Pinu; LoSG	Pinac; WL
P. torreyana	Torrey pine	Pine	Pinu; LoSG	Pinac; WL
P. cembroides	Mexican pinyon pine	Pine	Pinu; LoSG	Pinac; WL
P. remota	Papershell pinyon pine	Pine	Pinu; LoSG	Pinac; WL
P. longaeva	Great Basin bristlecone pine	Pine	Pinu; LoSG	Pinac; WL
P. monophylla var. fallax	Arizona pinyon pine	Pine	Pinu; LoSG	Pinac; WL

Table 2 continued

Scientific name	Common name	Jenkins group	Chojnacky et al. parameters when diameter is measured at	
			Breast height	**Root collar**
P. elliottii var. elliottii	Honduras pine	Pine	Pinu; LoSG	Pinac; WL
Pseudotsuga spp.	Douglas-fir spp.	Doug Fir	Pseud	Pinac; WL
P. macrocarpa	Bigcone Douglas-fir	Doug Fir	Pseud	Pinac; WL
P. menziesii	Douglas-fir	Doug Fir	Pseud	Pinac; WL
Sequoia sempervirens	Redwood	Cedar/Larch	Cupr; MedSG	Cupre; WL
Sequoiadendron giganteum	Giant sequoia	Cedar/Larch	Cupr; MedSG	Cupre; WL
Taxodium spp.	Baldcypress spp.	Cedar/Larch	Cupr; HiSG	Cupre; WL
T. distichum	Baldcypress	Cedar/Larch	Cupr; HiSG	Cupre; WL
T. ascendens	Pondcypress	Cedar/Larch	Cupr; HiSG	Cupre; WL
T. mucronatum	Montezuma baldcypress	Cedar/Larch	Cupr; HiSG	Cupre; WL
Taxus spp.	Yew spp.	T Fir/Hem	Pseud	
T. brevifolia	Pacific yew	T Fir/Hem	Pseud	
T. floridana	Florida yew	T Fir/Hem	Pseud	
Thuja spp.	Thuja spp.	Cedar/Larch	Cupr; MedSG	Cupre; WL
T. occidentalis	Northern white-cedar	Cedar/Larch	Cupr; LoSG	Cupre; WL
T. plicata	Western redcedar	Cedar/Larch	Cupr; MedSG	Cupre; WL
Torreya spp.	Torreya (nutmeg) spp.	T Fir/Hem	Pseud	
T. californica	California torreya	T Fir/Hem	Pseud	
T. taxifolia	Florida torreya	T Fir/Hem	Pseud	
Tsuga spp.	Hemlock spp.	T Fir/Hem	Tsug; HiSG	Pinac; WL
T. canadensis	Eastern hemlock	T Fir/Hem	Tsug; LoSG	Pinac; WL
T. caroliniana	Carolina hemlock	T Fir/Hem	Tsug; HiSG	Pinac; WL
T. heterophylla	Western hemlock	T Fir/Hem	Tsug; HiSG	Pinac; WL
T. mertensiana	Mountain hemlock	T Fir/Hem	Tsug; HiSG	Pinac; WL
Dead conifer	Unknown dead conifer	Pine	Pinu; LoSG	
Acacia spp.	Acacia spp.	Woodland	Fab/Jug	Fab/Ros; WL
A. farnesiana	Sweet acacia	Woodland	Fab/Jug	Fab/Ros; WL
A. greggii	Catclaw acacia	Woodland	Fab/Jug	Fab/Ros; WL
Acer spp.	Maple spp.	S Maple/Bir	Acer; LoSG	
A. barbatum	Florida maple	S Maple/Bir	Acer; HiSG	
A. macrophyllum	Bigleaf maple	S Maple/Bir	Acer; LoSG	
A. negundo	Boxelder	S Maple/Bir	Acer; LoSG	
A. nigrum	Black maple	H Maple/Oak	Acer; HiSG	
A. pensylvanicum	Striped maple	S Maple/Bir	Acer; LoSG	
A. rubrum	Red maple	S Maple/Bir	Acer; LoSG	
A. saccharinum	Silver maple	S Maple/Bir	Acer; LoSG	
A. saccharum	Sugar maple	H Maple/Oak	Acer; HiSG	
A. spicatum	Mountain maple	S Maple/Bir	Acer; LoSG	
A. platanoides	Norway maple	S Maple/Bir	Acer; LoSG	
A. glabrum	Rocky Mtn. maple	Woodland	Acer; LoSG	
A. grandidentatum	Bigtooth maple	Woodland	Acer; LoSG	
A. leucoderme	Chalk maple	Mixed HW	Acer; LoSG	
Aesculus spp.	Buckeye spp.	Mixed HW	Hip/Til	
A. glabra	Ohio buckeye	Mixed HW	Hip/Til	
A. flava	Yellow buckeye	Mixed HW	Hip/Til	

Table 2 continued

Scientific name	Common name	Jenkins group	Chojnacky et al. parameters when diameter is measured at	
			Breast height	Root collar
A.californica	California buckeye	Mixed HW	Hip/Til	
A.glabra var. arguta	Texas buckeye	Mixed HW	Hip/Til	
A.pavia	Red buckeye	Mixed HW	Hip/Til	
A.sylvatica	Painted buckeye	Mixed HW	Hip/Til	
Ailanthus altissima	Ailanthus	Mixed HW	Cor/Eri/Lau/Etc	
Albizia julibrissin	Mimosa/silktree	Mixed HW	Fab/Jug	Fab/Ros; WL
Alnus spp.	Alder spp.	Aspen/Alder	Betu; LoSG	
A. rubra	Red alder	Aspen/Alder	Betu; LoSG	
A. rhombifolia	White alder	Aspen/Alder	Betu; LoSG	
A. oblongifolia	Arizona alder	Aspen/Alder	Betu; LoSG	
A. glutinosa	European alder	Aspen/Alder	Betu; LoSG	
Amelanchier spp.	Serviceberry spp.	Mixed HW	Cor/Eri/Lau/Etc	Fab/Ros; WL
A. arborea	Common serviceberry	Mixed HW	Cor/Eri/Lau/Etc	Fab/Ros; WL
A. sanguinea	Roundleaf serviceberry	Mixed HW	Cor/Eri/Lau/Etc	Fab/Ros; WL
Arbutus spp.	Madrone spp.	Mixed HW	Cor/Eri/Lau/Etc	
A. menziesii	Pacific madrone	Mixed HW	Cor/Eri/Lau/Etc	
A. arizonica	Arizona madrone	Mixed HW	Cor/Eri/Lau/Etc	
A. xalapensis	Texas madrone	Mixed HW	Cor/Eri/Lau/Etc	
Asimina triloba	Pawpaw	Mixed HW	Cor/Eri/Lau/Etc	
Betula spp.	Birch spp.	S Maple/Bir	Betu; Med1SG	
B. alleghaniensis	Yellow birch	S Maple/Bir	Betu; Med2SG	
B. lenta	Sweet birch	S Maple/Bir	Betu; HiSG	
B. nigra	River birch	S Maple/Bir	Betu; Med1SG	
B. occidentalis	Water birch	S Maple/Bir	Betu; Med2SG	
B. papyrifera	Paper birch	S Maple/Bir	Betu; Med1SG	
B. uber	Virginia roundleaf birch	S Maple/Bir	Betu; Med2SG	
B. utahensis	Northwestern paper birch	S Maple/Bir	Betu; Med2SG	
B. populifolia	Gray birch	S Maple/Bir	Betu; Med1SG	
Sideroxylon lanuginosum	Chittamwood/gum bumelia	Mixed HW	Cor/Eri/Lau/Etc	
Carpinus caroliniana	American hornbeam	Mixed HW	Betu; Med2SG	
Carya spp.	Hickory spp.	H Maple/Oak	Fab/Jug/Carya	
C. aquatica	Water hickory	H Maple/Oak	Fab/Jug/Carya	
C. cordiformis	Bitternut hickory	H Maple/Oak	Fab/Jug/Carya	
C. glabra	Pignut hickory	H Maple/Oak	Fab/Jug/Carya	
C. illinoinensis	Pecan	H Maple/Oak	Fab/Jug/Carya	
C. laciniosa	Shellbark hickory	H Maple/Oak	Fab/Jug/Carya	
C. myristiciformis	Nutmeg hickory	H Maple/Oak	Fab/Jug/Carya	
C. ovata	Shagbark hickory	H Maple/Oak	Fab/Jug/Carya	
C. texana	Black hickory	H Maple/Oak	Fab/Jug/Carya	
C. alba	Mockernut hickory	H Maple/Oak	Fab/Jug/Carya	
C. pallida	Sand hickory	H Maple/Oak	Fab/Jug/Carya	
C. floridana	Scrub hickory	H Maple/Oak	Fab/Jug/Carya	
C. ovalis	Red hickory	H Maple/Oak	Fab/Jug/Carya	
C. carolinae-septentrionalis	Southern shagbark hickory	H Maple/Oak	Fab/Jug/Carya	

Table 2 continued

Scientific name	Common name	Jenkins group	Chojnacky et al. parameters when diameter is measured at	
			Breast height	Root collar
Castanea spp.	Chestnut spp.	Mixed HW	Faga; Decid	Fagac; WL
C. dentata	American chestnut	Mixed HW	Faga; Decid	Fagac; WL
C. pumila	Allegheny chinkapin	Mixed HW	Faga; Decid	Fagac; WL
C. pumila var. ozarkensis	Ozark chinkapin	Mixed HW	Faga; Decid	Fagac; WL
C. mollissima	Chinese chestnut	Mixed HW	Faga; Decid	Fagac; WL
Chrysolepis chrysophylla	Giant/golden chinkapin	Mixed HW	Faga; Evergrn	Fagac; WL
Catalpa spp.	Catalpa spp.	Mixed HW	Cor/Eri/Lau/Etc	
C. bignonioide	Southern catalpa	Mixed HW	Cor/Eri/Lau/Etc	
C. speciosa	Northern catalpa	Mixed HW	Cor/Eri/Lau/Etc	
Celtis	Hackberry spp.	Mixed HW	Cor/Eri/Lau/Etc	
C. laevigata	Sugarberry	Mixed HW	Cor/Eri/Lau/Etc	
C. occidentalis	Hackberry	Mixed HW	Cor/Eri/Lau/Etc	
C. laevigata var. reticulata	Netleaf hackberry	Mixed HW	Cor/Eri/Lau/Etc	
Cercis canadensis	Eastern redbud	Mixed HW	Fab/Jug	Fab/Ros; WL
Cercocarpus ledifoliu	Curlleaf mountain-mahogany	Woodland	Cor/Eri/Lau/Etc	Fab/Ros; WL
Cladrastis kentukea	Yellowwood	Mixed HW	Fab/Jug	Fab/Ros; WL
Cornus spp.	Dogwood spp.	Mixed HW	Cor/Eri/Lau/Etc	
C. florida	Flowering dogwood	Mixed HW	Cor/Eri/Lau/Etc	
C. nuttallii	Pacific dogwood	Mixed HW	Cor/Eri/Lau/Etc	
Crataegus spp.	Hawthorn spp.	Mixed HW	Cor/Eri/Lau/Etc	Fab/Ros; WL
C. crusgalli	Cockspur hawthorn	Mixed HW	Cor/Eri/Lau/Etc	Fab/Ros; WL
C. mollis	Downy hawthorn	Mixed HW	Cor/Eri/Lau/Etc	Fab/Ros; WL
C. brainerdii	Brainerd's hawthorn	Mixed HW	Cor/Eri/Lau/Etc	Fab/Ros; WL
C. calpodendron	Pear hawthorn	Mixed HW	Cor/Eri/Lau/Etc	Fab/Ros; WL
C. chrysocarpa	Fireberry hawthorn	Mixed HW	Cor/Eri/Lau/Etc	Fab/Ros; WL
C. dilatata	Broadleaf hawthorn	Mixed HW	Cor/Eri/Lau/Etc	Fab/Ros; WL
C. flabellata	Fanleaf hawthorn	Mixed HW	Cor/Eri/Lau/Etc	Fab/Ros; WL
C. monogyna	Oneseed hawthorn	Mixed HW	Cor/Eri/Lau/Etc	Fab/Ros; WL
C. pedicellata	Scarlet hawthorn	Mixed HW	Cor/Eri/Lau/Etc	Fab/Ros; WL
Eucalyptus spp.	Eucalyptus spp.	Mixed HW	Cor/Eri/Lau/Etc	
E. globulus	Tasmanian bluegum	Mixed HW	Cor/Eri/Lau/Etc	
E. camaldulensi	River redgum	Mixed HW	Cor/Eri/Lau/Etc	
E. grandis	Grand eucalyptus	Mixed HW	Cor/Eri/Lau/Etc	
E. robusta	Swamp mahogany	Mixed HW	Cor/Eri/Lau/Etc	
Diospyros spp.	Persimmon spp.	Mixed HW	Cor/Eri/Lau/Etc	
D. virginiana	Common persimmon	Mixed HW	Cor/Eri/Lau/Etc	
D. texana	Texas persimmon	Mixed HW	Cor/Eri/Lau/Etc	
Ehretia anacua	Anacua knockaway	Mixed HW	Cor/Eri/Lau/Etc	

Table 2 continued

Scientific name	Common name	Jenkins group	Chojnacky et al. parameters when diameter is measured at	
			Breast height	**Root collar**
Fagus grandifolia	American beech	H Maple/Oak	Faga; Decid	Fagac; WL
Fraxinus spp.	Ash spp.	Mixed HW	Olea; LoSG	
F. americana	White ash	Mixed HW	Olea; HiSG	
F. latifolia	Oregon ash	Mixed HW	Olea; LoSG	
F. nigra	Black ash	Mixed HW	Olea; LoSG	
F. pennsylvanica	Green ash	Mixed HW	Olea; LoSG	
F. profunda	Pumpkin ash	Mixed HW	Olea; LoSG	
F. quadrangulata	Blue ash	Mixed HW	Olea; LoSG	
F. velutina	Velvet ash	Mixed HW	Olea; LoSG	
F. caroliniana	Carolina ash	Mixed HW	Olea; LoSG	
F. texensis	Texas ash	Mixed HW	Olea; LoSG	
Gleditsia spp.	Honeylocust spp.	Mixed HW	Fab/Jug	Fab/Ros; WL
G. aquatica	Waterlocust	Mixed HW	Fab/Jug	Fab/Ros; WL
G. triacanthos	Honeylocust	Mixed HW	Fab/Jug	Fab/Ros; WL
Gordonia lasianthus	Loblolly-bay	Mixed HW	Cor/Eri/Lau/Etc	
Ginkgo biloba	Ginkgo	Mixed HW	Cor/Eri/Lau/Etc	
Gymnocluadus diocicus	Kentucky coffeetree	Mixed HW	Fab/Jug	Fab/Ros; WL
Halesia spp.	Silverbell spp.	Mixed HW	Cor/Eri/Lau/Etc	
H. carolina	Carolina silverbell	Mixed HW	Cor/Eri/Lau/Etc	
H. diptera	Two-wing silverbell	Mixed HW	Cor/Eri/Lau/Etc	
H. parviflora	Little silverbell	Mixed HW	Cor/Eri/Lau/Etc	
Ilex opaca	American holly	Mixed HW	Cor/Eri/Lau/Etc	
Juglans spp.	Walnut spp.	Mixed HW	Fab/Jug	
J. cinerea	Butternut	Mixed HW	Fab/Jug	
J. nigra	Black walnut	Mixed HW	Fab/Jug	
J. hindsii	No. California black walnut	Mixed HW	Fab/Jug	
J. californica	So. California black walnut	Mixed HW	Fab/Jug	
J. microcarpa	Texas walnut	Mixed HW	Fab/Jug	
J. major	Arizona walnut	Mixed HW	Fab/Jug	
Liquidambar styraciflua	Sweetgum	Mixed HW	Hama	
Liriodendron tulipifera	Yellow poplar	Mixed HW	Magno	
Lithocarpus densiflorus	Tanoak	Mixed HW	Faga; Evergrn	Fagac; WL
Maclura pomifera	Osage orange	Mixed HW	Cor/Eri/Lau/Etc	
Magnolia spp.	Magnolia spp.	Mixed HW	Magno	
M. acuminata	Cucumbertree	Mixed HW	Magno	
M. grandiflora	Southern magnolia	Mixed HW	Magno	
M. virginiana	Sweeetbay	Mixed HW	Magno	
M. macrophylla	Bigleaf magnolia	Mixed HW	Magno	
M. fraseri	Mountain/Frasier magnolia	Mixed HW	Magno	
M. pyramidata	Pyramid magnolia	Mixed HW	Magno	
M. tripetala	Umbrella magnolia	Mixed HW	Magno	
Malus spp.	Apple spp.	Mixed HW	Cor/Eri/Lau/Etc	Fab/Ros; WL
M. fusca	Oregon crab apple	Mixed HW	Cor/Eri/Lau/Etc	Fab/Ros; WL

Table 2 continued

Scientific name	Common name	Jenkins group	Chojnacky et al. parameters when diameter is measured at	
			Breast height	**Root collar**
M. angustifolia	Southern crabapple	Mixed HW	Cor/Eri/Lau/Etc	Fab/Ros; WL
M. coronaria	Sweet crabapple	Mixed HW	Cor/Eri/Lau/Etc	Fab/Ros; WL
M. ioensi	Prairie crabapple	Mixed HW	Cor/Eri/Lau/Etc	Fab/Ros; WL
Morus spp.	Mulberry spp.	Mixed HW	Cor/Eri/Lau/Etc	
M. alba	White mulberry	Mixed HW	Cor/Eri/Lau/Etc	
M. rubra	Red mulberry	Mixed HW	Cor/Eri/Lau/Etc	
M. microphyll	Texas mulberry	Mixed HW	Cor/Eri/Lau/Etc	
M. nigra	Black mulberry	Mixed HW	Cor/Eri/Lau/Etc	
Nyssa spp.	Tupelo spp.	Mixed HW	Cor/Eri/Lau/Etc	
N. aquatica	Water tupelo	Mixed HW	Cor/Eri/Lau/Etc	
N. ogeche	Ogeechee tupelo	Mixed HW	Cor/Eri/Lau/Etc	
N. sylvatica	Blackgum	Mixed HW	Cor/Eri/Lau/Etc	
N. biflora	Swamp tupelo	Mixed HW	Cor/Eri/Lau/Etc	
Ostrya virginiana	Eastern hophornbeam	Mixed HW	Betu; HiSG	
Oxydendrum arboreum	Sourwood	Mixed HW	Cor/Eri/Lau/Etc	
Paulownia tomentosa	Paulownia/empress tree	Mixed HW	Cor/Eri/Lau/Etc	
Persea spp.	Bay spp.	Mixed HW	Cor/Eri/Lau/Etc	
Persea borbonia	Redbay	Mixed HW	Cor/Eri/Lau/Etc	
Planera aquatica	Water elm/planetree	Mixed HW	Cor/Eri/Lau/Etc	
Platanus spp.	Sycamore spp.	Mixed HW	Cor/Eri/Lau/Etc	
P. racemosa	California sycamore	Mixed HW	Cor/Eri/Lau/Etc	
P. occidentalis	American sycamore	Mixed HW	Cor/Eri/Lau/Etc	
P. wrightii	Arizona sycamore	Mixed HW	Cor/Eri/Lau/Etc	
Populus spp.	Cottonwood/poplar spp.	Aspen/Alder	Sali; HiSG	
P. balsamifera	Balsam poplar	Aspen/Alder	Sali; LoSG	
P. deltoides	Eastern cottonwood	Aspen/Alder	Sali; HiSG	
P. grandidentata	Bigtooth aspen	Aspen/Alder	Sali; HiSG	
P. heterophylla	Swamp cottonwood	Aspen/Alder	Sali; HiSG	
P. deltoides	Plains cottonwood	Aspen/Alder	Sali; HiSG	
P. tremuloides	Quaking aspen	Aspen/Alder	Sali; HiSG	
P. balsamifera	Black cottonwood	Aspen/Alder	Sali; LoSG	
P. fremontii	Fremont cottonwood	Aspen/Alder	Sali; HiSG	
P. angustifolia	Narrlowleaf cottonwood	Aspen/Alder	Sali; HiSG	
P. alba	Silver poplar	Aspen/Alder	Sali; HiSG	
P. nigra	Lombardy poplar	Aspen/Alder	Sali; HiSG	
Prosopis spp.	Mesquite spp.	Woodland	Fab/Jug	Fab/Ros; WL
P. glandulosa	Honey mesquite	Woodland	Fab/Jug	Fab/Ros; WL
P. velutina	Velvet mesquite	Woodland	Fab/Jug	Fab/Ros; WL
P. pubescens	Screwbean mesquite	Woodland	Fab/Jug	Fab/Ros; WL
Prunus spp.	Cherry/plum spp.	Mixed HW	Cor/Eri/Lau/Etc	Fab/Ros; WL
P. pensylvanica	Pin cherry	Mixed HW	Cor/Eri/Lau/Etc	Fab/Ros; WL

Table 2 continued

Scientific name	Common name	Jenkins group	Chojnacky et al. parameters when diameter is measured at	
			Breast height	Root collar
P. serotina	Black cherry	Mixed HW	Cor/Eri/Lau/Etc	Fab/Ros; WL
P. virginiana	Chokecherry	Mixed HW	Cor/Eri/Lau/Etc	Fab/Ros; WL
P. persica	Peach	Mixed HW	Cor/Eri/Lau/Etc	Fab/Ros; WL
P. nigra	Canada plum	Mixed HW	Cor/Eri/Lau/Etc	Fab/Ros; WL
P. americana	American plum	Mixed HW	Cor/Eri/Lau/Etc	Fab/Ros; WL
P. emarginata	Bitter cherry	Woodland	Cor/Eri/Lau/Etc	Fab/Ros; WL
P. alleghaniensis	Allegheny plum	Mixed HW	Cor/Eri/Lau/Etc	Fab/Ros; WL
P. angustifolia	Chickasaw plum	Mixed HW	Cor/Eri/Lau/Etc	Fab/Ros; WL
P. avium	Sweet cherry (domestic)	Mixed HW	Cor/Eri/Lau/Etc	Fab/Ros; WL
P. cerasus	Sour cherry (domestic)	Mixed HW	Cor/Eri/Lau/Etc	Fab/Ros; WL
P. domestica	European plum (domestic)	Mixed HW	Cor/Eri/Lau/Etc	Fab/Ros; WL
P. mahaleb	Mahaleb cherry (domestic)	Mixed HW	Cor/Eri/Lau/Etc	Fab/Ros; WL
Quercus spp.	Oak spp.	H Maple/Oak	Faga; Decid	Fagac; WL
Q. agrifolia	California live oak	H Maple/Oak	Faga; Evergrn	Fagac; WL
Q. alba	White oak	H Maple/Oak	Faga; Decid	Fagac; WL
Q. arizonica	Arizona white oak	Woodland	Faga; Decid	Fagac; WL
Q. bicolor	Swamp white oak	H Maple/Oak	Faga; Decid	Fagac; WL
Q. chrysolepis	Canyon live oak	H Maple/Oak	Faga; Decid	Fagac; WL
Q. coccinea	Scarlet oak	H Maple/Oak	Faga; Decid	Fagac; WL
Q. douglasii	Blue oak	H Maple/Oak	Faga; Evergrn	Fagac; WL
Q. sinuata var. sinuata	Durand oak	H Maple/Oak	Faga; Decid	Fagac; WL
Q. ellipsoidalis	Northern pin oak	H Maple/Oak	Faga; Decid	Fagac; WL
Q. emoryi	Emory oak	Woodland	Faga; Decid	Fagac; WL
Q. engelmannii	Englemann oak	H Maple/Oak	Faga; Decid	Fagac; WL
Q. falcata	Southern red oak	H Maple/Oak	Faga; Decid	Fagac; WL
Q. pagoda	Cherrybark oak	H Maple/Oak	Faga; Decid	Fagac; WL
Q. gambelii	Gambel oak	Woodland	Faga; Decid	Fagac; WL
Q. garryana	Oregon white oak	H Maple/Oak	Faga; Decid	Fagac; WL
Q. ilicifolia	Scrub oak	H Maple/Oak	Faga; Decid	Fagac; WL
Q. imbricaria	Shingle oak	H Maple/Oak	Faga; Decid	Fagac; WL
Q. kelloggii	California black oak	H Maple/Oak	Faga; Decid	Fagac; WL
Q. laevis	Turkey oak	H Maple/Oak	Faga; Decid	Fagac; WL
Q. laurifolia	Laurel oak	H Maple/Oak	Faga; Evergrn	Fagac; WL
Q. lobata	California white oak	H Maple/Oak	Faga; Decid	Fagac; WL
Q. lyrata	Overcup oak	H Maple/Oak	Faga; Decid	Fagac; WL
Q. macrocarpa	Bur oak	H Maple/Oak	Faga; Decid	Fagac; WL
Q. marilandica	Blackjack oak	H Maple/Oak	Faga; Decid	Fagac; WL
Q. michauxi	Swamp chestnut oak	H Maple/Oak	Faga; Decid	Fagac; WL

Table 2 continued

Scientific name	Common name	Jenkins group	Chojnacky et al. parameters when diameter is measured at	
			Breast height	**Root collar**
Q. muehlenbergii	Chinkapin oak	H Maple/Oak	Faga; Decid	Fagac; WL
Q. nigra	Water oak	H Maple/Oak	Faga; Decid	Fagac; WL
Q. texana	Texas red oak	H Maple/Oak	Faga; Decid	Fagac; WL
Q. oblongifolia	Mexican blue oak	Woodland	Faga; Decid	Fagac; WL
Q. palustris	Pin oak	H Maple/Oak	Faga; Decid	Fagac; WL
Q. phellos	Willow oak	H Maple/Oak	Faga; Decid	Fagac; WL
Q. prinus	Chestnut oak	H Maple/Oak	Faga; Decid	Fagac; WL
Q. rubra	Northern red oak	H Maple/Oak	Faga; Decid	Fagac; WL
Q. shumardii	Shumard oak	H Maple/Oak	Faga; Decid	Fagac; WL
Q. stellata	Post oak	H Maple/Oak	Faga; Decid	Fagac; WL
Q. simili	Delta post oak	H Maple/Oak	Faga; Decid	Fagac; WL
Q. velutina	Black oak	H Maple/Oak	Faga; Decid	Fagac; WL
Q. virginiana	Live oak	H Maple/Oak	Faga; Evergrn	Fagac; WL
Q. wislizeni	Interier live oak	H Maple/Oak	Faga; Evergrn	Fagac; WL
Q. margarettiae	Dwarf post oak	H Maple/Oak	Faga; Evergrn	Fagac; WL
Q. minima	Dwarf live oak	H Maple/Oak	Faga; Evergrn	Fagac; WL
Q. incana	Bluejack oak	H Maple/Oak	Faga; Decid	Fagac; WL
Q. hypoleucoides	Silverleaf oak	Woodland	Faga; Decid	Fagac; WL
Q. oglethorpensis	Oglethorpe oak	H Maple/Oak	Faga; Decid	Fagac; WL
Q. prinoides	Dwarf chinkapin oak	H Maple/Oak	Faga; Decid	Fagac; WL
Q. grisea	Gray oak	Woodland	Faga; Decid	Fagac; WL
Q. rugosa	Netleaf oak	H Maple/Oak	Faga; Decid	Fagac; WL
Q. gracilliformis	Chisos oak	Woodland	Faga; Decid	Fagac; WL
Amyris elemifera	Sea torchwood	Mixed HW	Cor/Eri/Lau/Etc	
Annona glabra	Pond apple	Mixed HW	Cor/Eri/Lau/Etc	
Bursera simaruba	Gumbo limbo	Mixed HW	Cor/Eri/Lau/Etc	
Casuarina spp.	Sheoak spp.	Mixed HW	Cor/Eri/Lau/Etc	
C. glauca	Gray sheoak	Mixed HW	Cor/Eri/Lau/Etc	
C. lepidophloia	Belah	Mixed HW	Cor/Eri/Lau/Etc	
Cinnamomum camphora	Camphortree	Mixed HW	Cor/Eri/Lau/Etc	
Citharexylum fruticosum	Florida fiddlewood	Mixed HW	Cor/Eri/Lau/Etc	
Citrus spp.	Citrus spp.	Mixed HW	Cor/Eri/Lau/Etc	
Coccoloba diversifolia	Tietongue/pigeon plum	Mixed HW	Cor/Eri/Lau/Etc	
Colubrina elliptica	Soldierwood	Mixed HW	Cor/Eri/Lau/Etc	
Cordia sebestena	Longleaf geigertree	Mixed HW	Cor/Eri/Lau/Etc	
Cupaniopsis anacardioides	Carrotwood	Mixed HW	Cor/Eri/Lau/Etc	
Condalia hookeri	Bluewood	Woodland	Cor/Eri/Lau/Etc	
Ebenopsis ebano	Blackbead ebony	Woodland	Fab/Jug	Fab/Ros; WL
Leucaena pulverulenta	Great leadtree	Woodland	Fab/Jug	Fab/Ros; WL
Sophora affinis	Texas sophora	Woodland	Fab/Jug	Fab/Ros; WL
Eugenia rhombea	Red stopper	Mixed HW	Cor/Eri/Lau/Etc	
Exothea paniculata	Butterbough/inkwood	Mixed HW	Cor/Eri/Lau/Etc	
Ficus aurea	Florida strangler fig	Mixed HW	Cor/Eri/Lau/Etc	
Ficus citrifolia	Banyantree/shortleaf fig	Mixed HW	Cor/Eri/Lau/Etc	
Guapira discolo	Beeftree/longleaf blolly	Mixed HW	Cor/Eri/Lau/Etc	

Table 2 continued

Scientific name	Common name	Jenkins group	Chojnacky et al. parameters when diameter is measured at	
			Breast height	Root collar
Hippomane mancinella	Manchineel	Mixed HW	Cor/Eri/Lau/Etc	
Lysiloma latisiliquum	False tamarind	Mixed HW	Fab/Jug	Fab/Ros; WL
Mangifera indica	Mango	Mixed HW	Cor/Eri/Lau/Etc	
Metopium toxiferum	Florida poisontree	Mixed HW	Cor/Eri/Lau/Etc	
Piscidia piscipula	Fishpoison tree	Mixed HW	Fab/Jug	Fab/Ros; WL
Schefflera actinophylla	Octopus tree/schefflera	Mixed HW	Cor/Eri/Lau/Etc	
Sideroxylon foetidissimum	False mastic	Mixed HW	Cor/Eri/Lau/Etc	
Sideroxylon salicifolium	White bully/willow bustic	Mixed HW	Cor/Eri/Lau/Etc	
Simarouba glauca	Paradisetree	Mixed HW	Cor/Eri/Lau/Etc	
Syzygium cumini	Java plum	Mixed HW	Cor/Eri/Lau/Etc	
Tamarindus indica	Tamarind	Mixed HW	Fab/Jug	Fab/Ros; WL
Robinia pseudoacacia	Black locust	Mixed HW	Fab/Jug	Fab/Ros; WL
Robinia neomexicana	New Mexico locust	Woodland	Fab/Jug	Fab/Ros; WL
Acoelorraphe wrightii	Everglades palm	Mixed HW	Cor/Eri/Lau/Etc	
Coccothrinax argentata	Florida silver palm	Mixed HW	Cor/Eri/Lau/Etc	
Cocos nucifera	Coconut palm	Mixed HW	Cor/Eri/Lau/Etc	
Roystonea spp.	Royal palm spp.	Mixed HW	Cor/Eri/Lau/Etc	
Sabal Mexicana	Mexican palmetto	Mixed HW	Cor/Eri/Lau/Etc	
Sabal palmetto	Cabbage palmetto	Mixed HW	Cor/Eri/Lau/Etc	
Thrinax morrisii	Key thatch palm	Mixed HW	Cor/Eri/Lau/Etc	
Thrinax radiata	Florida thatch palm	Mixed HW	Cor/Eri/Lau/Etc	
Arecaceae	Other palms	Mixed HW	Cor/Eri/Lau/Etc	
Sapindus saponaria	Western soapberry	Mixed HW	Cor/Eri/Lau/Etc	
Salix spp.	Willow spp.	Aspen/Alder	Sali; HiSG	
S. amygdaloides	Peachleaf willow	Aspen/Alder	Sali; HiSG	
S. nigra	Black willow	Aspen/Alder	Sali; HiSG	
S. bebbiana	Bebb willow	Aspen/Alder	Sali; HiSG	
S. bonplandiana	Bonpland willow	Aspen/Alder	Sali; HiSG	
S. caroliniana	Coastal plain willow	Aspen/Alder	Sali; HiSG	
S. pyrifolia	Balsam willow	Aspen/Alder	Sali; HiSG	
S. alba	White willow	Aspen/Alder	Sali; HiSG	
S. scouleriana	Scouder's willow	Aspen/Alder	Sali; HiSG	
S. sepulcralis	Weeping willow	Aspen/Alder	Sali; HiSG	
Sassafras albidum	Sassafrass	Mixed HW	Cor/Eri/Lau/Etc	
Sorbus spp.	Mountain ash spp.	Mixed HW	Cor/Eri/Lau/Etc	Fab/Ros; WL
S. americana	American mountain ash	Mixed HW	Cor/Eri/Lau/Etc	Fab/Ros; WL
S. aucuparia	European mountain ash	Mixed HW	Cor/Eri/Lau/Etc	Fab/Ros; WL
S. decora	Northern mountain ash	Mixed HW	Cor/Eri/Lau/Etc	Fab/Ros; WL
Swietenia mahagoni	West Indian mahogany	Mixed HW	Cor/Eri/Lau/Etc	
Tilia spp.	Basswood spp.	Mixed HW	Hip/Til	
T. americana	American basswood	Mixed HW	Hip/Til	

Table 2 continued

Scientific name	Common name	Jenkins group	Chojnacky et al. parameters when diameter is measured at	
			Breast height	Root collar
T. americana var. heterophylla	White basswood	Mixed HW	Hip/Til	
T. americana var. caroliniana	Carolina basswood	Mixed HW	Hip/Til	
Ulmus spp.	Elm spp.	Mixed HW	Cor/Eri/Lau/Etc	
U. alata	Winged elm	Mixed HW	Cor/Eri/Lau/Etc	
U. americana	American elm	Mixed HW	Cor/Eri/Lau/Etc	
U. crassifolia	Cedar elm	Mixed HW	Cor/Eri/Lau/Etc	
U. pumila	Siberian elm	Mixed HW	Cor/Eri/Lau/Etc	
U. rubra	Slippery elm	Mixed HW	Cor/Eri/Lau/Etc	
U. serotina	September elm	Mixed HW	Cor/Eri/Lau/Etc	
U. thomasii	Rock elm	Mixed HW	Cor/Eri/Lau/Etc	
Umbellularia californica	California laurel	Mixed HW	Cor/Eri/Lau/Etc	
Yucca brevifolia	Joshua tree	Mixed HW	Cor/Eri/Lau/Etc	
Avicennia germinan	Black mangrove	Mixed HW	Cor/Eri/Lau/Etc	
Conocarpus erectus	Button mangrove	Mixed HW	Cor/Eri/Lau/Etc	
Laguncularia racemosa	White mangrove	Mixed HW	Cor/Eri/Lau/Etc	
Rhizophora mangle	American mangrove	Mixed HW	Cor/Eri/Lau/Etc	
Olneya tesota	Desert ironwood	Woodland	Fab/Jug	Fab/Ros; WL
Tamarix spp.	Saltcedar	Mixed HW	Cor/Eri/Lau/Etc	
Melaleuca quinquenervia	Melaleuca	Mixed HW	Cor/Eri/Lau/Etc	
Melia azedarach	Chinaberry	Mixed HW	Cor/Eri/Lau/Etc	
Triadica sebifera	Chinese tallowtree	Mixed HW	Cor/Eri/Lau/Etc	
Vernicia fordii	Tungoil tree	Mixed HW	Cor/Eri/Lau/Etc	
Cotinus obovatus	Smoketree	Mixed HW	Cor/Eri/Lau/Etc	
Elaeagnus angustifolia	Russian olive	Mixed HW	Cor/Eri/Lau/Etc	
Tree broadleaf	Unknown dead hardwood	Mixed HW	Cor/Eri/Lau/Etc	
Tree unknown	Unknown live tree	Mixed HW	Cor/Eri/Lau/Etc	
C. phaenopyrum	Washington hawthorn	Mixed HW	Cor/Eri/Lau/Etc	Fab/Ros; WL
C. succulenta	Fleshy hawthorn	Mixed HW	Cor/Eri/Lau/Etc	Fab/Ros; WL
C. uniflora	Dwarf hawthorn	Mixed HW	Cor/Eri/Lau/Etc	Fab/Ros; WL
F. berlandieriana	Berlandier ash	Mixed HW	Olea; LoSG	
Persea americana	Avocado	Mixed HW	Cor/Eri/Lau/Etc	
Ligustrum sinense	Chinese privet	Mixed HW	Olea; HiSG	
Q. gravesii	Graves oak	H Maple/Oak	Faga; Decid	Fagac; WL
Q. polymorpha	Mexican white oak	H Maple/Oak	Faga; Decid	Fagac; WL
Q. buckleyi	Buckley oak	H Maple/Oak	Faga; Decid	Fagac; WL
Q. laceyi	Lacey oak	H Maple/Oak	Faga; Decid	Fagac; WL
Cordia boissieri	Anacahuita Texas olive	Mixed HW	Cor/Eri/Lau/Etc	
Tamarix aphylla	Athel tamarisk	Mixed HW	Cor/Eri/Lau/Etc	

The first part of the Chojnacky parameter designator is the species group; text after a semicolon indicates the relevant category when more than one set of coefficients is given for a group

HiSG the coefficients given for the highest specific gravity in the designated species group, *LoSG* the lowest specific gravity given for a species group, *MedSG* select the coefficients given for the mid-range specific gravity. *WL* select the set of coefficients given for the woodland type. For example, Fagac; WL indicates that the second to the last line of Table 5, Woodland, Fagaceae should be used rather than the coefficients provided for Hardwood; Fagaceae

equivalent versus different is set by researchers and a conclusion of not-different, or equivalent, results from rejecting the null hypothesis (that the two are different).

Equivalence tests presented here are paired-sample tests [24, 25] because each sample is based on estimates from each of the Chojnacky and Jenkins groups. Our test statistic is the difference between estimates (Chojnacky minus Jenkins), and we set "equivalence" as a mean difference less than 5 % of the Jenkins-based estimate. Putting our test in terms of the null and alternative hypotheses following the format of publications describing this approach [22, 24], we have:

Null, H_0: (Chojnacky-Jenkins) <-5 % Jenkins or (Chojnacky-Jenkins) >5 Jenkins

and

Alternative, H_1: -5 % Jenkins $\leq$ (Chojnacky-Jenkins) ≤ 5 % Jenkins

We use the two one-sided tests (TOST) of our two-part null hypothesis that the plot-level difference was greater than 5 % of the Jenkins value and set $\alpha = 0.05$—one test that the mean difference is less than minus 5 % of the Jenkins estimate, and one test that the mean difference is greater than 5 % of the Jenkins estimate. Within an application of the TOST where α is set to 0.05, a one-step approach to accomplish the TOST result is establish a 2-sided 90 % confidence interval for the test statistic; if this falls entirely within the prescribed interval then the two populations can be considered equivalent [26]. We also extended the level of "equivalence" to within 10 % of the Jenkins-based estimates for some analyses in order to look for more general trends, or broad agreement between the two approaches.

Our equivalence tests are based on the paired estimates of carbon tonnes per hectare on the single-condition forested plots variously classified according to regions described in Fig. 1, forest type-groups listed in Table 1, or stand size class as in Fig. 3 (see [13] for additional details about these classifications). The distribution of the test statistic (mean difference) was obtained from resampling with replacement [27] ten thousand times, with a mean value determined for each sample. The number of plots available varied depending on the classification (Table 1; Fig. 3). We did not test for equivalence if fewer than 30 plots were available, and if over 2000 plots were available we randomly selected 2000 for resampling. The choice of 2000 is based on preliminary analysis of these data that showed the confidence interval from resampling converge with percentiles obtained directly from the distribution of the large number of sample plots, usually well below 1000; the 2000 is simply a round number well beyond this convergence without getting too computationally intense. The 90 % confidence interval (the same as the 95 % interval of TOST) obtained for the

distribution of the mean difference is according to a bias corrected and accelerated percentile method [28, 29]. Note that our tests for equivalence are based on comparing this confidence interval to the ± 5 % of the corresponding Jenkins based estimate. Table 1 provides the estimates from the two approaches, with the equivalence test results indicated with asterisks. Similarly, the equivalence test results in Fig. 3 are not in the tonnes per hectare of the resampled values and the confidence interval, they are represented as percentage of Jenkins estimates— for this, equivalence is established if the entire confidence interval is within the zero side of the respective 5 %.

Authors' contributions

Design and analysis was split equally between JS and CH; JS was responsible for coding and calculations, CH developed the figures and tables, and writing was equally divided between JS and CH. All authors read and approved the final manuscript.

Acknowledgements

The authors gratefully acknowledge the helpful feedback from Linda Heath and William Leak on the draft manuscript. We would also like to thank the anonymous reviewers for their time and comments.

Competing interests

The authors declare that they have no competing interests.

References

1. Jenkins JC, Chojnacky DC, Heath LS, Birdsey RA. National-scale biomass estimators for United States tree species. For Sci. 2003;49:12–34.
2. Hoover CM, Rebain SA. Forest carbon estimation using the forest vegetation simulator: seven things you need to know. Gen. Tech. Rep. NRS-77. Newtown Square, PA: U.S. Department of Agriculture, Forest Service, Northern Research Station; 2011.
3. Twery MJ, Knopp, PD, Thomasma, SA, Nute, DE. NED-2 reference guide. Gen. Tech. Rep. NRS-86. Newtown Square, PA: U.S. Department of Agriculture, Forest Service, Northern Research Station; 2012.
4. US EPA. Inventory of U.S. Greenhouse gas emissions and sinks: 1990–2009. EPA 430-R-11-005. U.S. Environmental Protection Agency, Office of Atmospheric Programs, Washington, DC; 2011. http://www.epa.gov/climatechange/ghgemissions/usinventoryreport/archive.html.
5. Liénard JF, Gravel D, Strigul NS. Data-intensive modeling of forest dynamics. Environ Model Softw. 2015;67:138–48.
6. Ziter C, Bennett EM, Gonzalez A. Temperate forest fragments maintain aboveground carbon stocks out to the forest edge despite changes in community composition. Oecologia. 2014;176:893–902.
7. Carter DR, Tahey RT, Dreisilker K, Bialecki MB, Bowles ML. Assessing patterns of oak regeneration and C storage in relation to restoration-focused management, historical land use, and potential trade-offs. For Ecol Manage. 2015;343:53–62.
8. Reinikainen M, D'Amato AW, Bradford JB, Fraver S. Influence of stocking, site quality, stand age, low-severity canopy disturbance, and forest composition on sub-boreal aspen mixedwood carbon stocks. Can J For Res. 2014;44:230–42.
9. DeSiervo MH, Jules ES, Safford HD. Disturbance response across a productivity gradient: postfire vegetation in serpentine and nonserpentine forests. Ecosphere. 2015;6(4):60. doi:10.1890/ES14-00431.1.
10. Dore S, Kolb TE, Montes-Helu M, et al. Carbon and water fluxes from ponderosa pine forests disturbed by wildfire and thinning. Ecol Appl. 2010;20:663–83.

11. Magruder M, Chhin S, Palik B, Bradford JB. Thinning increases climatic resilience of red pine. Can J For Res. 2013;43:878–89.

12. Chojnacky DC, Heath LS, Jenkins JC. Updated generalized biomass equations for North American tree species. Forestry. 2014;87:129–51.

13. USDA Forest Service. Forest Inventory and Analysis National Program, FIA library: Database Documentation. U.S. Department of Agriculture, Forest Service, Washington Office; 2015. http://www.fia.fs.fed.us/library/database-documentation/.

14. Jenkins JC, Chojnacky DC, Heath LS, Birdsey RA. Comprehensive database of diameter-based biomass regressions for North American tree species. Gen. Tech. Rep. NE-319. Newtown Square, PA: U.S. Department of Agriculture, Forest Service, Northeastern Research Station; 2004.

15. Bragg DC, McElligott KM. Comparing aboveground biomass predictions for an uneven-aged pine-dominated stand using local, regional, and national models. J Ark Acad Sci. 2013;67:34–41.

16. Melson SL, Harmon ME, Fried JS, Domingo JB. Estimates of live-tree carbon stores in the Pacific Northwest are sensitive to model selection. Carbon Balance Manag. 2011;6:2.

17. Bragg DC. Modeling loblolly pine aboveground live biomass in a mature pine-hardwood stand: a cautionary tale. J. Ark. Acad. Sci. 2011;65:31–8.

18. USDA Forest Service. Forest Inventory and Analysis National Program: FIA Data Mart. U.S. Department of Agriculture Forest Service. Washington, DC; 2015. http://apps.fs.fed.us/fiadb-downloads/datamart.html. Accessed 2 June 2015.

19. USDA, NRCS. The PLANTS Database. National Plant Data Team, Greensboro, NC, USA; 2015. http://plants.usda.gov. Accessed 23 September 2015.

20. Robinson AP, Duursma RA, Marshall JD. A regression-based equivalence test for model validation: shifting the burden of proof. Tree Phys. 2005;25:903–13.

21. MacLean RG, Ducey MJ, Hoover CM. A comparison of carbon stock estimates and projections for the northeastern United States. For Sci. 2014;60(2):206–13.

22. Parkhurst DF. Statistical significance tests: equivalence and reverse tests should reduce misinterpretation. BioSci. 2001;51:1051–7.

23. Brosi BJ, Biber EG. Statistical inference, Type II error, and decision making under the US Endangered Species Act. Front Ecol Environ. 2009;7(9):487–94.

24. Feng S, Liang Q, Kinser RD, Newland K, Guilbaud R. Testing equivalence between two laboratories or two methods using paired-sample analysis and interval hypothesis testing. Anal Bioanal Chem. 2006;385:975–81.

25. Mara CA, Cribbie RA. Paired-samples tests of equivalence. Commun Stat Simulat. 2012;41:1928–43.

26. Berger RL, Hsu JC. Bioequivalence trials, intersection-union tests and equivalence confidence sets. Stat Sci. 1996;11(4):283–319.

27. Efron B, Tibshirani RJ. An introduction to the bootstrap. New York: Chapman and Hall; 1993

28. Carpenter J, Bithell J. Bootstrap confidence intervals: when, which, what? A practical guide for medical statisticians. Statis Med. 2000;19:1141–64.

29. Fox J. Bootstrapping regression models. In: Applied Regression Analysis and Generalized Linear Models, 2nd edition. Thousand Oaks: Sage, Inc; 2008. pp 587–606.

## 6

Carbon dioxide fluxes from contrasting ecosystems in the Sudanian Savanna in West Africa

Emmanuel Quansah[1,3], Matthias Mauder[2*], Ahmed A Balogun[1], Leonard K Amekudzi[3], Luitpold Hingerl[4], Jan Bliefernicht[4] and Harald Kunstmann[2,5]

Abstract

Background: The terrestrial land surface in West Africa is made up of several types of savanna ecosystems differing in land use changes which modulate gas exchanges between their vegetation and the overlying atmosphere. This study compares diurnal and seasonal estimates of CO_2 fluxes from three contrasting ecosystems, a grassland, a mixture of fallow and cropland, and nature reserve in the Sudanian Savanna and relate them to water availability and land use characteristics.

Results: Over the study period, and for the three study sites, low soil moisture availability, high vapour pressure deficit and low ecosystem respiration were prevalent during the dry season (November to March), but the contrary occurred during the rainy season (May to October). Carbon uptake predominantly took place in the rainy season, while net carbon efflux occurred in the dry season as well as the dry to wet and wet to dry transition periods (AM and ND) respectively. Carbon uptake decreased in the order of the nature reserve, a mixture of fallow and cropland, and grassland. Only the nature reserve ecosystem at the Nazinga Park served as a net sink of CO_2, mostly by virtue of a several times larger carbon uptake and ecosystem water use efficiency during the rainy season than at the other sites. These differences were influenced by albedo, LAI, EWUE, PPFD and climatology during the period of study.

Conclusion: These results suggest that land use characteristics affect plant physiological processes that lead to flux exchanges over the Sudanian Savanna ecosystems. It affects the diurnal, seasonal and annual changes in NEE and its composite signals, GPP and RE. GPP and NEE were generally related as NEE scaled with photosynthesis with higher CO_2 assimilation leading to higher GPP. However, CO_2 effluxes over the study period suggest that besides biomass regrowth, other processes, most likely from the soil might have also contributed to the enhancement of ecosystem respiration.

Keywords: West Africa; Sudanian Savanna; Carbon fluxes; Net ecosystem exchange

Background

Globally, the terrestrial biosphere has been recognised to have the capacity to partially offset anthropogenic CO_2 by carbon assimilation into the biomass through the process of photosynthesis [1,2]. The African ecosystems contribute significantly to the inter-annual variability in the global atmospheric CO_2, mainly through the emission of trace gases from land use changes and bush burning [3,4]. The terrestrial ecosystem in West Africa is made up of several types of savannas which have exhibited strong diurnal and seasonal variability in their CO_2 fluxes, an indication of how the carbon budget can be affected by

the ecosystem's response to meteorological forcing [5]. For example, Brümmer et al. [6] showed that a Southern Sudanian Savanna ecosystem in Burkina Faso was a source of CO_2 into the atmosphere in the dry season, but only marginally, whereas appreciable CO_2 uptake was observed in the rainy season. In addition, Ardö et al. [7] found that for a sparse savanna in the semi-arid Sudan, during two short investigation periods, in February 2005 (dry season), and September 2005 (rainy season), the studied ecosystem was a sink of carbon during both periods.

The contributions of the West African savanna ecosystems to the global carbon budgets as well as factors influencing their impact on the temporal and spatial variation of the terrestrial carbon uptake and emission are still highly uncertain. An insufficient network of sites using measurement equipment to determine CO_2

* Correspondence: matthias.mauder@kit.edu
[2]Institute of Meteorology and Climate Research, Karlsruhe Institute of Technology, Garmisch-Partenkirchen, Germany
Full list of author information is available at the end of the article

fluxes such as eddy covariance (EC) stations have been identified as one of the main reasons [6]. Although over the years many EC experiments have been performed in West Africa within various research projects such as the Sahelian Energy Balance EXperiment (SEBEX, [8]), Hydrological and Atmospheric Pilot Experiment-Sahel in Niger (HAPEX-Sahel, [9-11]), the African Monsoon Multidisciplinary Analyses (AMMA) project [12], and the CARBOAFRICA project [13], the network of eddy covariance stations in West Africa is still very sparse compared to the network in North America, Europe and Asia, [14]. A situation which has resulted in a limited scope of comprehensive climatic information essential for the development of regional climate adaptation policies especially in setting CO_2 reduction targets. Therefore, attempts to generalise or extrapolate outcomes of flux measurements from one area to another area of a given target region based on results from former flux measurements [15] are associated with high uncertainties. The main reason is that carbon dynamics vary for different ecosystems, depending on the climate, land management and further factors. Another reason is that many former measurement experiments were mostly performed for short periods, e.g. for a specific season or for several months and for a few selected sites [15,16].

As part of the efforts to develop effective adaptation and mitigation measures to climate change and land use in West Africa, the meteorological network has been refined in three regions of the Sudanian Savanna of West Africa by installing further climate stations and three eddy covariance (EC) stations. The establishment of the EC and climate stations has been realised within the framework of the West African Science Service Centre on Climate Change and Adapted Land Use (WASCAL) project. The EC stations have been established in October 2012 and January 2013 in the West African Sudanian Savanna along a gradient of changing land use (cover) characteristics (grassland, a mixture of fallow and cropland and nature reserve), close to the border between Ghana and Burkina Faso [17].

The objective of this study was to perform an intercomparison of CO_2 flux exchanges across three contrasting ecosystems, and to quantify their magnitude and temporal variability in net ecosystem exchange (NEE), gross primary production (GPP) and ecosystem respiration (ER) in responses to changing vegetation characteristics and different soil moisture conditions. This will improve our understanding of the impact of physiological and environmental factors on CO_2 efflux and sequestration patterns over the region. It will also contribute to our knowledge regarding how NEE and its composite signals (GPP and ER) vary with land use change.

Results and discussions
General meteorology
The meteorological conditions (air temperature and relative humidity) over the study areas were evaluated at each individual EC site. The results (Figure not shown) revealed that the variability in the daily averages of air temperature (Tair), relative humidity (RH) and precipitation during the study period was typical of the Sudanian Savanna. The highest daily mean Tair of 34.49, 34.73 and 34.88°C, were recorded in March in the dry season (November – April) for the Nazinga Park, Sumbrungu and Kayoro respectively. While the minimum daily mean Tair of 22.04, 22.58 and 23.27°C occurred in August for Kayoro and in September, the rainy season (May–October) for the Nazinga Park and Sumbrungu respectively. The daily averaged RH ranged between 8.64 to 94.38%, 7.86 to 93.28 and 7.80 to 95.13% for the Nazinga Park, Sumbrungu and Kayoro respectively. The minimum values of RH were recorded in February, while the maximum values were obtained in August for the Nazinga Park and Kayoro, and in September for Sumbrungu.

Annual rainfall obtained from Sumbrungu was approximately 375 mm yr^{-1}. In this station, gaps (N = 188) in the dry season (December) were assumed to be zero. Results from a 'gap-free' data at a nearby climate station in Bongo Soe (Table 1) with similar vegetation characteristics as Sumbrungu was 542 mm yr^{-1} (Figure 1). This suggest rainfall over the region was sporadic during the period. Nevertheless, our values were still within what had been reported (320–1100 mm) over the Southern Sudanian Savanna [7,18-21]. Annual rainfall from the remaining EC stations could not be determined because of several gaps in the rainfall data, especial during the rainy season.

Volumetric soil moisture content
The daily averages of the volumetric water content measured from the topsoil at 0.03 m from each EC sites were converted into water-filled pore space (WFPS) for the one-year study period based on the formula described in Brümmer et al. [6]. The results during the period ranged between 3.90 to 34.05% for

Table 1 Coordinates of the climate stations surrounding the three eddy covariance sites

Name of climate station	Latitude (°)	Longitude (°)	Elevation (m)
Bongo Soe	10.973	-0.783	200
Doninga	10.617	-1.420	162
Nabuobelle	10.703	-1.869	328
Oualem	11.205	-1.309	295
Nebou	11.305	-1.879	310
Tabou	11.367	-2.169	312
Gwasi	10.478	-1.648	284

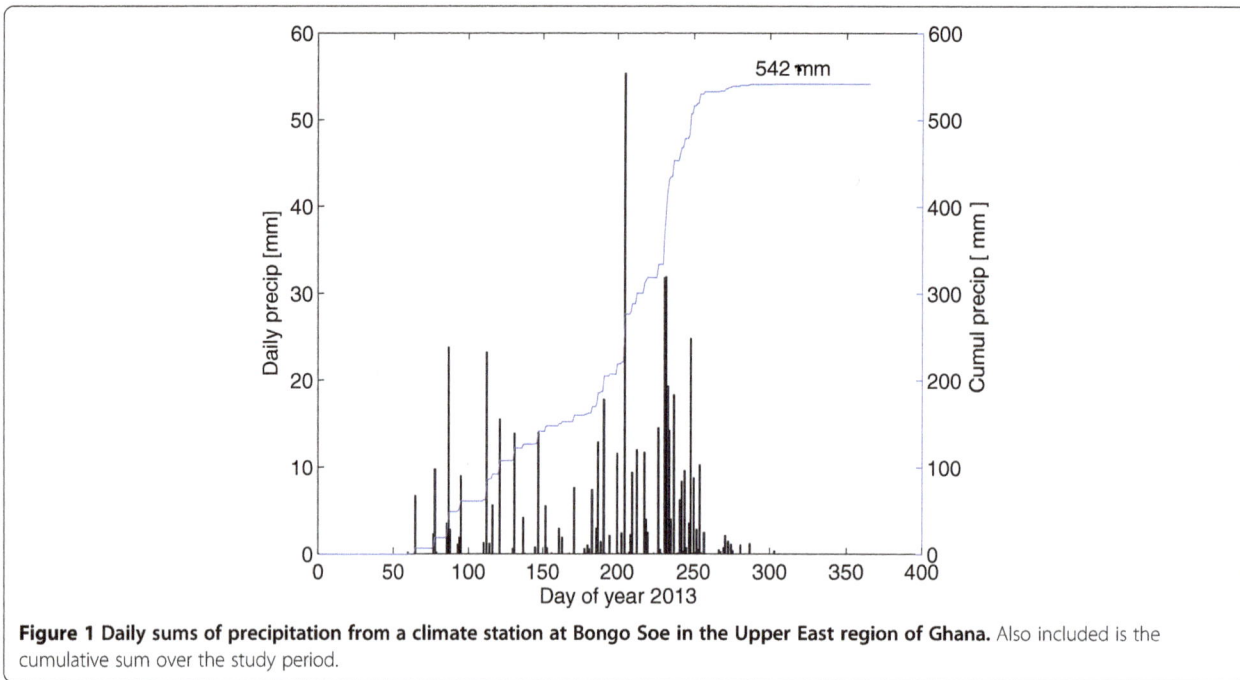

Figure 1 Daily sums of precipitation from a climate station at Bongo Soe in the Upper East region of Ghana. Also included is the cumulative sum over the study period.

Sumbrungu, 2.69 to 30.24% for Kayoro and 3.76 to 43.25% for the station in the Nazinga (Figure 2), which suggest that the top few centimetres of the soil (5 cm) was not entirely dried (not below 3%) even in the dry season (November to April). The soil moisture increased after the onset of the rain in March in the dry season. Afterwards, the soil water content decreased

drastically again until the beginning of the rainy season in May. There was strong variability in the WFPS during the rainy season (May to October), but no clear indication of the soil moisture reaching saturation condition. The soil moisture begun to dry up at the end of the rainy season in October. However, the WFPS was never close to zero. There was a strong correlation between the near surface

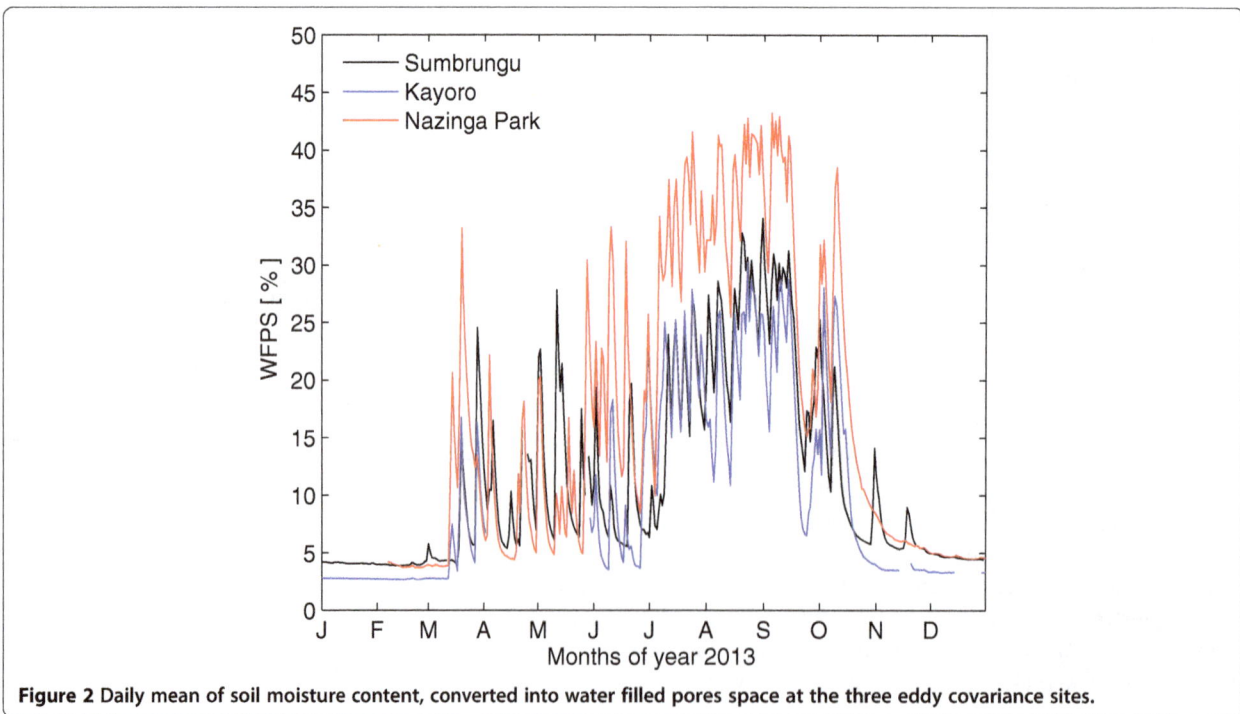

Figure 2 Daily mean of soil moisture content, converted into water filled pores space at the three eddy covariance sites.

soil moisture and the precipitation, an indication of a quick response to rain events by the topsoil.

Evaluation of the energy balance closure

The half-hourly averaged flux and meteorological data set of G, H, λE and R_n (number of data points (N) = 12240, 11470 and 10291) for Sumbrungu, Kayoro and Nazinga respectively, were used to evaluate the performance of the eddy covariance measurements for the entire study period, based on the linear regression relation between the turbulent fluxes ($H + \lambda E$) and available energy ($R_n - G$). The regression fits between ($R_n - G$) and ($H + \lambda E$) resulted in a coefficient of determination of 0.89 for Sumbrungu, 0.90 for Kayoro and 0.92 for Nazinga. The slopes and intercepts were 0.67 and 32.60 W m^{-2}, 0.67 and 11.34 W m^{-2}, and 0.89 and 9.85 W m^{-2} for the three sites accordingly (Figure 3). The slopes of our regression indicated that the energy balance closures (EBC) were moderate for Sumbrungu and Kayoro. Our results were within the typical closure since the sum of the turbulence fluxes ($H + \lambda E$) is usually found to be 10 to 30% less than the available energy (Rn − G) due to large-scale transport not captured by regular eddy covariance tower measurements and measurement errors associated with individual instruments [22-24].

The EC station at Kayoro was selected to determine the impact of meteorological conditions on the EBC for four different periods within the year (JFM, AM, JJASO and ND), that corresponded with the complete dry season, dry to wet transition, rainy season, and wet to dry transition respectively. The results revealed that the EBC was variable with slopes of 0.66, 0.50, 0.65 and 0.58 according to the analysed periods. In addition, analyses of the EBC at Kayoro for two selected days, one without rain and the other with total rain of 8.6 mm day^{-1} in the rainy season, gave a slight reduction in the EBC for the latter. These suggested that over the studied

period the EBC could have been affected by the changes in the meteorological conditions prevalent during the dry and rainy seasons. It is therefore likely that besides the possible causes of the non-closure of the EBC as mentioned above, the scattered feature of the points observed over the one-year study period was due to the sensitivity of the measurement EC device to rain drops and dust. As we focused on the carbon dioxide fluxes, no adjustments were made to compensate for the imbalances in the EBC in this study, which is in accordance with Foken et al. [25].

Seasonal changes in daily NEE

The seasonal changes in daily sums of NEE for the three study sites are illustrated in Figure 4. There were effluxes between March and April, the transition between the dry to rainy season. And these were prominent for Sumbrungu and Kayoro, peaking at 4.97, 5.55 and 2.62 g C m^{-2} d^{-1} for Sumbrungu, Kayoro and Nazinga respectively. This was followed predominantly by net uptake for Nazinga between May and peaking at -6.78 g C m^{-2} d^{-1} at the end of August, and decreasing until the end of November, the transition period between the rainy and dry season. However for Sumbrungu and Kayoro both uptakes and effluxes were observed between the rainy season (May to October), with maximum uptakes in August and September at -2.93 and -3.51 g C m^{-2} d^{-1} for Sumbrungu and Kayoro respectively.

The occurrence of the maximum daily NEE uptake in the rainy season over the one-year investigated period, revealed the important roles precipitation and soil moisture play on the rates of CO_2 uptakes and effluxes at the studied ecosystems.

The evolution of the albedo, a function comprising of the surface and radiation characteristics, such as the land cover type, vegetation phenology, soil moisture, incident angle, and wavelength [26-28], was assessed at

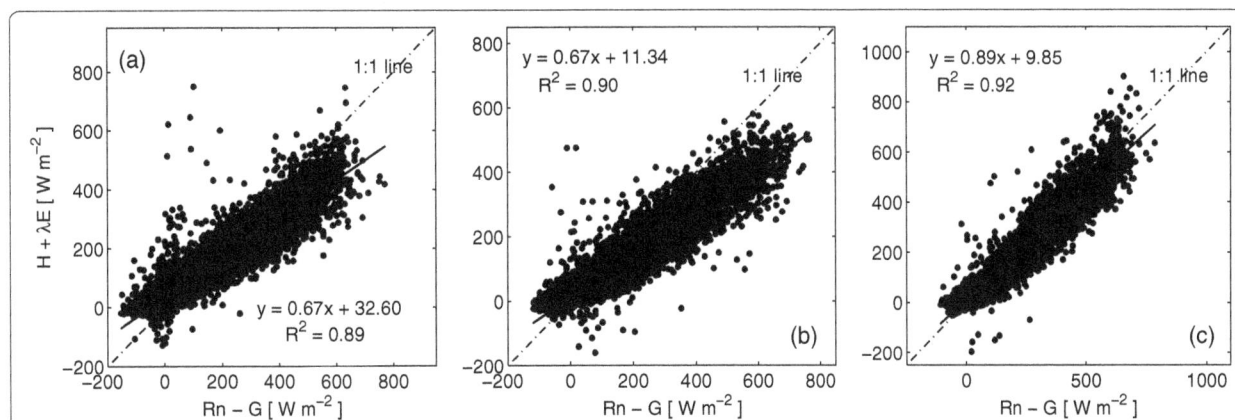

Figure 3 Energy balance closure for the study period at (a) Sumbrungu, (b) Kayoro, all in the Upper West Region of Ghana, and (c) the Nazinga Park near Pô in the Nahouri province in Burkina Faso.

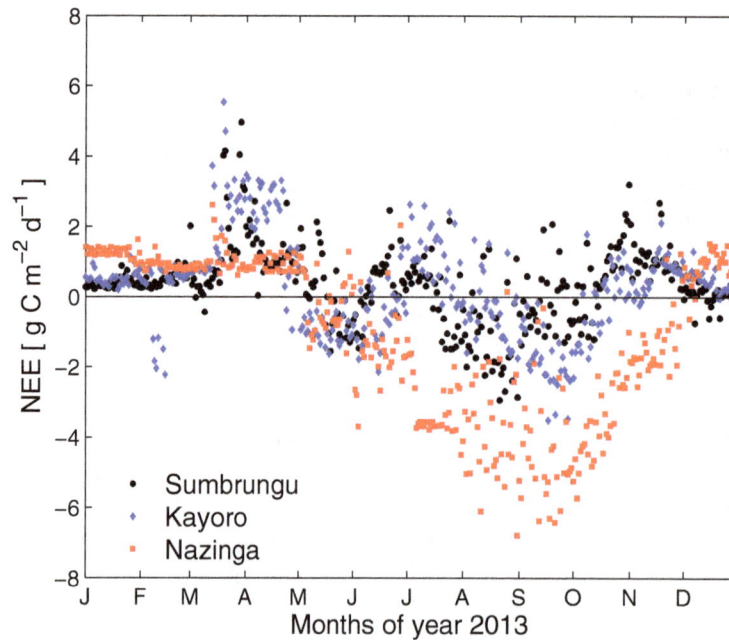

Figure 4 Seasonal changes in daily sums of net ecosystem exchange at the eddy covariance sites.

the study sites over the period. The results suggested a decrease in albedo from the onset of the rain in March until the end of November, before it started to rise again until the end of December. Figure 5 shows the results of the daily mean albedo and the daily mean energy fluxes (sensible and latent heat) during the period for Sumbrungu, which ranged between 0.20–0.38. Similar patterns were observed for Kayoro and Nazinga (Figures not shown), between 0.16–0.29 and 0.12–0.27 respectively. The high albedo values were observed in the dry season when most of the leaves on the ecosystem had withered and sensible heat dominated the energy exchanges. The lower values occurred in the rainy season when an enhancement in soil moisture (high latent heat)

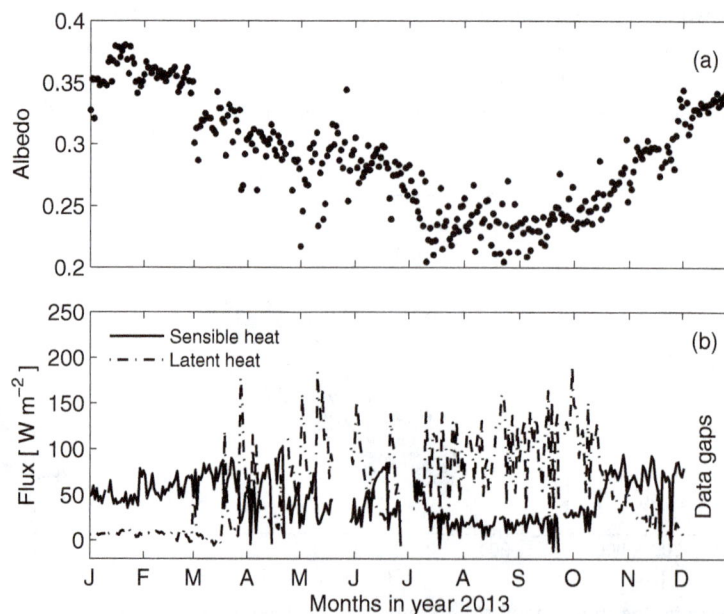

Figure 5 Seasonal changes in daily mean of (a) albedo, and (b) energy fluxes (sensible and latent heat) at Sumbrungu.

led to productivity pulsation, followed by the accumulation of biomass. Leaf area index measured for Sumbrungu and Kayoro between June and October 2013 ranged between 1.3–2.5 and 3.0–5.3 m^2 m^{-2} respectively.

The ecosystem water use efficiency (EWUE) was simply calculated as the ratio of the rate of net CO_2 uptake (GPP) to the water vapour flux (WVF) according to Law et al. [29]. However, no filtering was applied to remove overestimation as a result of precipitation events as described in Grelle et al. [30]. The linear regression between the daily integrated GPP and WVF (Figure 6), revealed that the slopes (EWUE) were 1.35, 1.46 and 2.05 μmol CO_2 m^{-2} $mm(H_2O)^{-1}$ for Sumbrungu, Kayoro and the Nazinga Park respectively, during the entire period of the study. This suggest evapotranspiration and gross primary production were strongly related and that EWUE showed daily variability throughout the year. However, the values of EWUE obtained for the three sites during the rainy season (May to October) were 1.68, 1.71 and 2.17 μmol CO_2 m^{-2} $mm(H_2O)^{-1}$ for Kayoro, Sumbrungu and Nazinga Park respectively. This was an indication that the difference in the climatology between the dry and rainy season as well as the land surface characteristics and the morphology of the vegetation influenced the NEE and its composite fluxes.

Increase in soil moisture during the rainy season, enhanced the processes leading to both autotrophic and heterotrophic respiration, which resulted in increasing rates in the daily sums of GPP and ER in the increasing order, from Kayoro to Sumbrungu and to Nazinga

(Figure 7), which agreed well with the EWUE values obtained in the rainy season for the respective stations. Figure 8 illustrates the disparities in how the NEE scaled with photosynthesis. Higher carbon dioxide assimilations (GPP) coincided with higher uptake rates of NEE decreasing in the order of Nazinga, Kayoro and Sumbrungu accordingly. The higher values of NEE during the day for all the three sites coupled with photosynthetic photon flux density (PPFD, Figure 9), with maximum NEE occurring at PPFD of 2000 μmol m^{-2} s^{-1}.

Seasonal variability of monthly carbon fluxes

The seasonality of the carbon fluxes including the GPP and ER derived from the half-hour measurements of the gap-filled-NEE at the study areas was investigated using their monthly sums (Figure 10). The outcomes illustrate strong variability in the monthly fluxes for the three sites over the study period, mostly influenced by the soil moisture. The NEE for the complete dry season (JFM) for Sumbrungu, Kayoro and Nazinga were respectively, 68.58, 83.25 and 97.52 g C m^{-2}. In contrast, during the rainy season (JJASO) NEE values were -14.29, -51.02 and -494.68 g C m^{-2} for Sumbrungu, Kayoro and Nazinga respectively. Analyses for the transition periods, i.e. the dry to wet transition (AM) showed NEE rates at 39.56, 39.35 and 21.90 g C m^{-2} for Sumbrungu, Kayoro and Nazinga. While the wet to dry transition (ND), also showed NEE rates at 1.96, 15.46 and 27.32 g C m^{-2}. The results suggest that the ecosystems served as sources of CO_2 during the dry season (November – April), as well as the transition periods AM and ND when soil

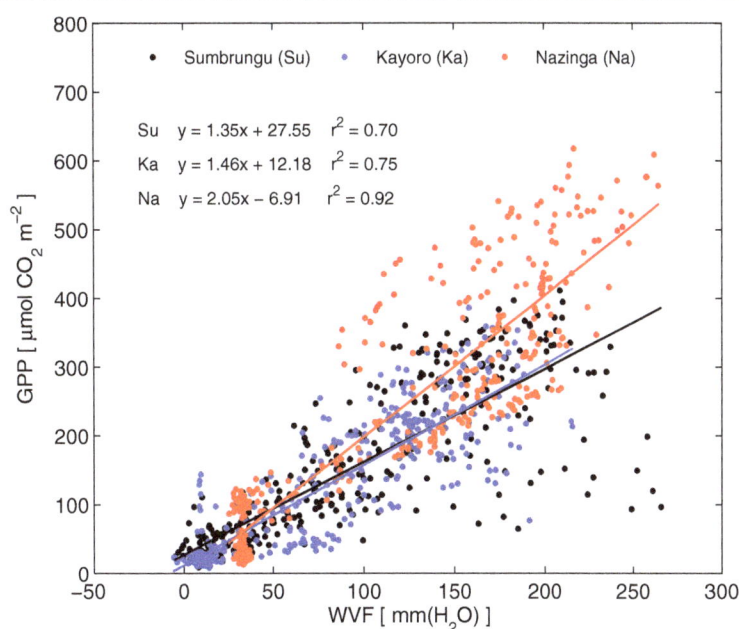

Figure 6 Daily sums of gross primary production against daily sums of water vapour flux.

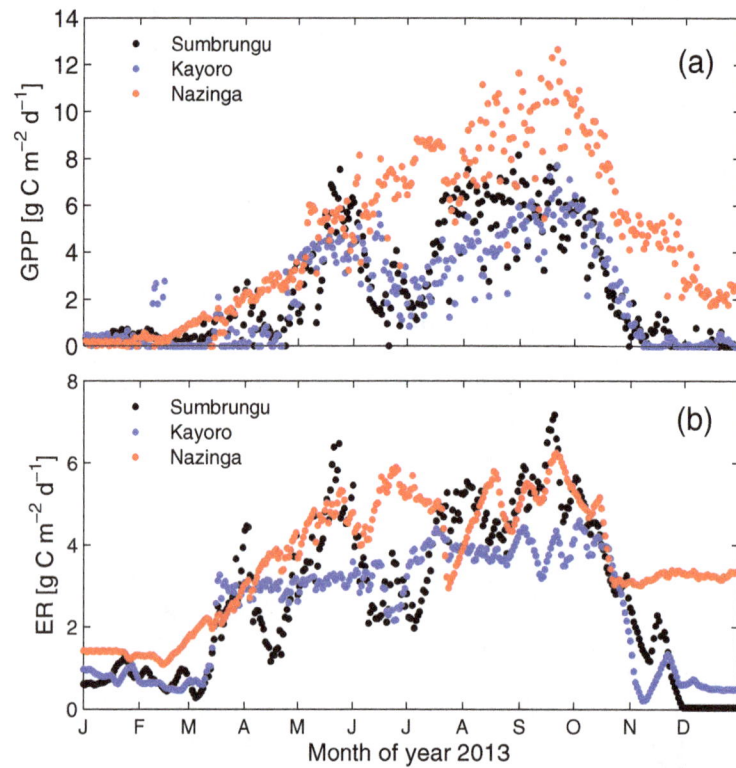

Figure 7 Seasonal changes in (a) daily sums of gross primary production, and (b) daily sums of ecosystem respiration.

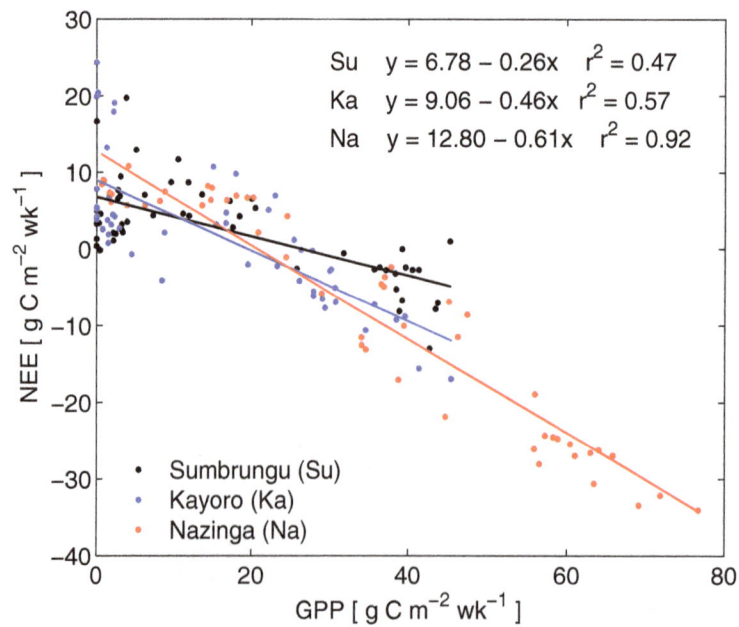

Figure 8 Scaling of weekly sums of ecosystem carbon exchange (NEE = net ecosystem exchange) with weekly sums of ecosystem photosynthesis (GPP = gross primary production).

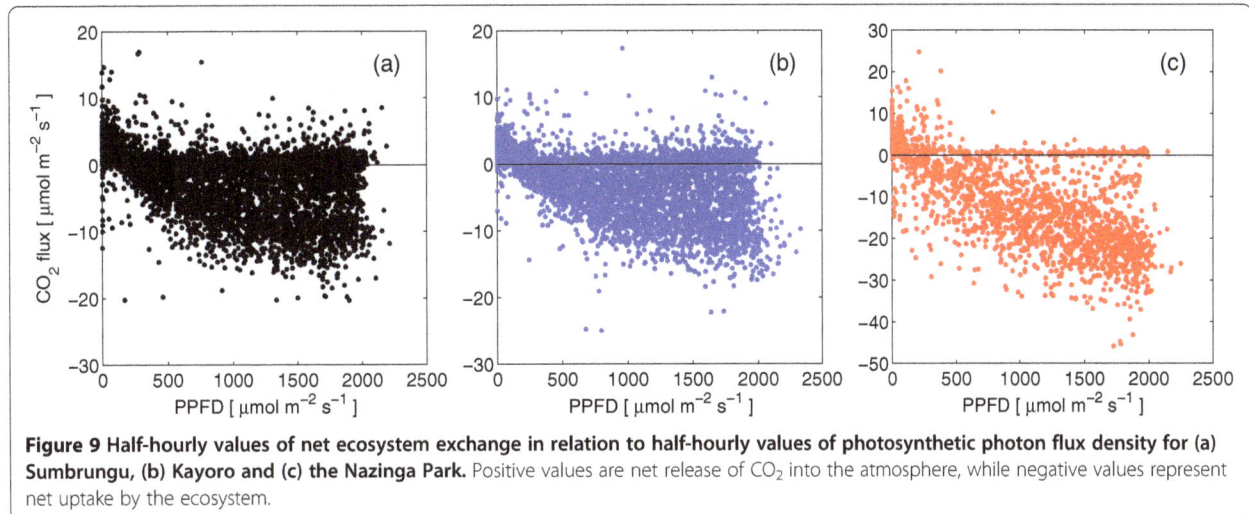

Figure 9 Half-hourly values of net ecosystem exchange in relation to half-hourly values of photosynthetic photon flux density for (a) Sumbrungu, (b) Kayoro and (c) the Nazinga Park. Positive values are net release of CO_2 into the atmosphere, while negative values represent net uptake by the ecosystem.

moisture was low. However, during the rainy season (May–October), the ecosystems served as sinks of CO_2 at all the three sites. During this period the NEE in Kayoro was more than three times that at Sumbrungu, while that of Nazinga was almost ten times that of Kayoro. While all three ecosystems had a similar NEE during the dry season, the nature reserve ecosystem was a much larger carbon sink in the rainy season.

Furthermore, the results for ND revealed a quick switch from net uptake to net release during the year, but the same cannot be said for AM.

In July 2013, during one of the routine visits to the field site in Kayoro, it was observed that a residence in a nearby community had planted cowpea (*Vigna unguiculata*) in the vicinity of the EC setup that extended into our fetch. Although we removed the portion that

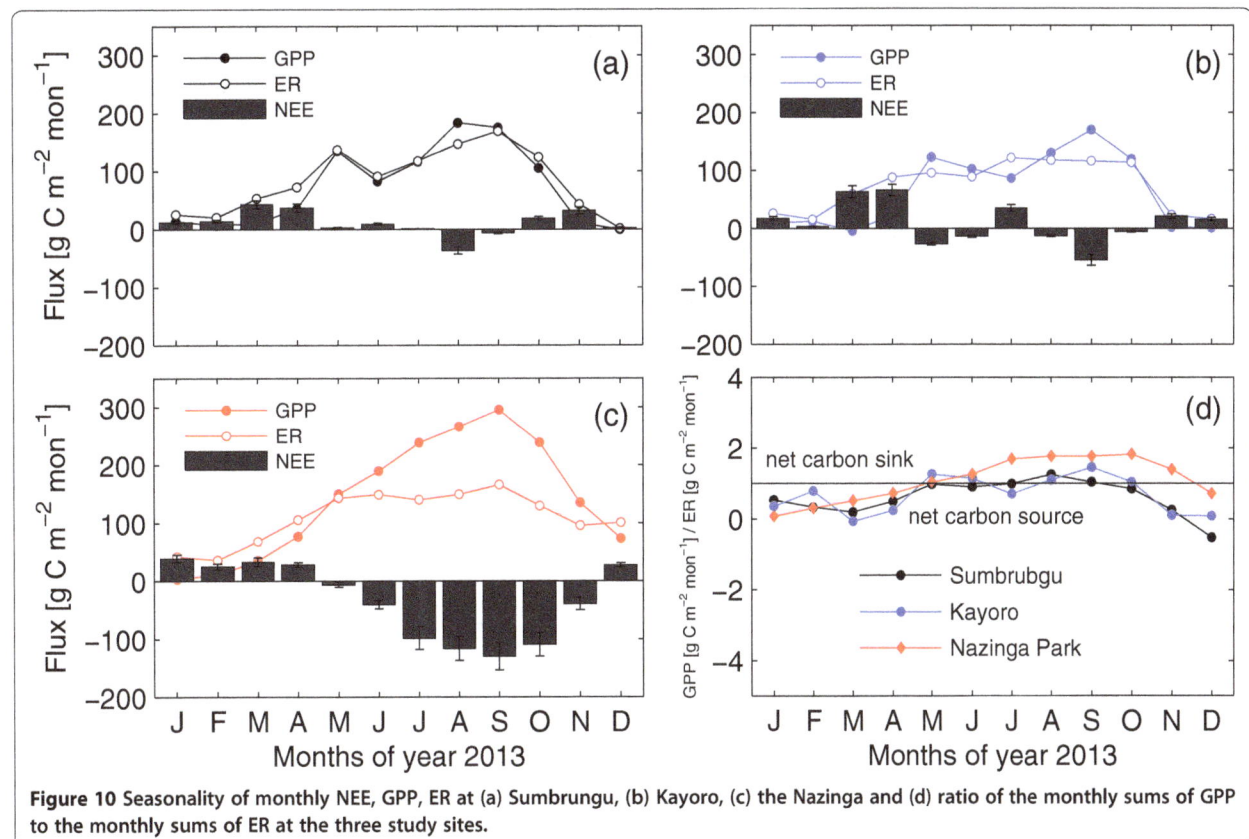

Figure 10 Seasonality of monthly NEE, GPP, ER at (a) Sumbrungu, (b) Kayoro, (c) the Nazinga and (d) ratio of the monthly sums of GPP to the monthly sums of ER at the three study sites.

affected our fetch to allow for regrowth of grass, it is likely that the preparations for the cowpea plantation caused the sudden release of CO_2 in July (Figure 10b). If this is the case, then considering the fact that cowpea is one of the principal crops cultivated in the region, land management over the region could have considerable impact on the diurnal, seasonal and annual variability of CO_2 uptake and release across the region.

The outcomes of the monthly (seasonal) integrated total amount of productivity in the ecosystem, GPP and ER derived from the gap-filled NEE over the period for all the study sites showed that, in all cases, the assimilation of carbon (GPP) was mostly limited to the rainy season when soil moisture content was increased. Furthermore, ER rates were clearly higher in the rainy season compared to the complete dry season as well as the transitions periods (Figure 10a,b and c). This may be attributed to the enhancement of the processes that lead to both autotrophic and heterotrophic respirations 'fuelled' by the general plant growth driving forces (precipitation and soil moisture). The higher carbon assimilation (GPP) rates corresponded with higher NEE uptakes in the rainy season. The GPP to ER flux

partitioning ratios (Figure 10d) revealed a net carbon uptake (GPP/ER > 1) for months from May to November and peaking in October, the end of the rainy season, especially for Nazinga. The ratios decreased after November as soil water content declined (see Figure 2) until the beginning of the rainy season in May. In the dry season, high air temperature and low relative humidity values led to a high vapour pressure deficit. This resulted in a stronger driving forces on evapotranspiration and hence low ER rates during the complete dry season (JFM) and the transition periods (AM and ND), compared to the rainy season.

The estimated annual (January to December) carbon fluxes (NEE, GPP and ER) derived from their cumulative daily sums over the period showed that only the nature reserve ecosystem at Nazinga served as a net sink of CO_2 at -387.3 ± 23.1 g C m^{-2} yr^{-1}, while the remaining two sites were net sources of CO_2 into the atmosphere at 127.8 ± 7.2 and 108.0 ± 5.5 g C m^{-2} yr^{-1} for Sumbrungu and Kayoro respectively (Figure 11a). The corresponding values of the annual cumulative GPP and RE over the period (Figures 11b and c) are provided in Table 2.

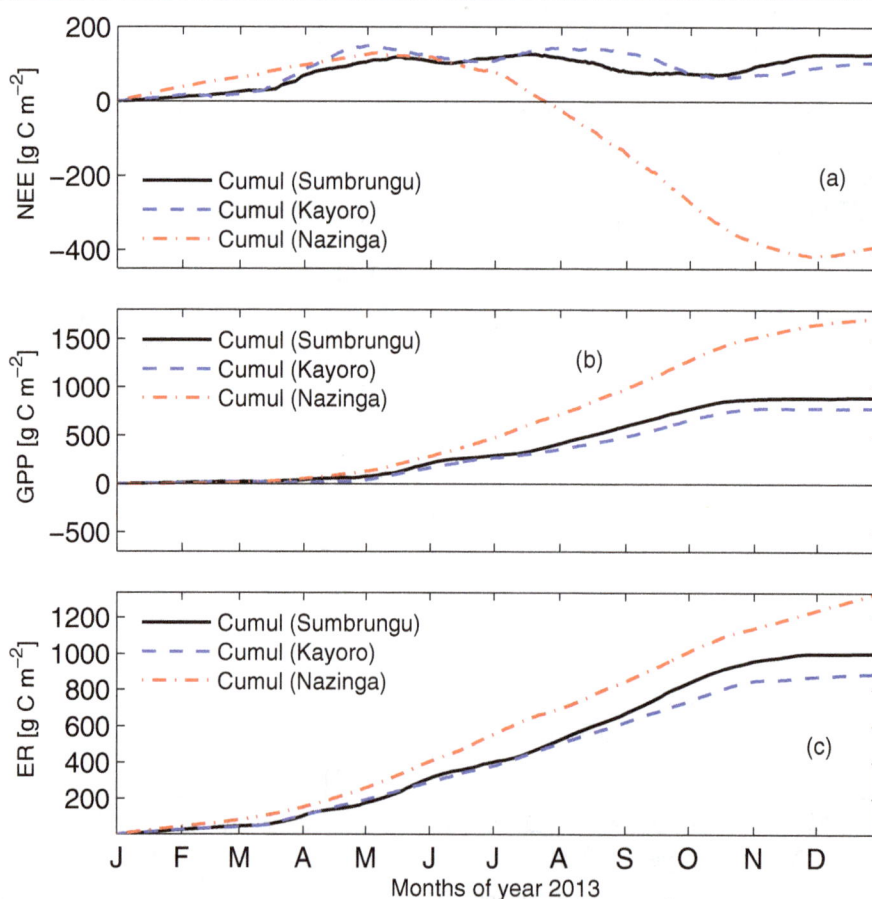

Figure 11 Annual aggregation of (a) measured NEE, (b) estimated GPP, and (c) estimated ER over the study sites.

Table 2 Summary of the seasonal and annual integrated carbon fluxes (NEE = net ecosystem exchange, GPP = gross primary production, ER = respiration; [g C m^{-2} (time unit)$^{-1}$]) over the study sites, with their corresponding standard deviations (std)

Variable	Sumbrungu		Kayoro		Nazinga Park	
	Value	std	Value	std	Value	std
Rainy season NEE	-11.8	0.5	-77.6	4.6	-501.2	38.9
Dry season NEE	139.5	10.8	185.6	12.3	113.9	6.4
Rainy season GPP	798.0	108.3	734.4	95.5	1385.0	193.4
Dry season GPP	76.0	11.8	46.9	6.4	340.6	57.1
Rainy season RE	786.3	102.8	656.8	87.1	883.4	114.0
Dry season RE	215.5	34.2	232.5	36.7	454.4	65.8
Annual NEE	127.8	7.2	108.0	5.5	-387.3	23.1
Annual GPP	874.0	17.8	781.3	15.8	1725.1	32.6
Annual ER	1001.8	19.0	889.3	16.5	1337.8	23.0

Comparison of our GPP results with the alternative flux partitioning approach of Lasslop et al. [31] resulted in coefficients of determinations of 0.74, 0.89 and 0.92 and regression slopes of 0.80, 0.92 and 0.96 for Nazinga, Kayoro and Sumbrungu respectively. Similarly, the ER comparison gave coefficient of determination of 0.36, 0.79 and 0.83 and regression slopes of 0.65, 0.97 and 0.70 for Nazinga, Kayoro and Sumbrungu respectively. While the NEE comparison resulted in slopes of 0.84, 0.70 and 0.70 and coefficient of determination of 0.84, 0.71 and 0.67 for Nazinga, Kayoro and Sumbrungu respectively. This relatively good agreement shows that the MDS gap-filling technique is quite suitable for our study sites within the typical uncertainty range of such estimates of 20 to 30% [32]. Although the MDS approach resulted in larger estimates of both NEE, GPP and ER, the estimate for the cumulative NEE agreed quite well for all three sites (Table 2).

Comparison with selected former flux measurements

Our daily values of photosynthetically driven carbon uptake of -2.93, -3.51 and -6.78 g C m^{-1} d^{-1} for the grassland, mixture of fallow and cropland and nature reserve respectively compared well with values of -0.48, -1.8, -2.4 -3.6 and -5.9 g C m^{-1} d^{-1} obtained for some former flux measurements (Table 3). Similarly, the rates of daily release of CO_2 from our study sites at 4.97, 5.55 and 2.62 g C m^{-2} d^{-1} for Sumbrungu, Kayoro and Nazinga respectively, were comparable with values of 0.4, 0.6 and 2.4 g C m^{-1} s^{-1}, from the former flux measurements (Table 3) [7]. The monthly and annual fluxes also compared well. Although the comparison revealed some slight variability in CO_2 efflux and uptake values for the variety of ecosystems presented (Table 3), the general observation revealed that soil moisture as well as other physiological properties of the ecosystems were very significant in the CO_2 sequestration and efflux patterns

over the regions. Example, a nature reserve near Bontioli in Burkina Faso, which shared similar characteristics with our third EC site, the Nazinga Park, produced total net ecosystem CO_2 uptake of -179 ± 98 g C m^{-2} yr^{-1} in the first year and -429 ± 100 g C m^{-2} yr^{-1} in the second year of investigations. And the large difference in values between the two years was attributed to the increment in the rainfall amount, 785 mm, in the first year as against 919 mm in the second year [6]. Comparing those daily, monthly and annual estimates with values obtained from our study sites suggest that ecosystems that shared similar characteristics were more comparable with each other.

The annually estimated GPP values obtained from our studies were 874.0, 781.3 and 1725.1 g C m^{-2} yr^{-1} (Table 2). These looked similar to values obtained from studies summarised in Sjöström et al. [33] for a variety of ecosystems in Africa that ranged approximately between 50 to 2050 g C m^{-2} yr^{-1}.

Summary and conclusions

The carbon fluxes of three contrasting ecosystem in the Sudanian Savanna in West Africa have been estimated using one year of eddy covariance CO_2 flux data. Three eddy covariance stations were built close to the Ghanaian-Burkinabe border in October 2012 and January 2013. The first EC site represents a grassland ecosystem, which served as a pasture occasionally to the cattle and sheep owned by the inhabitants living in the nearby communities. The natural vegetation in the area is characterised by grasses of average height, 0.10 m when fully grown after the rainy season in October. The second site is dominated by tall grasses with an average height of 1 m after the rainy season in October and represents a mixture of fallow and cropland, and the third EC site is a representation of a near-natural Sudanian Savanna with no agricultural activities. The latter is located in a nature reserve area and

Table 3 Net ecosystem exchange from former flux measurements

Site	Vegetation	MAP	Annual NEE	Monthly NEE	Daily NEE
Baja	Desert shrub	174	-39, -52	$0.7 - 25^D$	0.4^D
				-12 to -41^R	-0.48^R
Demokeya	Sparse savanna	320	-	-	-0.2^D
					-1.8^R
Maun	Savanna woodland	464	12	-	$1.2 - 2.4^D$
					-0.6 to -2.4^R
Hapex	Sahelian fallow savanna	495	-32	-	$\sim0.3^D$
					0.6 to -3.6^R
Virgina Park	Open woodland savanna	667	44	$\sim6^D$	-
				14 to -52^R	-
Bontioli	Shrub dominated savanna	926	-179, -429	$5 -20^D$	$0.2 - 0.4^D$
				-35 to -175^R	-1.4 to -5.9^R
Aguas Emendadas	Cerrado	1500	-		0.6^D
				-	-1.2^R
Sumbrungu	Short grassland savanna	375	127.8	$2 - 43^D$	4.97^D
				-6 to -37^R	-2.93^R
Kayoro	Fallow land	-	108	$3 - 66^D$	5.55^D
				-6 to -54^R	-3.51^R
Nazinga Park	Nature reserve	-	-387.3	$25 - 39^D$	2.62^D
				-6 to -130^R	-6.78^R

$[NEE] = g\ C\ m^{-2}\ (time\ unit)^{-1}$.
MAP – Mean Annual Precipitation, mm yr^{-1}.
NEE – negative values denote uptake.
[R] – Rainy season.
[D] – Dry season.

characterised by tall grass of approximately 3 m when fully grown after the rainy season in October.

Carbon uptake predominantly took place during the rainy season, while a net carbon release was observed in the dry season as well as during the transition periods between the dry and rainy seasons (i.e. AM and ND). The seasonal trends in NEE varied substantially among all the three sites. Climatology and biophysical factors including, LAI, albedo, surface roughness length modulated the phenological processes leading to biomass regrowth and CO_2 assimilation at the study sites. It appeared that soil moisture availability played a significant role in the variability of NEE during the transition periods.

Our study revealed that land use and management had a large impact on the disparities in NEE values at the study sites. Generally, the carbon uptake decreased in the order from nature reserve over mixture of fallow and cropland to grassland. Only the nature reserve ecosystem at Nazinga served as a net sink of CO_2, mostly by virtue of a much larger carbon uptake and ecosystem water use efficiency during the rainy season than at the other sites.

In order to improve our knowledge about the impact of the plant-physiological processes on flux exchanges over the three contrasting ecosystems, numerical simulations of energy and CO_2 fluxes as well as a gap-filling model with a special parameterisation adapted to such rain driven ecosystems over the Sudanian Savanna are warranted.

Materials and methods
Study site description

Three eddy covariance stations were built close to the Ghanaian-Burkinabe border (Figure 12) in October 2012 and January 2013. The first EC station was located in Sumbrugu Aguusi (10.846° N, 0.917° W, 200 m a.s.l.) within the catchment of the river Vea, about 35 km from Bolgatanga in the Upper East Region of Ghana. This site was representative of a grassland ecosystem which occasionally served as pasture for some livestock (e.g. cattle and sheep) owned by inhabitants of the nearby communities. The stocking density of the grazing animals ranged between 6–8 animals per hectare. The natural vegetation in the area was characterised by grasses (e.g. *Brachiaria lata, Chloris piloasa and Cassia mimosoides*) of average height, 0.10 m when fully grown after the peak of the rainy season in October. The leaf area index (LAI) measurements undertaking between June and

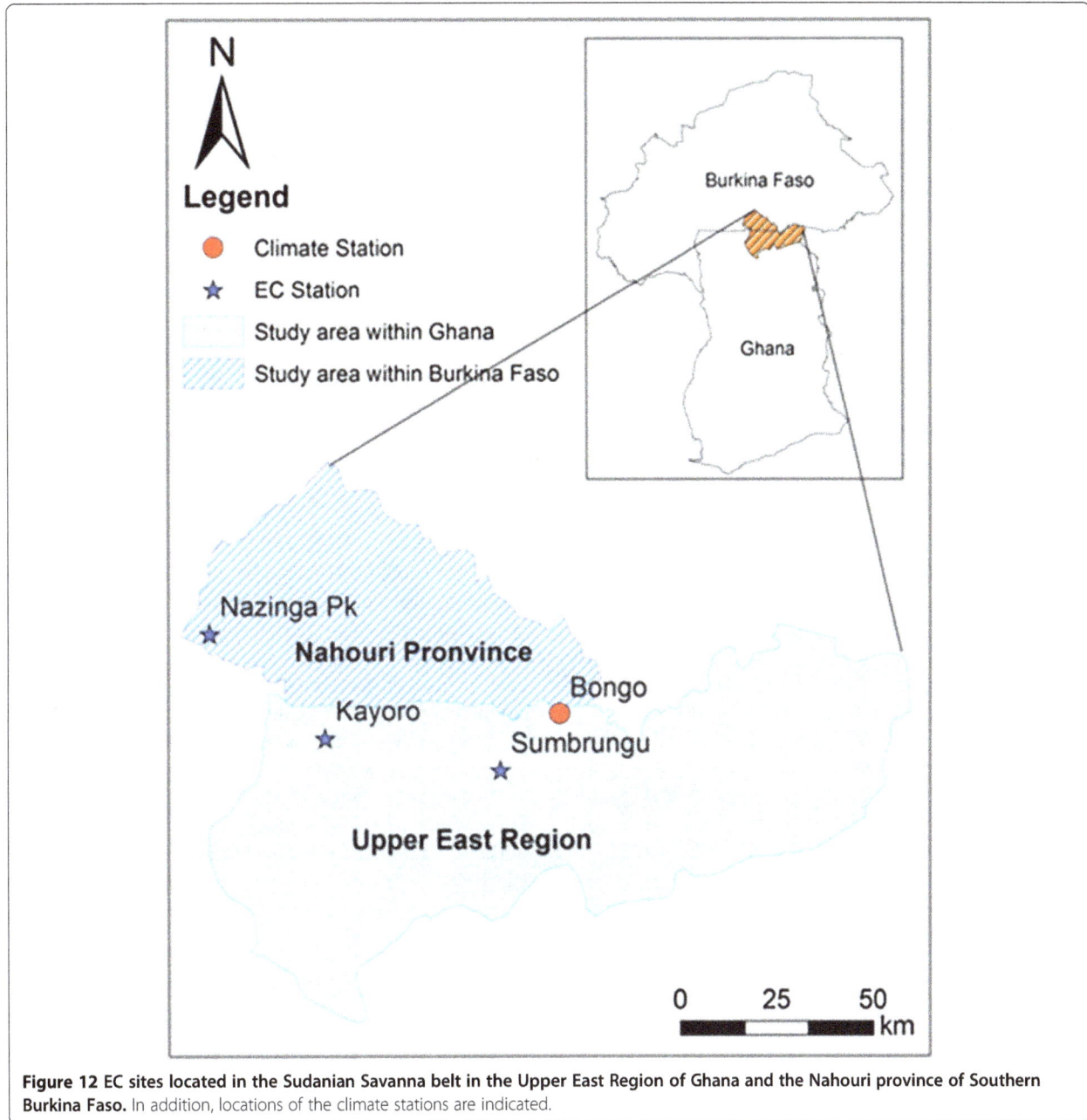

Figure 12 EC sites located in the Sudanian Savanna belt in the Upper East Region of Ghana and the Nahouri province of Southern Burkina Faso. In addition, locations of the climate stations are indicated.

October 2013 gave values ranging between 1.3–2.5 m^2 m^{-2}. The area was interspersed with trees of heights between 3–5 m and around 30–40 m apart. *Parkia biglobosa, Adansonia* and *Lannea microcarpa* were some of the dominant tree species in this area. The measurements were carried out in an area dominated by grass. The location of the eddy covariance system was selected in a manner to allow for the required fetch. The tower position was chosen in order that all the trees were at a distance of more than 30 m away and were not within the two main wind directions (easterly and westerly) during the study period. The micrometeorological

instruments were installed in a 2 m high wired-fence surrounding an area of 7 m × 7 m to protect the devices from destruction by the livestock.

The second EC station was established at Kayoro Dakorenia (10.918° N, 1.321° W, 292 m a.s.l.) within the catchment of the Tono river, approximately 100 km away from Bolgatanga in the Upper East Region of Ghana. The vegetation of this site was dominated by tall grasses (e.g. *Andropogon and Cenchrus*) with an average height of 1 m after the rainy season in October. The LAI measured between June and October 2013 were between 3.0–5.3 m^2 m^{-2}. The dominant tree species in the area included: *Entada africana*, *Acacia dudgeoni* and *Vitellaria paradoxa*. This site represented a mixture of fallow and cropland, occasionally used for cropping. Similarly at this site, the tower location allowed for sampling from grass dominated areas and all the trees were at distances of more than 200 m away and were also not within the dominant wind directions during the period of the study.

The third eddy covariance station was located in the Nazinga Park (11.152° N, 1.586° W, 293 m a.s.l.), one of the biggest natural reserves in Burkina Faso, about 50 km from Pô in the Nahouri province (Figure 12). It was located in the Sissili river basin. The site represented a near-natural Sudanian Savanna with no agricultural activities. It was located at a specific research area of the natural reserve close to the Nazinga Ranch, and endowed with rivers with many small dams close to the ranch with rich biological diversity (mammals, birds, and flora). The eddy covariance instrument was located in a protected area of the nature reserve reserved for research purposes. The vegetation of the site was made up of a mosaic of shrubs and tree savannas (e.g. *Daniellia oliveri*, *Burkea Africana and Isoberlinia doka*), between 4–5 m tall with averaged canopy height of about 4.5 m. While the grasses were mainly of annual and perennial Graminae, with average height of about 2.5 m at the end of the rainy season in October. The LAI was not determined for this site.

Climate of the study areas

All the three study sites are representative of the Sudanian savanna climate and vegetation, with a mono-modal rainfall pattern, mainly between the months of May and October, while the dry season begins from November and ends in April each year [34,35]. The mean annual precipitation is between 320 and 1100 mm [7,18-21]. The general climate is characterised by the seasonal changes in water availability and fires which are influenced by the dry and rainy seasons, also linked to the movement of the Inter-Tropical Convergence Zone (ITCZ, [36]).

The dry season is associated with dry and dust-laden 'Harmattan' winds with low relative humidity and low night temperatures, while the contrary prevails during the rainy season. The horizontal wind characteristics are predominantly north easterly in the dry season, but south westerly in the rainy season. The observed temperatures over the area were low during the month of August with an average of 22°C but the high values were recorded between March and April, with a peak value of about 40°C, while the annual relative humidity were between 6 to 95% [15,37,38].

Instrumentation

All three EC stations were equipped with the same measurement devices. The EC sensors included CSAT3 3D sonic anemometer (Campbell Scientific Inc., USA) to measure wind speed and direction as well as the sonic temperature, and a Licor 7500A open-path infrared gas analyser (LI-COR, Biosciences Inc., USA) to measure CO_2 and H_2O in the atmosphere. In addition, each station was equipped with a tipping bucket, model 52293 (R. M. Young, USA) and a weighing gauge pluviometer (Ott, Germany) to measure rainfall, the HMP155A (Campbell Scientific Inc., USA) to measure air temperature and relative humidity and the CNR4 net radiometer (Kipp & Zonen, Netherlands) to measure the incoming and outgoing shortwave and longwave radiations. Furthermore, the CS616 soil moisture probes (Campbell Scientific Inc., USA) were employed to measure the volumetric soil moisture content, at the depth of 0.03 m at each site, while the TCAV thermal sensors (Campbell Scientific Inc., USA), and the HFP01SC self-calibrating heat flux plates (Hukseflux, Netherlands) were used to measure soil temperature at three different depths (0.03, 0.10, and 0.30 m) and soil heat fluxes at 0.08 m respectively. Also installed were MX-Q24M-Sec-D11 web cameras (Mobotix, Germany) to take daily pictures from the surrounding areas for studying the vegetation phenology. In addition, the multi-sensor WXT 520 (Vaisala, Finland) was installed to measure rainfall, air temperature, horizontal wind speed and direction, relative humidity and air pressure. Data from each station were automated and recorded onto the CR3000 and CR1000 data loggers (Campbell Scientific Inc., USA). Two solar panels were built at each station for the power supply of the EC stations and the data transmission unit for an automatic transfer of the measurements on the daily basis.

Determination of Carbon dioxide, water vapour and energy fluxes

The carbon dioxide and the energy flux data were recorded at the three study sites at high-frequency of 20 Hz. The flux and meteorological data from each study site were stored every 30 minutes. The turbulence (vertical) fluxes of the carbon dioxide, F_c (mmol m^{-2} s^{-1}), the sensible heat flux, H (W m^{-2}) and the latent heat flux, λE (W m^{-2}) for each time step of the investigated period were calculated as described in [39,40]:

$$F_c = \rho_a \overline{w'\rho'_c} \tag{1}$$

$$H = \rho_a c_p \overline{w' \, T'_a} \tag{2}$$

$$\lambda E = \lambda \overline{w'\rho'_v} \tag{3}$$

where ρ_a is the density of dry air (kg m^{-3}) at a given air temperature, c_p is the specific heat capacity of dry air at constant pressure (J kg^{-1} K^{-1}), λ is the latent heat of vapourisation (J kg^{-1}), ρ_c is the molar density of CO_2 gas (mol m^{-3}) and ρ_v is the molar density of water vapour (mol m^{-3}). T_a is the air temperature derived from the sonic anemometer (K) and w is the vertical wind velocity component (m s^{-1}). Over bars denote time averages and primes indicate fluctuations about the averages.

Measurement and parameterisation of ground heat flux

The soil heat flux is often assumed to be negligible in many studies. The heat fluxes are often estimated from in situ soil measurements which are affected by measurement errors. This can lead to large uncertainties regarding the estimation of the heat fluxes [41], which also strongly influences the energy balance closure.

In this paper, the soil heat fluxes were measured at each station using three heat flux plates, each buried at 0.08 m depth and 0.05 m apart within the soil. The average heat fluxes from these plates at each time intervals (30 minutes) were used as the measured soil heat fluxes. However, the change of the heat storage (G_s) above the plates at 0.03 m depth was calculated based on a simplified equation proposed by Liebethal et al. [41], which uses the measured soil temperature at 0.03 m at each time step. Hence the soil or ground heat flux was calculated based on the combination method, the sum of the measured heat flux and the calculated heat storage (G_s):

$$G = G_m + G_s \tag{4}$$

where G is the ground heat flux (W m^{-2}), G_m is the measured ground heat flux (W m^{-2}) using the heat flux plates and G_s is the calculated heat storage (W m^{-2}) above the heat flux plates. G_s was calculated based on the following formula:

$$G_s = \int_0^z C_v \frac{\partial T_s}{\partial t} dz \tag{5}$$

where z is the depth (m) above the heat flux plate from the soil surface, T_s is the soil temperature (°C) measurements at 0.03 m at each time step (30 minutes) and t is the time intervals (s). C_v, the volumetric heat capacity (J m^{-3} K^{-1}), was estimated from the composition of the soil following De Vries [42]:

$$C_v = (1.90 \cdot f_m + 2.47 \cdot f_o + 4.12 \cdot V_s) \cdot 10^6 \tag{6}$$

where f_m is the volumetric fraction of minerals (%) in the soil, f_o is the fraction of organic content in the soil (%) and V_s is the time dependent (30 minute) volumetric soil moisture content (m^3 m^{-3}). The values of f_m and f_o (see Table 4) were kept constant throughout the investigated period.

Data acquisition and quality control

The CO_2 and the energy flux data were obtained from the three study sites using the eddy covariance method [43,44]. The raw turbulence data including the mixing ratios of CO_2, H$_2$O, air temperature and pressure, as well as the three-dimensional wind speeds were processed using the software TK3.1, which had the capability to perform all post field processing of turbulence measurements and to produce statistically quality assured turbulence fluxes for a station automatically [40]. Within the TK3.1 software, a number of calculations and standard corrections for open-path sensors to produce quality controlled (QC) and quality assured (QA) data were performed after the raw data such as air temperature, water vapour, CO_2 and the three components of the wind velocity (u, v and w) had been retrieved from the CR3000 and CR1000 data loggers. The programme applied standardised quality assessment routines and user specified consistency limits [40,45,46], to detect and reject physically or electronically not possible values of the EC data including CO_2, sensible heat (H) and the latent heat (λE) fluxes, emanating from unfavourable micrometeorological conditions, such as strong non-stationary and non-turbulence as well as low-quality data caused by precipitation, dust, or other contamination on the sensor optics. The study state quality test of the mean and variance of the data set at 20 Hz scanning frequency and 5 minutes interval was performed based on Foken et al. [47]. The spikes were detected according to Vickers and Mahrt [48], based on Højstrup [49], and these were values which exceeded 4.5 times standard deviations in a window of 15 values. However, in a situation where this criterion was fulfilled by 4 or more values in a row, they were not considered as such. They were supposed to be 'real' in this case. The Planar Fit method after Wilczak et al. [50] was applied for the coordinate transformation. Moreover, the Schotanus et al. [51] correction was applied for obtaining the sensible heat flux from the sonic measurements and the Webb et al. [52] correction was applied to account for density fluctuations in the fluxes of carbon dioxide and water vapour.

All the climatic and eddy covariance variables used in this work, except the gross primary production and ecosystem respiration were determined using the aforementioned eddy covariance setups, meteorological instruments and equations.

Energy balance closure

The performance of the eddy covariance measurements was evaluated based on the linear regression relation between the half-hourly averaged turbulent fluxes ($H + \lambda E$) and the available energy ($R_n - G$):

$$H + \lambda E = m(R_n - G) + b \tag{7}$$

where m is the slope and b (W m^{-2}) is the intercept respectively. Under an ideal condition the m should be equal to 1. However, the sum of the turbulence fluxes is usually found to be between 10 to 30% less than the available energy due to several factors such as large-scale transport not captured by regular eddy covariance tower

Table 4 Summary of the characteristics of the eddy covariance sites, including the measurement height of the open-path gas analyser and selected soil physio-chemical properties taking from soil samples at 0.10 m

	Sumbrungu	Kayoro	Nazinga Park
Location (Lat/Lon)	10.846° N	10.918° N	11.152° N
	0.917° W	1.321° W	1.586° W
Elevation (m)	200.00	292.00	293.00
Management	Highly degraded used for grazing	Mixture of fallow and cropland	Nature reserve
EC height (m)	2.65	3.15	7.19
Average height of grass when fully grown after the rainy season (m)	0.10	1.00	3.00
Average canopy height of trees (m)	4.50	3.00	4.50
Land cover type	Short grass savanna	Tall grass savanna	Tall grass/shrub savanna
Soil texture	Loamy sandy	Loamy sandy	Sandy loam
Soil bulk density (g cm^{-3})	1.41	1.42	1.42
Organic carbon (%)	1.60	0.62	3.30
Soil mineral (%)	51.74	52.84	50.30
C:N ratio	11.07	11.00	11.28
pH	5.79	6.13	6.37
Sand (%)	77.99	77.09	51.52
Silt (%)	16.34	20.05	38.93
Clay (%)	5.67	2.87	9.53

measurements and measurement errors associated with individual instruments [23,45,53].

Data gap-filling and flux partitioning schemes

The flux of CO_2 across the interface between the vegetation and the atmosphere, the net ecosystem exchange (NEE), at the study sites can directly be calculated via the eddy covariance approach. The sign convention of NEE is with respect to the atmosphere, i.e. a negative sign means the ecosystem captures CO_2 from the atmosphere, while a positive sign means the ecosystem is losing CO_2 to the atmosphere. The NEE is comprised of signals made up of photosynthetic uptake, gross primary production (GPP), as well as autotrophic and heterotrophic respiration, called the ecosystem respiration (ER). The gaps in NEE data that emanated from strong non-stationary and non-turbulence, power failures, spikes, precipitation and dust among others, accounted for 33, 39 and 43% for Sumbrungu, Kayoro and the Nazinga Park respectively, during the study periods. These gaps were consistent with data gaps of between 30 to 40%, usually found in flux measurements on an annual time scale which were required to be gap-filled [2,54,55], in order to obtained continuous data for the calculations of the daily, monthly (seasonal) or annual integrated carbon balances.

In this paper, the NEE was gap-filled as well as decomposed into its constituent signals (GPP and RE) using the online gap filling and flux partitioning tool publicly available at: (http://www.bgc-jena.mpg.de/~MDIwork/

eddyproc/: [56]. This Marginal Distribution Sampling (MDS) technique has a similar performance compared to other techniques [2]. The gap filling process using the MDS technique requires meteorological data (30 minutes averages) such as solar radiations, air temperature, relative humidity, vapour pressure deficit and the soil temperature. The technique employs temporal autocorrelation of fluxes, and estimate the missing data based on the mean value of available data under similar meteorological conditions within an averaging window of 7 days. In the case of larger data gaps with no similar meteorological conditions within the 7-day window, the averaging window is increased by 14 days until similar meteorological conditions are met. The flux partitioning component of the on line tool estimates the ecosystem respiration (ER) by employing the regression model proposed by Lloyd and Taylor [57]:

$$\mathrm{ER}(T_s) = \mathrm{ER}_{ref} \cdot e^{E_o\left(\frac{1}{T_{ref}-T_o} - \frac{1}{T_s-T_o}\right)} \quad (8)$$

where T_s (°C) is the soil temperature, T_o (°C) is the regression parameter, E_o (K^{-1}) is the activation-energy that determines temperature sensitivity, T_{ref} (°C) is the reference temperature and ER_{ref} (μmol m^2 s^{-1}) is the ecosystem respiration at the reference temperature. The regression parameter E_o and the reference temperature T_{ref} were kept constant as in the original model by Lloyd and Taylor [57]. GPP was finally derived from the difference between NEE and ER. In order for easy comparison of our results with those from the former flux measurements, e.g. Brümmer et al. [6], Ardö et al. [7] and others, the daily, monthly (seasonal) and annual courses of the carbon fluxes (NEE, GPP and ER) were expressed as g C m^{-2} (time unit)$^{-1}$. Results of former measurements in other units were also converted into g C m^{-2} (time unit)$^{-1}$ for easy comparability.

Abbreviations
AM: April-May; AMMA: African Monsoon Multidisciplinary Analyses; BMBF: German Federal Ministry of Education and Research; CARBOAFRICA: Project to quantify, understand and predict carbon cycle and other greenhouse gases in Sub-Saharan Africa; EBC: Energy balance closure; EC: Eddy covariance; ER: Ecosystem respiration; EWUE: Ecosystem water use efficiency; GPP: Gross primary production; HAPEX-Sahel: Hydrological and Atmospheric Pilot Experiment-Sahel; ITCZ: Inter-Tropical Convergence Zone; JFM: January-February-March; JJASO: June-July-August-September-October; LAI: Leaf area index; MDS: Marginal Distribution Sampling; NEE: Net ecosystem exchange; ND: November-December; PPFD: Photosynthetic photon flux density; QA: Quality assurance; QC: Quality control; RH: Relative humidity; SEBEX: Sahelian Energy Balance Experiment; Tair: Air temperature; TK3.1: 'Turbulentzknecht' version 3.1 (software for processing of eddy fluxes); VPD: Vapour pressure deficit; WASCAL: West African Science Service Centre on Climate Change and Adapted Land Use; WFPS: Water-filled pore space; WVF: Water vapour flux; ZEF: Centre for development research.

Competing interests
The authors declare that they have no competing interests.

Authors' contributions
EQ prepared the manuscript with contributions from all co-authors. MM, AAB and LKA assisted in the data analyses. LH and JB designed the experiments and carried out the installations of the EC and climate stations. HK is the coordinator of the climate research activities within the Core Research Program of WASCAL and he also coordinated and assisted in the installation of the EC and climate stations. All authors read and approved the final manuscript.

Acknowledgements
This work is part of the research program of the West African Science Service Centre on Climate Change and Adapted Land Use (WASCAL) funded by the German Federal Ministry of Education and Research (BMBF). The authors wish to acknowledge BMBF for providing the financial support for the installation and maintenance of the EC and automatic weather stations at the study areas. M. Mauder's contribution was partly funded by the Helmholtz-Association through the President's Initiative and Networking Fund. We are grateful for the very helpful comments by Lutz Merbold and one anonymous reviewer. We also wish to thank Samuel Guug who provided the field site technical support, and finally the authority of the Nazinga Park (the Ministry of the Environment of Burkina Faso) for allowing us to build the EC station in the nature reserve.

Author details
[1]Federal University of Technology, Akure, Nigeria. [2]Institute of Meteorology and Climate Research, Karlsruhe Institute of Technology, Garmisch-Partenkirchen, Germany. [3]Kwame Nkrumah University of Science and Technology, Kumasi, Ghana. [4]Chair for Regional Climate and Hydrology, University of Augsburg, Augsburg, Germany. [5]Head of Chair for Regional Climate and Hydrology, University of Augsburg, Augsburg, Germany.

References
1. Janssens IA, Freibauer A, Ciais P, Smith P, Nabuurs G-J, Folberth G, et al. Europe's terrestrial biosphere absorbs 7 to 12% of European anthropogenic CO_2 emissions. Science. 2003;300:1538–42.

2. Thomas MV, Malhi Y, Fenn KM, Fisher JB, Morecroft MD, Lloyd CR, et al. Carbon dioxide fluxes over an ancient broadleaved deciduous woodland in southern England. Biogeosciences Discuss. 2010;7:3765–814.

3. Bombelli A, Valentini R, editors. Africa and Carbon Cycle. World Soil Resources Reports No. 105. Rome: FAO; 2011.

4. Ciais P, Bombelli A, Williams M, Piao SL, Chave J, Ryan CM, et al. The carbon balance of Africa: synthesis of recent research studies. Phil Trans R Soc. 2011;A369:1–20.

5. Pilegaard K, Hummelshoej P, Jensen NO, Chen Z. Two years of continuous CO_2 eddy-flux measurements over a Danish beech forest. Agric Forest Meteorol. 2001;107(1):29–41.

6. Brümmer C, Falk U, Papen H, Szarzynski J, Wassmann R, Brüggemann N. Diurnal, seasonal, and interannual variation in carbon dioxide and energy exchange in shrub savanna in Burkina Faso (West Africa). J Geophys Res. 2008;113:G02030. doi:10.1029/2007JG000583.

7. Ardö J, Mölder M, El-Tahir BA, Elkhidir HAM. Seasonal variation of carbon fluxes in a sparse savanna in semi-arid Sudan. Carbon Balance Manage. 2008;3(7):1–18.

8. Wallace JS, Wright IR, Steward JB, Holwill CJ. The Sahelian Energy Balance Experiment (SEBEX): Ground based measurements and their potential for spatial extrapolation using satellite data. Adv Space Res. 1991;11(3):131–41.

9. Verhoef A, Allen S, De Bruin H, Jacobs C, Heusinkveld B. Fluxes of carbon dioxide and water vapour from a Sahelian savanna. Agric For Meteorol. 1996;80(2–4):231–48.

10. Friborg T, Boegh E, Soegaard H. Carbon dioxide flux, transpiration and light response of millet in the Sahel. J Hydrol. 1997;188–189:633–50.

11. Hanan NP, Kabat P, Dolman JA, Elbers JA. Photosynthesis and carbon balance of a Sahelian fallow savanna. Glob Chang Biol. 1998;4:523–38.

12. Redelsperger J-L, Thorncroft CD, Diedhiou A, Lebel T, Parker DJ, Polcher J. African monsoon multidisciplinary analysis: an international research project and field campaign. Bull Am Meteorol Soc. 2006;87:1739–46.

13. Bombelli A, Henry M, Castaldi S, Adu-Bredu S, Arneth A, de Grandcourt A, et al. An outlook on the Sub-Saharan Africa carbon balance. Biogeosciences. 2009;6(10):2193–205.

14. A Global Network – Historical Site Status [http://fluxnet.ornl.gov/site_status; 13-5-2014]

15. Bagayoko F. Impact of land-use intensity on evaporation and surface runoff: Processes and parameters for eastern Burkina Faso, West Africa, PhD thesis. Germany: University of Bonn; 2006.

16. Schüttemeyer D. The Surface Energy Balance Over Drying Semi-Arid Terrain in West Africa, PhD thesis. University of Wageningen, Netherlands; 2005.

17. Bliefernicht J, Kunstmann H, Hingerl L, Rummler T, Andresen S, Mauder M, et al. Field and simulation experiments for investigating regional land-atmosphere interactions in West Africa: experimental set-up and first results, Climate and Land Surface Changes in Hydrology. Gothenburg, Sweden: IAHS Publ.; 2013. p. 359p.

18. Kpongor D. Spatially Explicit Modeling of Sorghum (Sorghum bicolor L) Production on Complex Terrain of a Semi-arid Region of Ghana using APSIM, PhD thesis. University of Bonn, Germany; 2007.

19. Brümmer C, Papen H, Wassmann R, Brüggemann N. Fluxes of CH_4 and CO_2 from soil and termite mounds in south Sudanian savanna of Burkina Faso (West Africa). Global Biogeochem Cycles. 2009;23:GB1001. doi:10.1029/2008GB003237.

20. Grote R, Lehmann E, Brümmer C, Brüggemann N, Szarzynski J, Kunstmann H. Modelling and observation of biosphere atmosphere interactions in natural savannah in Burkina Faso, West Africa. Phys Chem Earth. 2009;34(4–5):251–60.

21. Ibrahim B, Polcher J, Karambiri H, Rockel B. Characterization of the rainy season in Burkina Faso and it's representation by regional climate models. Clim Dyn. 2012;39:1287–302.

22. Foken T. The energy balance closure problem - An overview. Ecolog Appl. 2008;18:1351–67.

23. Ingwersena J, Steffens K, Högy P, Warrach-Sagi K, Zhunusbayeva D, Poltoradnev M, et al. Comparison of Noah simulations with eddy covariance and soil water measurements at a winter wheat stand. Agric For Meteorol. 2011;151:345–55.

24. Leuning R, Eva van G, Massman WJ, Isaac PR. Reflections on the surface energy imbalance problem. Agric For Meteorol. 2012;156:65–74.

25. Foken T, Leuning R, Oncley SP, Mauder M, Aubinet M. Corrections and data quality. In: Aubinet M, Vesala T, Papale D, editors. Eddy Covariance: A Practical Guide to Measurement and Data Analysis, vol. 4. Dordrecht: Springer; 2012. p. 85–132.

26.   Henderson-Sellers A, Wilson MF. Albedo observations of the Earth's surface for climate research. Phil Trans Roy Soc London. 1983;309(1508):285–94.

27.   Wenge N, Woodcock C. "Surface albedo of boreal conifer forests: Modeling and measurements," Geoscience and Remote Sensing Symposium, IGARSS '99 Proceedings. IEEE 1999 International. 1999;2:1068–70.

28.   Schmitt M, Bahn M, Wohlfahrt G, Tappeiner U, Cernusca A. Land use affects the net ecosystem CO_2 exchange and its components in mountain grasslands. Biogeosciences. 2010;7(8):2297–309. doi:10.5194/bg-7-2297-2010.

29.   Law BE, Falge E, Gu L, Baldocchi DD, Bakwin P, Berbigier P, et al. Environmental controls over carbon dioxide and water vapor exchange of terrestrial vegetation. Agr Forest Meteorol. 2002;113:97–120.

30.   Grelle A, Lundberg A, Lindroth A, Moren A-S, Cienciala E. Evaporation components of a boreal forest: variations during the growing season. J Hydrol. 1997;197:70–87.

31.   Lasslop G, Reichstein M, Papale D, Richardson AD, Arneth A, Barr AG, et al. Separation of net ecosystem exchange into assimilation and respiration using a light response curve approach: critical issues and global evaluation. Glob Chang Biol. 2010;16:187–208.

32.   Schulze ED, Ciais P, Luyssaert S, Schrumpf M, Janssens IA, Thiruchittampalam B, et al. The European carbon balance. Part 4: integration of carbon and other trace-gas fluxes. Glob Chang Biol. 2010;16(5):1451–69.

33.   Sjöström M, Zhao M, Archibald S, Arneth A, Cappelaere B, Falk U, et al. Evaluation of MODIS gross primary productivity for Africa using eddy covariance data. Remote Sens Environ. 2013;131:275–86.

34.   Ofori-Sarpong E. Impact of climate change on agriculture and farmers coping strategies in the upper east region of Ghana. West Afr J Appl Ecol. 2001;2:21–35.

35.   Callo-Concha D, Gaiser T, Ewert F. Farming and cropping systems in the West African Sudanian Savanna. WASCAL research area: Northern Ghana, Southwest Burkina Faso and Northern Benin, ZEF Working Paper Series, No. 100, 2012. Available at: http://hdl.handle.net/10419/88290.

36.   Sultan B, Janicot S. The West African monsoon dynamics, Part II: The "pre-onset" and the "onset" of the summer monsoon. J Clim. 2003;16:3407–27.

37.   Oguntunde PG. Evapotranspiration and Complimentarity Relations in the Water Balance of the Volta Basin: Field Measurements and GIS-based regional Estimate. In: Vlek PLG, Denich M, Martius C, van de Giesen N, editors. Ecology and Development Series, vol. 22. Cuvillier Verlag, Göttingen: 2004. p. 1–170.

38.   Sandwidi JP. Groundwater Potential to Supply Population Demand within the Kompienga dam basin in Burkina Faso. In: Vlek PLG, Denich M, Martius C, Rodgers C, van de Giesen N, editors. Ecology and Development Series, vol. 55. Cuvillier Verlag, Göttingen: 2007. p. 1–157.

39.   Grünwald T, Bernhofer C. A decade of carbon, water and energy flux measurements of an old spruce forest at the Anchor Station Tharandt. Tellus. 2007;59B:387–96.

40.   Mauder M, Foken T. Documentation and Instruction Manual of the Eddy-Covariance software Package TK3. Arbeitsergebnisse 46, Bayreuth, Germany: Universität Bayreuth, Abteilung Mikrometeorologie; 2011. p. 60p. ISSN: 1614-8924.

41.   Liebethal C, Huwe B, Foken T. Sensitivity analysis for two ground heat flux calculation approaches. Agric For Meteorol. 2005;132:253–62.

42.   De Vries DA. Thermal Properties of Soils. In Physics of Plant Environment. Edited by VanWijk WR. Amsterdam: North-Holland Publishing Company; 1963:210 – 235.

43.   Kaimal JC, Finnigan JJ. Atmospheric Boundary Layer Flows: Their Structure and Measurement. New York: Oxford University Press; 1994.

44.   Aubinet M, Grelle A, Ibrom A, Rannik Ü, Moncrieff J, Foken T, et al. Estimates of the annual net carbon and water exchange of European forests: the EUROFLUX methodology. Adv Ecol Res. 2000;30:113–75.

45.   Mauder M, Jegede OO, Okogbue EC, Wimmer F, Foken T. Surface energy balance measurements at a tropical site in West Africa during the transition from dry to wet season. Theor Appl Climatol. 2007;89:171–83.

46.   Mauder M, Cuntz M, Drüe C, Graf A, Rebmann C, Schmid HP, et al. A strategy for quality and uncertainty assessment of long-term eddy-covariance measurements. Agric For Meteorol. 2013;169:122–35.

47.   Foken T, Wichura B. Tools for quality assessment of surface-based flux measurements. Agric For Meteorol. 1996;78:83–105.

48.   Vickers D, Mahrt L. Quality control and flux sampling problems for tower and aircraft data. J Atmos Ocean Technol. 1997;14:512–26.

49.   Højstrup J. A statistical data screening procedure. Meas Sci Technol. 1993;4:153–7.

50.   Wilczak JM, Oncley SP, Stage SA. Sonic anemometer tilt correction algorithms. Bound-Layer Meteor. 2001;99:127–50.

51.   Schotanus P, Nieuwstadt FTM, De Bruin HAR. Temperature measurement with a sonic anemometer and its application to heat and moisture fluxes. Bound-Layer Meteorol. 1983;26:81–93.

52.   Webb EK, Pearman GI, Leuning R. Correction of the flux measurements for density effects due to heat and water vapour transfer. Quart J Roy Meteorol Soc. 1980;106:85–100.

53.   Stoy PC, Mauder M, Foken T, Marcolla B, Boegh E, Ibrom A, et al. A data-driven analysis of energy balance closure across FLUXNET research sites: The role of landscape scale heterogeneity. Agric For Meteorol. 2013;171–172:137–52.

54.   Moffat AM, Papale D, Reichstein M, Hollinger DY, Richardson AD, Barr AG, et al. Comprehensive comparison of gap-filling techniques for eddy covariance net carbon fluxes. Agric For Meteorol. 2007;147:209–32.

55.   Baldocchi D. TURNER REVIEW No. 15, "Breathing" of the terrestrial biosphere: lessons learned from a global network of carbon dioxide flux measurement systems. Aust J Bot. 2008;56:1–26.

56.   Reichstein M, Falge E, Baldocchi D, Papale D, Valentini R, Aubinet M, et al. On the separation of net ecosystem exchange into assimilation and ecosystem respiration: review and improved algorithm. Global Change Biol. 2005;11:1–16.

57.   Lloyd J, Taylor JA. On the temperature dependence of soil respiration. Funct Ecol. 1994;8:315–23.

## 7

Airborne lidar-based estimates of tropical forest structure in complex terrain: opportunities and trade-offs for REDD+

Veronika Leitold[1,2*], Michael Keller[3,4], Douglas C Morton[2], Bruce D Cook[2] and Yosio E Shimabukuro[1]

Abstract

Background: Carbon stocks and fluxes in tropical forests remain large sources of uncertainty in the global carbon budget. Airborne lidar remote sensing is a powerful tool for estimating aboveground biomass, provided that lidar measurements penetrate dense forest vegetation to generate accurate estimates of surface topography and canopy heights. Tropical forest areas with complex topography present a challenge for lidar remote sensing.

Results: We compared digital terrain models (DTM) derived from airborne lidar data from a mountainous region of the Atlantic Forest in Brazil to 35 ground control points measured with survey grade GNSS receivers. The terrain model generated from full-density (~20 returns m^{-2}) data was highly accurate (mean signed error of 0.19 ± 0.97 m), while those derived from reduced-density datasets (8 m^{-2}, 4 m^{-2}, 2 m^{-2} and 1 m^{-2}) were increasingly less accurate. Canopy heights calculated from reduced-density lidar data declined as data density decreased due to the inability to accurately model the terrain surface. For lidar return densities below 4 m^{-2}, the bias in height estimates translated into errors of 80–125 Mg ha^{-1} in predicted aboveground biomass.

Conclusions: Given the growing emphasis on the use of airborne lidar for forest management, carbon monitoring, and conservation efforts, the results of this study highlight the importance of careful survey planning and consistent sampling for accurate quantification of aboveground biomass stocks and dynamics. Approaches that rely primarily on canopy height to estimate aboveground biomass are sensitive to DTM errors from variability in lidar sampling density.

Keywords: Tropical montane forest; Airborne lidar; Digital Terrain Model; Elevation accuracy; Data thinning; Canopy height; Biomass estimation; REDD+

Background

Tropical forests are important reservoirs of carbon and biodiversity. Characterizing the spatial distribution of aboveground biomass (AGB) is a prerequisite for understanding carbon cycle dynamics in tropical forests over time. Precise estimates of AGB and changes in carbon stocks from human activities are also required for ongoing climate mitigation efforts to Reduce Emissions from Deforestation and Forest Degradation (REDD+) [1].

Airborne lidar has been successfully used to estimate aboveground biomass in a range of forest ecosystems [2-9]. Typical approaches to predict AGB with lidar data are based on regression models linking lidar metrics to biomass estimates from forest inventory plots. The model is then used to estimate AGB over larger areas. Lidar-derived metrics most frequently used to predict biomass include mean or maximum canopy height [10-13] and vertical canopy profile measures, such as height percentiles and variance of heights [14,15]. Airborne lidar remote sensing supports high-resolution carbon mapping across broad spatial scales and a range of ecosystems [16-18], with great potential to aid carbon monitoring and climate change mitigation efforts (e.g. REDD+).

Estimation of forest canopy height using lidar data depends upon an accurate representation of the ground surface in digital terrain models (DTMs). For forestry studies in particular, lidar is capable of characterizing both terrain and vegetation structure effectively. However, any error in

* Correspondence: veronika.leitold@nasa.gov
[1]Remote Sensing Division, National Institute for Space Research (INPE), São José dos Campos, SP CEP 12201-970, Brazil
[2]Biospheric Sciences Laboratory, NASA Goddard Space Flight Center, Greenbelt, MD 20771, USA
Full list of author information is available at the end of the article

the DTM will propagate to affect the accuracy of the derived vegetation metrics [19] and canopy height models (CHM). Therefore, it is necessary to characterize uncertainties associated with lidar-derived DTMs in order to accurately quantify uncertainties in the overlying vegetation heights.

Ground data are the most common method for estimating the accuracy of lidar-derived elevation estimates. Control points are collected using an independent method with higher accuracy, assuming that the calculated height differences or elevation errors are normally distributed [20]. In this context, and for the purposes of the present study, the quality of the DTM is expressed in terms of vertical accuracy, i.e., how close the lidar-measured terrain elevation is to the *reference value* established from *in-situ* GNSS observations.

The accuracy of lidar-derived DTMs can differ significantly across topographic and land cover gradients. Uncertainty in lidar-derived DTMs encompasses three sources of error: (1) sensor-specific uncertainties associated with the navigation, positioning and lidar systems during data acquisition; (2) geometric uncertainties related to the flight altitude and ranging distance, scan angle, or the local topography; and (3) uncertainties arising during the post-processing steps, such as point classification or surface interpolation [21]. Over open areas with relatively flat terrain, it is common to achieve elevation accuracies below 0.15 m root mean square error (RMSE) [22-24]. In a study evaluating DTM accuracy for six different land-cover types, Hodgson and Bresnahan [25] observed RMSE values ranging from a low of 0.17 to 0.19 m in pavement and low grass classes to a high of 0.26 m in a deciduous forest. In areas covered by dense vegetation, DTM elevation errors tend to increase because less energy reaches the ground, resulting in fewer ground points for DTM surface interpolation [26]. Several studies have assessed lidar-derived DTM accuracy in temperate coniferous, deciduous and mixed forests, reporting RMSE values that range between 0.32 m and 1.22 m [27-29]. However, there have been relatively few studies of elevation accuracy under complex, multilayered tropical rain forest canopies [26] where REDD+ efforts are concentrated.

In this study, we analyzed 1000 hectares of high-density lidar data collected along a steep elevational gradient (100 m to 1100 m a.s.l.) with coastal Atlantic Forest in Southeast Brazil. Lidar data collection covered nine 1-ha permanent field plots divided between submontane and montane forest areas (Figure 1). We evaluated the accuracy of a DTM derived from the airborne lidar data for the topographically complex study area of the Serra do Mar and assessed the impact of variable survey conditions (i.e. changes in flying height, ranging distance and footprint

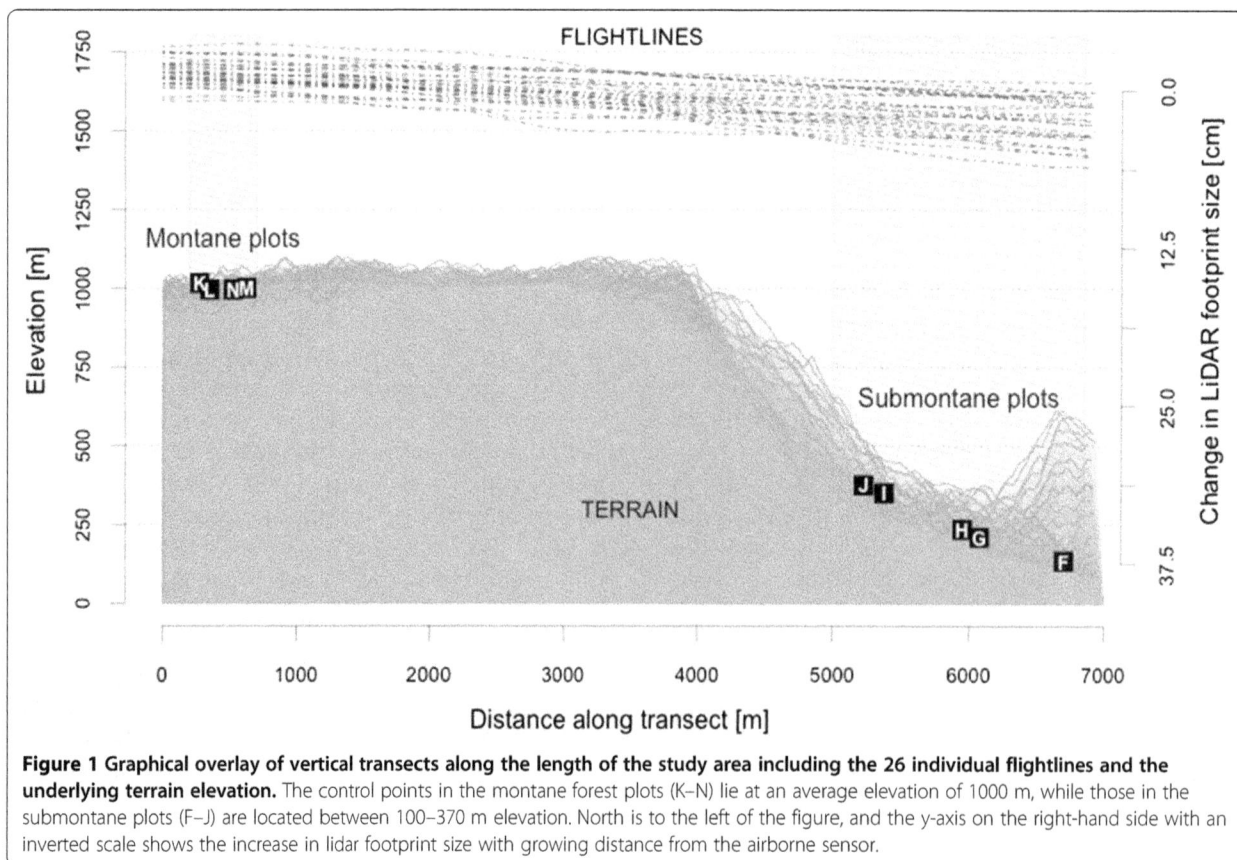

Figure 1 Graphical overlay of vertical transects along the length of the study area including the 26 individual flightlines and the underlying terrain elevation. The control points in the montane forest plots (K–N) lie at an average elevation of 1000 m, while those in the submontane plots (F–J) are located between 100–370 m elevation. North is to the left of the figure, and the y-axis on the right-hand side with an inverted scale shows the increase in lidar footprint size with growing distance from the airborne sensor.

size) on the characterization of the ground surface. We then assessed how changes in lidar data density influenced DTM accuracy, and examined how DTM uncertainty propagated into lidar-derived canopy height metrics. Our study targeted two main objectives: 1) to provide guidance regarding the minimum lidar point density required to generate DTM accuracies needed for lidar-based studies of forest biomass, and 2) to quantify the impacts of DTM errors on estimates of aboveground biomass. With its complex terrain, steep elevational gradient, and dense multilayered tropical forest canopy, the study site is unlike most of the areas considered in previous lidar forestry studies, but similar to fragments of Brazil's Atlantic Forest and other tropical forests.

Results

Full-density lidar data

Field GNSS elevations and lidar-derived DTM values (1m resolution, full-density data) showed excellent agreement. The error analysis of elevations using all 35 valid control points resulted in a mean signed error of 0.19 ± 0.97 m ($\mu \pm \sigma$), and the calculated RMSE value was 0.97 m. DTM elevations were higher on average than the corresponding GNSS elevations. Considering the uncertainty in calculating the lidar DTM (vertical $1\sigma = 0.15$ m on flat terrain) and error in the GNSS measurements, this 0.19 m elevation difference indicates a very good agreement between field data and the terrain model. Moreover, using only the 30 most accurate control points ($\sigma < 1$ m) for comparison, the mean signed error dropped by 63% to 0.07 ± 0.89 m difference of terrain elevations. Based on a one-sided t-test performed with the 30 most accurate control points, the DTM errors were not significantly different from zero (95% confidence level, p-value = 0.662).

DTM accuracy did not differ significantly by forest type or elevation for ground control points collected in submontane and montane forests. Differences between elevation errors associated with submontane and montane areas were evaluated assuming a normal distribution of the errors (Kolmogorov-Smirnov test, p-value = 0.923). Using the 30 most accurate control points for comparison, calculated mean signed errors for submontane vs. montane areas revealed positive differences between DTM and GNSS elevations at lower altitudes (0.23 ± 0.88 m), indicating a slight overestimate from lidar-derived terrain elevations in this area. For montane sites, the difference between DTM and GNSS values was smaller in magnitude and negative (-0.14 ± 0.90 m). However, mean signed errors were not significantly different from each other, based on a two-sided t-test performed with the two sets of errors (95% confidence level, p-value = 0.139). Thus, variability in flying height (footprint size) did not result in a statistically significant difference in DTM accuracy across the study area.

Reduced-density lidar data

Lower point densities in the thinned lidar data resulted in less accurate DTMs. Five data density levels were analyzed: the original density of 20 returns m^{-2} (D20) and the thinned return densities of 8, 4, 2 and 1 m^{-2} (denoted D8, D4, D2 and D1, respectively). When compared with GNSS control points, mean signed errors of the thinned DTM elevations increased as data density was reduced from D20 (0.19 ± 0.97 m) to D1 data (3.21 ± 3.12 m) (Figure 2). DTM elevations were higher than the GNSS elevations in all cases, with increasing error magnitudes as data were thinned. Calculated RMSE values showed a similar increasing trend with decreasing data density, ranging from a low of 0.97 m for the D20 DTM to a high of 4.45 m for the D1 data (Table 1).

Elevation errors in the thinned DTMs were larger in the submontane region than in the montane area for all the data densities. This observed difference between elevation classes became larger with increased levels of data thinning; the mean signed error difference between submontane and montane areas with 20 returns m^{-2} (0.31 m) increased to 2.64 m when data density dropped to 1 return m^{-2}. The trend in RMSE values also followed a similar pattern, with growing differences between submontane and montane DTM accuracy as data were thinned (Figure 3). With the highest data density, submontane and montane RMSE values were nearly identical (<0.1 m difference), while with lower data densities, montane RMSE values remained low while submontane RMSE values increased rapidly (0.76 to 3.08 m difference). The elevation error statistics based on thinned data are summarized in Table 1. These differences likely reflect the combined influence of greater ranging distance and topographic complexity in submontane areas.

To illustrate the spatial variability of DTM elevation errors across the landscape, a transect line was drawn along the center of the study area and DTM elevations were sampled from the 1-meter raster grids for all data densities. The difference between the cell values of the full-density DTM extracted along the transect line and the corresponding cell values of each thinned DTM was calculated and the elevation differences plotted (Figure 4). In general, the elevation difference between full-density and thinned DTMs was larger at lower altitudes, along the hillslope and in the valley, and smaller on top of the plateau. The magnitude of the difference increased with increased data thinning throughout the whole area, and the spatial distribution of the errors was associated with the level of complexity of the terrain in all DTMs examined. Where the terrain surface was more accentuated (i.e. greater rate of change of elevation), the corresponding difference in full-density vs. thinned DTM values was also larger, while with a smoother terrain surface, the associated DTM differences were smaller in magnitude.

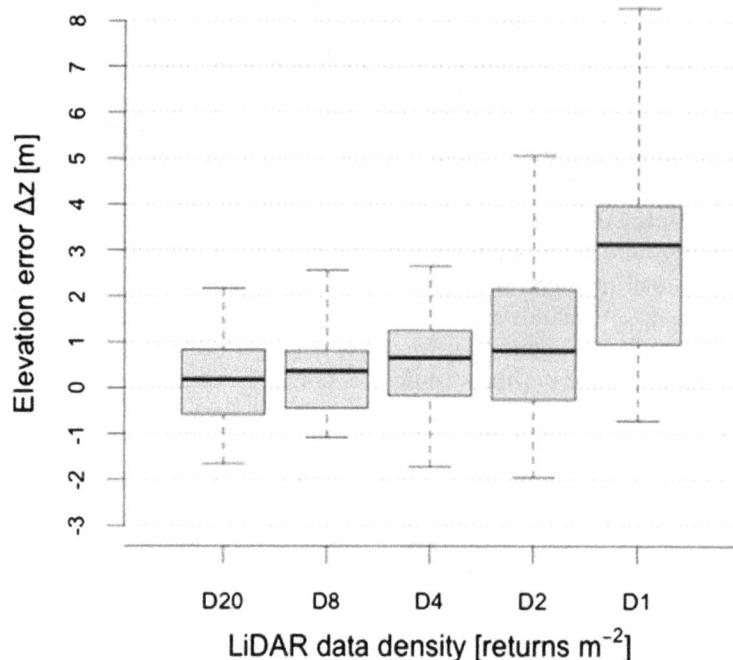

Figure 2 Distribution of the errors between GNSS and DTM elevations with data density levels of 20, 8, 4, 2 and 1 returns m^{-2} (D20, D8, D4, D2 and D1, respectively).

Effects of data thinning on estimated canopy height

Thinned lidar data consistently underestimated canopy heights in the 1-ha plots (Figure 5). With the full D20 data, mean canopy heights for the nine inventory plots ranged between 19.52 and 22.91 m (Plots F-N). With increasing levels of data thinning, the mean canopy

Table 1 Summary statistics from the DTM error analysis after data thinning

| Data type | | Error statistics (Δz) in meters | | | | |
		Min	Max	Mean	Stdev	RMSE
D20	submontane	−1.23	2.18	0.33	0.92	0.95
	montane	−1.65	1.86	0.02	1.02	0.99
	ALL	**−1.65**	**2.18**	**0.19**	**0.97**	**0.97**
D8	submontane	−2.88	4.51	0.54	1.60	1.65
	montane	−1.07	1.85	0.19	0.90	0.89
	ALL	**−2.88**	**4.51**	**0.38**	**1.32**	**1.35**
D4	submontane	−1.72	6.98	1.81	2.45	2.99
	montane	−0.94	2.25	0.30	0.97	0.99
	ALL	**−1.72**	**6.98**	**1.12**	**2.04**	**2.30**
D2	submontane	−1.96	14.62	2.36	3.96	4.52
	montane	−1.39	3.33	0.66	1.32	1.44
	ALL	**−1.96**	**14.62**	**1.59**	**3.13**	**3.47**
D1	submontane	0.46	14.05	4.42	3.24	5.43
	montane	−0.72	7.49	1.78	2.32	2.87
	ALL	**−0.72**	**14.05**	**3.21**	**3.12**	**4.45**

heights decreased on average by 0.70 m (3%), 1.75 m (8%), 3.40 m (16%) and 5.26 m (25%) for return densities of D8, D4, D2 and D1, respectively. The magnitude of canopy height changes was generally larger for the submontane plots (F, G, H, I and J), resulting in mean decreases of 0.79 m, 1.99 m, 3.93 m and 6.08 m with increasing thinning levels. In comparison, the mean canopy height changes in the montane plots (K, L, M and N) was 0.60 m, 1.45 m, 2.73 m and 4.24 m for the return densities of D8, D4, D2 and D1, respectively.

Lidar-derived canopy surfaces (digital surface models, DSMs) at the field plot locations showed little variation with the different levels of data thinning. A visual assessment of the DSM for each plot indicated that the canopy surface became slightly more rugged with increased data thinning, but the overall canopy surface elevation and shape did not change. In comparison, the terrain surface showed larger changes with increased levels of thinning. DTM errors in the thinned datasets resulted from an incorrect classification of vegetation features as ground. The overall effect of thinning was a positive bias in the ground elevation, which translated into lower canopy heights with decreasing data density.

Underestimation of mean canopy height (MCH) in the thinned lidar data had a significant impact on modeled aboveground biomass (AGB). We developed a simple regression model for the nine plot locations based on MCH: AGB = 24.13 × MCH - 204.76; r^2 = 0.43; RMSE = 30.0 Mg ha^{-1}. Aboveground biomass predictions (mean ± standard

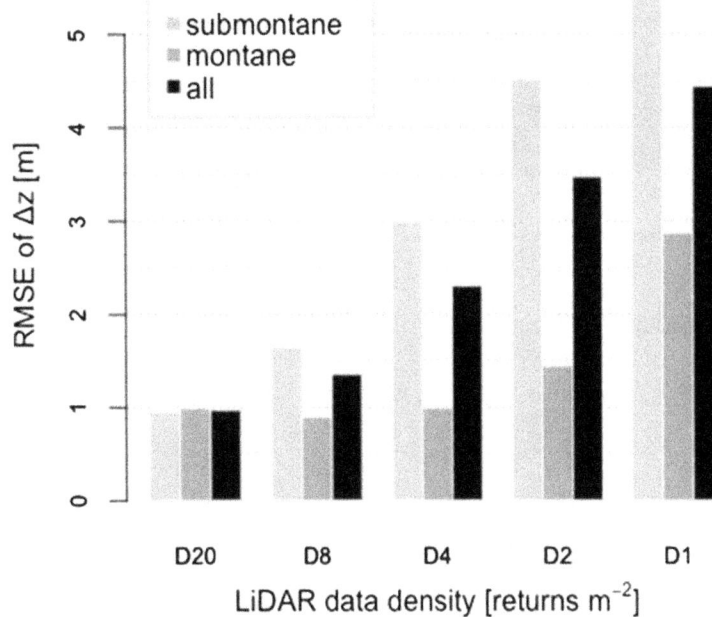

Figure 3 Comparison of RMSE values in the DTMs based on the five data density levels of 20, 8, 4, 2 and 1 returns m^{-2} (D20, D8, D4, D2 and D1, respectively).

deviation across nine permanent plots) for the different thinning levels ranged from 295.3 (±27.9) Mg ha^{-1} with full-density lidar data to 168.2 (±31.5) Mg ha^{-1} with the lowest data density of 1 return m^{-1} (Figure 6). In this study, a 1–5 m bias in MCH from incorrect ground detection may lead to errors in AGB estimates on the order of 15–125 Mg ha^{-1}. For lidar return densities below 4 m^{-2}, the bias in height estimates translated into aboveground biomass errors substantially greater than the model error of ~30 Mg ha^{-1}. These findings illustrate how approaches that rely on mean canopy height to estimate aboveground biomass are sensitive to DTM errors that arise from variability in lidar sampling density.

Discussion

Lidar-derived ground topography

Lidar coverage at the Atlantic Forest study site resulted in a very accurate DTM, despite large elevation differences, steep slopes, and closed canopy tropical forest cover. The ability to generate a highly accurate terrain model in such a challenging environment can be attributed, in part, to the high lidar point density (20 returns m^{-2} on average). Typical lidar data densities used for forest research and management purposes have been within the range of 0.5 - 4 returns m^{-2} [30-32], occasionally reaching a higher value of 10 to 12 returns m^{-2} [33,34]. Our approach to test the impact of data density on DTM accuracy highlights the potential variability in terrain elevations (and therefore canopy characterization) from low-density lidar coverage in regions with complex topography. Thinning of the

point cloud below 4 returns m^{-2} led to elevation errors that rendered the resulting DTM inadequate for consistent retrievals of vegetation heights. We therefore recommend a minimum lidar point density of 4 m^{-2} for studies of dense forest vegetation in complex terrain. Dense lidar data coverage is also critical for REDD+ and related applications that require repeat acquisitions to monitor changes in forest structure and aboveground carbon stocks; accurate DTMs are critical for change detection in regions with complex topography.

The results of this study are consistent with previous efforts to validate DTM products from small-footprint lidar systems [27-29], including an exponential increase in errors as data density decreases [35]. Clark and collaborators [26] reported a DTM accuracy of 0.58 m RMSE in open-canopy flat areas of an old-growth Costa Rican rain forest, and overall RMSE of 2.29 m when steep slopes and multilayered dense vegetation areas were also considered. In our study, the lidar-derived DTM consistently overestimated the ground elevation compared to the reference points, likely due to the incorrect classification of vegetation features as ground by the point-filtering algorithm. This overestimation of ground elevation was small in the full density data, but increased with successive thinning of the data.

Importantly, submontane areas consistently showed larger changes in DTM accuracy than montane areas after data thinning – consistent with longer ranging distances, larger lidar footprints, and more complex topography at lower elevations in the study site. Consistent flying altitude

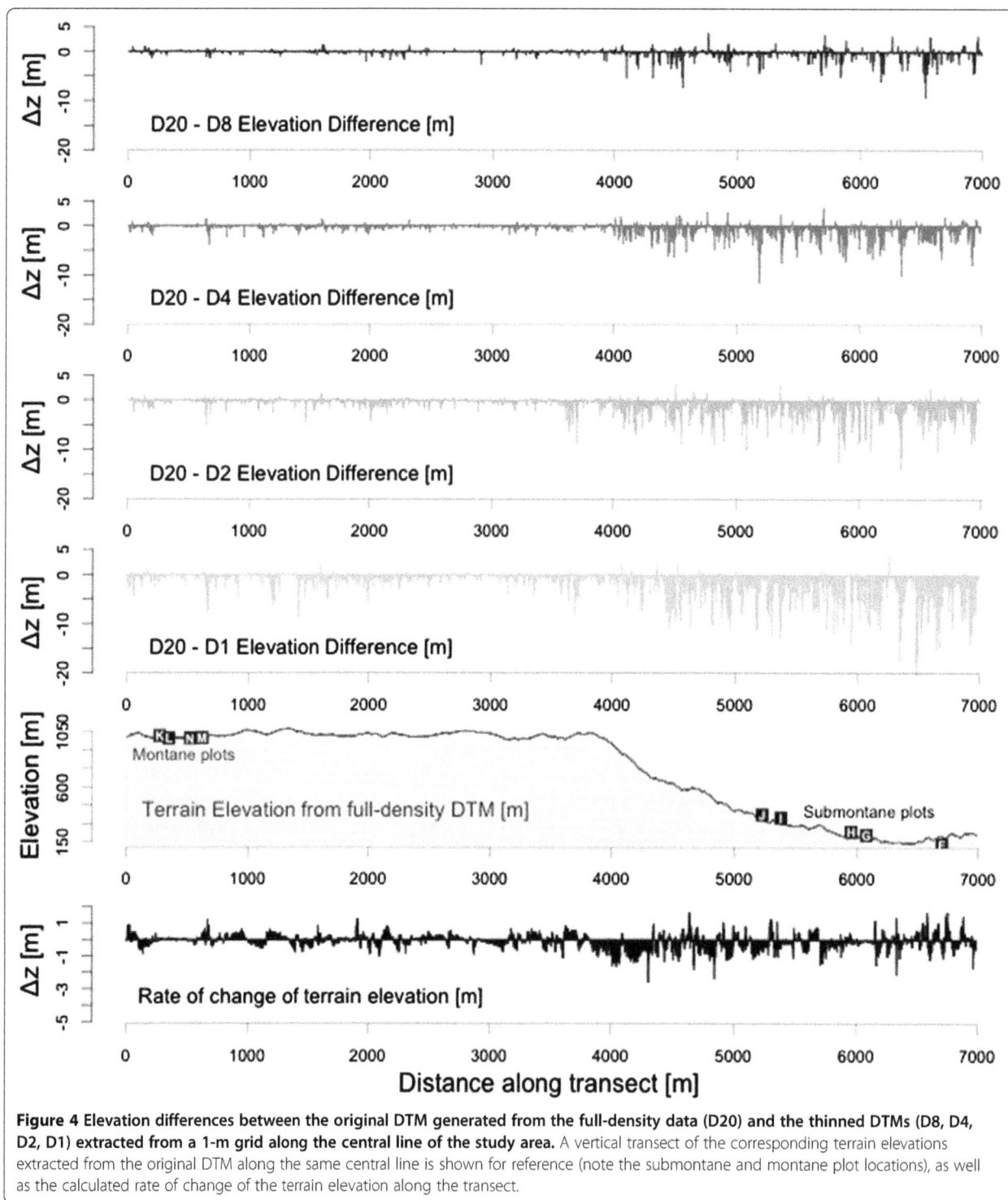

Figure 4 Elevation differences between the original DTM generated from the full-density data (D20) and the thinned DTMs (D8, D4, D2, D1) extracted from a 1-m grid along the central line of the study area. A vertical transect of the corresponding terrain elevations extracted from the original DTM along the same central line is shown for reference (note the submontane and montane plot locations), as well as the calculated rate of change of the terrain elevation along the transect.

for data collection resulted in a change in footprint size as ranging distances increased between montane and submontane areas (Figure 1). Longer ranging distances lower the proportion of pulses that penetrate the forest canopy to generate a return from the ground surface [19,36]. Terrain complexity has been identified as a cause for the variation in DTM accuracy across landscapes [37].

The steeper slopes and more variable topography in the submontane region might be harder to capture by the lidar system than the generally more homogeneous terrain on top of the plateau in the montane forest. Optimization of the flight line configuration at the time of data collection (e.g. constant flying height above ground, even point distribution) could potentially minimize

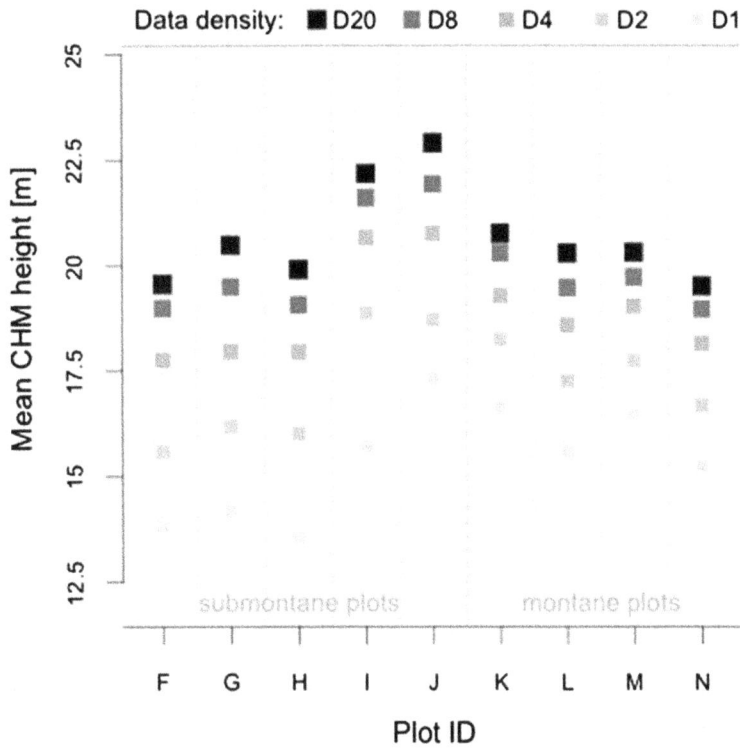

Figure 5 Mean canopy surface heights associated with the field plot locations (submontane Plots F - J and montane Plots K - L) based on CHMs generated from original and thinned lidar data (D20, D8, D4, D2 and D1 indicate the different data density levels).

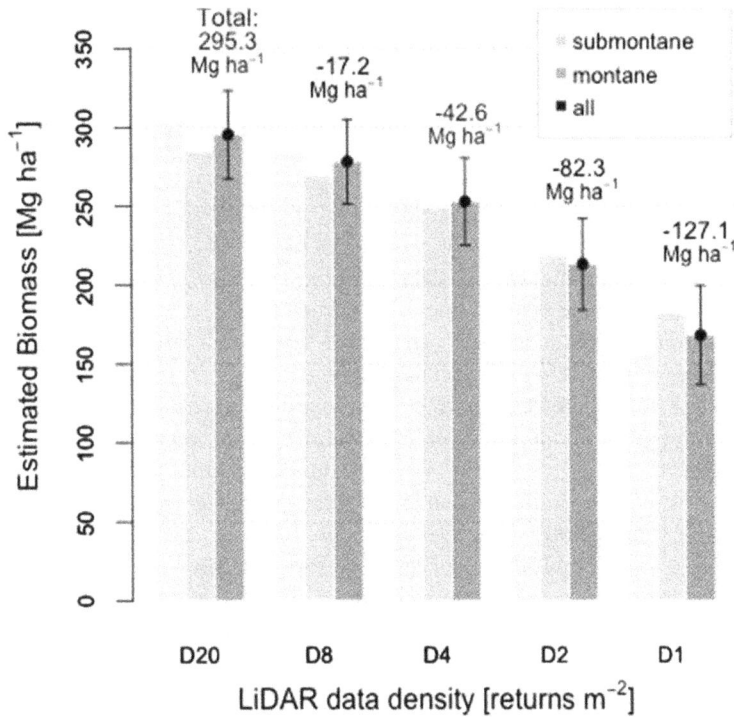

Figure 6 Aboveground biomass estimates in submontane and montane classes and across all nine permanent plots (mean ± standard deviation) for different data densities predicted with a linear model based on mean canopy surface height.

the observed difference between DTM accuracy for areas with different elevations, with important implications for data quality and forest applications.

Lidar-derived canopy height

The results of this study illustrate how errors in DTM accuracy propagate into estimates of forest canopy structure. Accurate characterization of the ground surface is a prerequisite for lidar vegetation studies because vegetation heights are calculated relative to the associated bare earth surface. Variability in DTM accuracy introduces error in the canopy height calculations, ultimately leading to erroneous estimation of related forest metrics or modeled aboveground biomass. Careful attention to lidar collection and analysis is particularly important in regions with complex topography, given previous issues with large-footprint lidar data in sloped terrain [38] and the relative inaccessibility for field measurements in these sites. Biases that propagate from lidar-derived canopy structure to estimates of aboveground forest biomass on sloped terrain would therefore be less likely to be detected by field validation efforts.

Lidar-based biomass estimates that rely on mean canopy height may be particularly sensitive to height biases from sampling issues that influence the accuracy of the DTM. The consistent overestimation of ground elevation in the analysis of thinned lidar data for Serra do Mar highlights the potential for a directional bias (underestimation of canopy heights) in regions with more sparse lidar sampling. No significant change was observed in the DSM heights with data thinning, suggesting that even with low point density, it is possible to capture the highest points of tree crowns and generate a canopy surface model representative of the true outer vegetation surface. Mountainous areas and other regions with complex topography present unique challenges for uniform lidar sampling. REDD+ efforts at the subnational or national scale will confront these sampling and analytic challenges for forests on steep slopes or other complex terrain. This study provides important guidance on the trade-offs associated with sampling density and biomass estimation over large regions with complex topography.

Lidar-based biomass estimates

Detailed knowledge of the spatial distribution of aboveground forest biomass is critical to improve estimates of carbon sources and sinks over time. Tropical forest biomass estimates are limited by knowledge of the allometry of tropical trees. The extreme diversity of tree species in tropical forests generally precludes species-specific allometries and instead general relations are applied [39,40]. As in other biomes, quantification of biomass depends on relations between lidar metrics (mainly mean or total canopy height) and estimates of plot biomass from field measurements and allometric equations [41,42]. Lidar-based estimates of forest biomass could greatly improve mapping of aboveground carbon stocks and monitoring carbon emissions over large areas for tropical forests. However, this study suggests caution when applying generalized biomass models based on a single lidar metric (MCH or TCH) across a heterogeneous landscape with both flat and sloped terrain and dense vegetation, like the Serra do Mar, especially at low lidar return densities. Increasing point density mitigates the problem of accurate canopy height (and DTM) generation but increases costs.

Our study points to the need for careful attention to lidar data acquisition parameters to assess aboveground biomass in tropical forests with complex topography. Mascaro and colleagues [43] have called for a global airborne lidar campaign to cover tropical forests. We endorse this proposal but add two important caveats. First, some tropical forest environments will be more costly and complex than others for airborne lidar data acquisition. Additional costs reflect the need to adapt data collection parameters to provide equivalent lidar sampling in domains of simple and complex topography. Wall-to-wall mapping with consistent data collection in montane environments drastically reduces the efficiency of the *fly high and fast* strategy advocated by Mascaro and colleagues. Second, and perhaps more importantly, the legislation controlling airborne lidar survey varies across tropical nations. Brazil contains the largest area of tropical forests of any nation. However, because Brazil has a highly regulated market for aerial survey, achieving pricing as low as estimated by Mascaro and colleagues would be difficult or impossible at present. Regardless of these difficulties, airborne lidar offers a promising avenue for more detailed characterization of the world's tropical forests – with unique advantages for assessing the spatial and structural complexity of tropical forests in addition to benchmarking forest carbon stocks.

Conclusions

We found that small-footprint lidar data can be used to characterize the sub-canopy terrain elevation with high vertical accuracy (<1 m) in the topographically complex Serra do Mar region. The accuracy of the lidar-derived ground elevations was more strongly influenced by sampling point density than either the ranging distance or complexity of the terrain features. From the perspective of forest carbon monitoring and REDD+, return densities above 4 m^{-2} are recommended for generating forest structure data for biomass estimation. In addition, we recommend a constant flying height above ground (i.e. equal lidar footprint size), and careful flight planning to generate uniform data density throughout the lidar coverage. A consistent sampling frame is prerequisite for improved lidar-based estimates of aboveground biomass and consistent long-term monitoring under REDD+ and related activities. For dense tropical forests on steep terrain, variability in sampling density and footprint characteristics can introduce large biases in lidar-based estimates of aboveground biomass (up to 80–125 Mg ha^{-1} error in estimated biomass vs. 30 Mg ha^{-1} model error in our case), based on the underestimation of canopy height in areas with low sampling density.

Methods

Study area

The study area is located within the São Paulo State Park of Serra do Mar (PESM) (23°34′S and 45°02′W; 23°17′S and 45°11′W) in Southeast Brazil. It is characterized by complex terrain along an altitudinal gradient (0–1200 m a.s.l.) and is covered by the dense vegetation of the Atlantic Forest. The humid tropical forest in this area is subdivided into vegetation types by altitude – lowland, submontane and montane forests – from sea level up to 1200 m elevation [44]. Terrain slope at the study site is steepest at intermediate elevations in the submontane forest areas (200–900 m a.s.l; ~37° average slope), which account for approximately 37% of the study area. The remaining 63% of the study area consists of the relatively flat lowland forests (4.9%) just above sea level (~21° mean slope) and the montane forest region (58.1%) on flatter sites atop the plateau (900–1100 m a.s.l; ~24° mean slope). Our study included nine permanent forest inventory plots that were established along an altitudinal transect in the PESM [45,46]. One plot is located in the lowland forest at an elevation of 100 m (Plot F), four plots in the submontane forest between 180–370 m (Plots G, H, I and J), and four plots in the montane forest at about 1000 m a.s.l. (Plots K, L, M and N). The permanent plots each have a projected area of 1 ha.

Lidar dataset

Lidar data were collected by the GEOID Ltda. (Belo Horizonte, MG) in April 2012 as part of the Sustainable Landscapes Brazil joint project of the Brazilian Corporation of Agricultural Research (EMBRAPA) and the United States Forest Service (USFS). The study area was overflown with an Optech ALTM 3100 laser scanner instrument at an average flying altitude of 1600 m a.s.l., covering a rectangular strip of the surface (about 1.5 km × 7 km) with a total area of approximately 1000 ha (Table 2). Average pulse density was 12 m^{-2}, resulting in an average return density of 20 m^{-2}. Aircraft position information for individual flight lines was used to characterize changes in footprint characteristics across the study site. The original lidar data and associated metadata are freely available on the Sustainable Landscapes Brazil Project's website: http://mapas.cnpm.embrapa.br/paisagenssustentaveis/.

Table 2 Laser system parameters

Parameter	Specification
Positioning system	POS AV™ 510 (OEM) - GNSS/L-Band receiver
Horizontal accuracy	≤50 cm (1:1000 scale; PEC "A"); 1σ
Vertical accuracy	≤15 cm; 1σ
System frequency (PRF)	50 kHz
Scan frequency	25 Hz
Scan angle (FOV)	≤20°
Data recording	first/last mode (up to 2 returns per pulse)
Average flight altitude	1600 m a.s.l.
Beam divergence	0.25 mrad (1/e)
Overlap between flight lines	30%

Lidar processing

Flight line calibration to adjust variables such as heading, roll, pitch and height was performed by the data provider, and the lidar point cloud was processed using the methodology developed by the G-LiHT research group at NASA Goddard Space Flight Center [47]. Height filtering was carried out using a progressive morphological (PM) filter to select ground points from the data set – a critical step for DTM generation from lidar data [37]. The PM filter is used to identify objects in grayscale images based on spatial structure, and works with dilation and erosion in combination with opening and closing operators to separate ground points from non-ground ones [48]. Point classification was followed by Delaunay triangulation to create a triangular irregular network (TIN) of the filtered ground returns, and the TIN was used to interpolate the ground elevations onto a 1-meter raster grid, thus obtaining the DTM [47].

Lidar thinning

The original lidar point cloud consisted of multiple return data. Data were thinned from the original point density (~20 m^{-2}) to four predefined return densities (8, 4, 2 and 1 m^{-2}). Thinning was done randomly at 10 × 10 m resolution to achieve the desired point densities. The resulting datasets simulate lower-density lidar coverage. Random thinning reduced the density of returns classified as ground from the full density dataset (D20, 0.289 m^{-2}) to 0.113, 0.058, 0.033, and 0.023 for D8, D4, D2, and D1, respectively. Reclassification of ground and canopy returns in the thinned datasets resulted in a larger fraction of points being classified as ground returns after thinning. On average, montane plots had a higher ground point density than submontane plots, but this difference became less apparent with increased levels of thinning.

Full and reduced-density datasets were processed to generate three different data products representing the terrain surface, the canopy heights above ground, and the outer surface of the forest vegetation: Digital Terrain

Model (DTM), Canopy Height Model (CHM) and Digital Surface Model (DSM) raster layers at 1-meter resolution. DTM raster grids were created using the G-LiHT methodology, described above. CHM products were also generated using the G-LiHT algorithm by selecting the highest lidar return in every 1-meter grid cell, building a TIN based on these points, and interpolating the canopy heights on a 1-meter raster grid [47]. The DSMs of the outer canopy were produced from only the first-return points in the lidar point cloud using the BCAL LIDAR Tools open-source software package [49].

Ground data acquisition

Ground survey data collected in June 2013 within the study area were treated as a reference dataset for lidar DTM validation. A total of 36 points were measured under closed forest canopy in the hilly terrain along the altitudinal transect, marking the corner points of the nine permanent forest inventory plots located within the lidar coverage. We used two Topcon HiPer (L1/L2) GNSS receivers, one used as a rover and a second as a base for subsequent differential corrections. These receivers are survey-grade dual-frequency units capable of receiving both NAVSTAR and GLONASS signals. Raw data at the unknown points were collected for 20–35 minutes on average and up to 60 minutes when reception was poor. Base measurements were made at a survey marker (INCRA "ABE M0693") located at the Santa Virginia station in the PESM, in an open area less than 10 km of the forest plots. Post-processing of the GNSS data was performed to produce the estimated position of the unknown points. Out of the 36 control points, 35 were measured with success, and 30 points had sub-meter accuracy ($\sigma < 1$m) in all three coordinates x, y, z (UTM easting, northing and elevation). The remaining 5 points were less accurate ($\sigma < 2.2$ m). The GNSS system parameters and measurement conditions during the survey are summarized in Table 3.

Statistical analysis of the datasets

We compared GNSS and lidar DTM elevations for ground reference locations using mean signed error, absolute error, and root mean square error (RMSE) [21]. Mean signed error can be useful to identify the tendency for under- or over-estimation of elevations (i.e. bias), while RMSE represents the overall mean elevation accuracy of a DTM. We note that RMSE has been criticized as a metric for evaluation of DTMs and other map position data [50-52]. However, the criticisms relate to data distributions that deviate strongly from normality. Inspection of Q-Q plots showed no outliers and no obvious deviation from normality, therefore we had no reason to employ alternative metrics. To determine if the difference between the two sets of height points (DTM vs. GNSS elevations) is

Table 3 GNSS system parameters, survey conditions and control points

Parameter	Specification
GNSS system	Topcon HiPer L1/L2 receiver
Horizontal accuracy	3 mm + 0.5 PPM
Vertical accuracy	5 mm + 0.5 PPM
System frequency	20 Hz
Linear units	meters
Angular units	degrees
Datum	WGS84
Projection	UTM Zone 23 South
Geoid	MAPGEO 2010
Base Reference Point	INCRA "ABE M0693"
Number of points measured with success	35 (out of 36 total)
Points with $\sigma < 1$ m (x,y,z)	30
Points with $1 \leq \sigma < 2.2$ m (x/y/z)	5
Accuracy (RMSE) Easting	0.006 - 2.130 m; mean = 0.473 m
Accuracy (RMSE) Northing	0.006 - 1.876 m; mean = 0.225 m
Accuracy (RMSE) Elevation	0.019 - 2.195 m; mean = 0.469 m

statistically significant, a two-sided t-test was performed with a confidence level of 95% and assuming a normal distribution of the errors (Kolmogorov-Smirnov test for normality, p-value = 0.923).

Given the significant variation in terrain elevation across the study area (from about 100 m a.s.l. up to 1100 m a.s.l.) and the relatively constant flying altitude during the lidar survey (~1600 m a.s.l.), the sensor height above the ground varied substantially across the 1000 ha lidar coverage (Figure 1). The mean ranging distance between the sensor and the ground surface was ~660 m for the montane region on top of the plateau, while it was about twice as large (~1320 m) for the submontane region. Because of beam divergence, increasing lidar ranging distance results in a larger footprint on the ground. Variation of sensor height above the ground can influence the measurement results, such as laser point density, penetration, ground detection, and calculated metrics [53]. In this study, the lidar footprint diameter doubled between the montane and submontane regions, from ~0.16 m to ~0.33 m, based on the 0.25 mrad beam divergence. To assess the effect of different ranging distances (i.e. variable footprint size) on DTM error across the study area, the control points were grouped into submontane and montane elevational classes, and the error distributions between the groups were compared. To test if the means of the errors associated with the specified elevation classes are statistically different, a two-sided t-test was performed with a confidence level of 95%.

The accuracy of the DTMs generated after data thinning was evaluated using the same approach as with the full-density DTM. Additionally, we assessed the total number of lidar returns and the number of ground returns in the reduced-density point clouds for each permanent field plot location. The ground point density (points m^{-2}) and the fraction of ground returns out of all returns (%) was calculated for each thinning level to quantify the change in commission errors resulting from the ground classification algorithm.

Plot-scale lidar metrics and forest inventory data were used to establish lidar-biomass relationships following standard methods. The goal of this effort was to assess the impact of DTM errors from variability in sampling density on predicted aboveground biomass. We used forest inventory data from the Serra do Mar permanent plot network (Biota Project, see [46]) to calculate field-based AGB estimates in the nine plots following the methodology applied by Alves and colleagues [45]. A linear model was developed to predict AGB based on plot-level mean canopy surface heights derived from the full-density lidar data. We used this regression equation to generate biomass estimates based on the thinned lidar datasets with mean canopy surface height as the predictor, and compared the resulting values across the different data densities.

Abbreviations
AGB: Aboveground biomass; CHM: Canopy height model; DSM: Digital surface model; DTM: Digital terrain model; LiDAR: Light detection and ranging; MCH: Mean canopy height; TCH: Top-of-canopy height; TIN: Triangular irregular network; REDD: Reduced emissions from deforestation and forest degradation; RMSE: Root mean square error.

Competing interests
The authors declare that they have no competing interests.

Authors' contributions
VL conducted field reference data collection, carried out all analyses and drafted the manuscript. MK and DCM helped design the study, guided the research, and assisted with the writing. BDC provided technical assistance with lidar data processing. YES and all other authors read and approved the final manuscript.

Acknowledgments
This research was supported by NASA's Terrestrial Ecology and Carbon Monitoring System Programs (NASA NNH13AW64I) and CAPES (Brazilian Federal Agency for the Support and Evaluation of Graduate Education) graduate scholarship offered through the Brazilian National Institute for Space Research (INPE). Lidar data were acquired with support from USAID and the US Department of State with the technical assistance of the Brazilian Corporation for Agricultural Research (EMBRAPA) and the US Forest Service Office of International Programs. We thank Luciana F. Alves for sharing the biomass data of the Atlantic forest sites. Forest inventory work was supported by USAID and the State of São Paulo Research Foundation (FAPESP 03/12595-7 to C. A. Joly and L. A. Martinelli), within the BIOTA/FAPESP Program – The Biodiversity Virtual Institute (http://www.biota.org.br). COTEC/IF 41.065/2005, COTEC/IF 663/2012 and IBAMA/CGEN 093/2005 permits.

Author details
[1]Remote Sensing Division, National Institute for Space Research (INPE), São José dos Campos, SP CEP 12201-970, Brazil. [2]Biospheric Sciences Laboratory, NASA Goddard Space Flight Center, Greenbelt, MD 20771, USA. [3]International Institute of Tropical Forestry, USDA Forest Service, San Juan 00926, Puerto Rico. [4]EMBRAPA Satellite Monitoring, Campinas SP CEP 13070-115, Brazil.

References
1. Angelsen A, editor. Moving ahead with REDD: issues, options and implications. Bogor, Indonesia: Center for International Forestry Research (CIFOR); 2008. p. 156.
2. Naesset E. Estimating timber volume of forest stands using airborne laser scanner data. Remote Sens Environ. 1997;51:246–53.
3. Lefsky MA, Harding DJ, Cohen WB, Parker GG. Surface lidar remote sensing of basal area and biomass in deciduous forests of eastern Maryland, USA. Remote Sens Environ. 1999;67:83–98.
4. Lefsky MA, Cohen WB, Harding DJ, Parker GG, Acker SA, Gower ST. Lidar remote sensing of aboveground biomass in three biomes. Glob Ecol Biogeogr. 2002;11:393–400.
5. Drake JB, Knox RG, Dubayah RO, Clark DB, Condit R, Blair JB, et al. Aboveground biomass estimation in closed canopy Neotropical forests using lidar remote sensing: factors affecting the generality of relationships. Glob Ecol Biogeogr. 2003;12:147–59.
6. Lim K, Treitz P, Wulder MA, St-Onge B, Flood M. Lidar remote sensing of forest structure. Prog Phys Geogr. 2003;27:88–106.
7. Naesset E, Gobakken T, Holmgren J, Hyyppä H, Hyyppä J, Maltamo M, et al. Laser scanning of forest resources: the Nordic experience. Scand J For Res. 2004;19:482–99.
8. Naesset E, Gobakken T. Estimation of above- and below-ground biomass across regions of the boreal forest zone using airborne laser. Remote Sens Environ. 2008;112:3079–90.
9. Asner GP, Hughes RF, Varga TA, Knapp DE, Kennedy-Bowdoin T. Environmental and biotic controls over aboveground biomass throughout a tropical rain forest. Ecosystems. 2009;12:261–78.
10. Clark ML, Roberts DA, Ewel JJ, Clark DB. Estimation of tropical rain forest aboveground biomass with small-footprint lidar and hyperspectral sensors. Remote Sens Environ. 2011;115:2931–42.
11. Mascaro J, Detto M, Asner GP, Muller-Landau H. Evaluating uncertainty in mapping forest carbon with airborne LiDAR. Remote Sens Environ. 2011;115:3770–4.
12. Asner GP, Mascaro J, Muller-Landau HC, Vieilledent G, Vaudry R, Rasamoelina M, et al. A universal airborne LiDAR approach for tropical forest carbon mapping. Oecologia. 2011;168:1147–60.
13. Asner GP, Mascaro J. Mapping tropical forest carbon: calibrating plot estimates to a simple LiDAR metric. Remote Sens Environ. 2014;140:614–24.
14. Ni-Meister W, Lee S, Strahler AH, Woodcock CE, Schaaf C, Yao T, et al. Assessing general relationships between aboveground biomass and vegetation structure parameters for improved carbon estimate from lidar remote sensing. J Geophys Res. 2010;115:G00E11.
15. D'Oliveira MVN, Reutebuch SE, McGaughey RJ, Andersen H-E. Estimating forest biomass and identifying low-intensity logging areas using airborne scanning lidar in Antimary State Forest, Acre State, Western Brazilian Amazon. Remote Sens Environ. 2012;124:479–91.
16. Asner GP, Powell GVN, Mascaro J, Knapp DE, Clark JK, Jacobson J, et al. High-resolution forest carbon stocks and emissions in the Amazon. Proc Natl Acad Sci U S A. 2010;107:16738–42.
17. Asner GP, Hughes RF, Mascaro J, Uowolo AL, Knapp DE, Jacobson J, et al. High-resolution carbon mapping on the million-hectare Island of Hawaii. Front Ecol Environ. 2011;9:434–9.
18. Asner GP, Mascaro J, Anderson C, Knapp DE, Martin RE, Kennedy-Bowdoin T, et al. High-fidelity national carbon mapping for resource management and REDD+. Carbon Bal Manage. 2013;8:7.
19. Tinkham WT, Smith AMS, Hoffman C, Hudak AT, Falkowski MJ, Swanson ME, et al. Investigating the influence of LiDAR ground surface errors on the utility of derived forest inventories. Can J For Res. 2012;42:413–22.
20. Aguilar FJ, Mills JP. Accuracy assessment of LiDAR-derived digital elevation models. Photogramm Rec. 2008;23:148–69.
21. Su J, Bork E. Influence of vegetation, slope, and LiDAR sampling angle on DEM accuracy. Photogramm Eng Remote Sens. 2006;72:1265–74.
22. Cobby DM, Mason DC, Davenport IJ. Image processing of airborne scanning laser altimetry data for improved river flood modelling. ISPRS J Photogramm Remote Sens. 2001;56:121–38.

23. Hodgson ME, Jensen J, Raber G, Tullis J, Davis BA, Thompson G, et al. Evaluation of lidar-derived elevation and terrain slope in leaf-off conditions. Photogramm Eng Remote Sens. 2005;71:817–23.

24. Spaete LP, Glenn NF, Derryberry DR, Sankey TT, Mitchell JJ, Hardegree SP. Vegetation and slope effects on accuracy of a LiDAR-derived DEM in the sagebrush steppe. Remote Sens Lett. 2010;2:317–26.

25. Hodgson ME, Bresnahan P. Accuracy of airborne LiDAR-derived elevation: empirical assessment and error budget. Photogramm Eng Remote Sens. 2004;70:331–9.

26. Clark ML, Clark DB, Roberts DA. Small-footprint lidar estimation of sub-canopy elevation and tree height in a tropical rain forest landscape. Remote Sens Environ. 2004;91:68–89.

27. Reutebuch SE, McGaughey RJ, Anderson HE, Carson WW. Accuracy of a high-resolution lidar terrain model under a conifer forest canopy. Can J Remote Sens. 2003;29:527–35.

28. Kraus K, Pfeifer N. Determination of terrain models in wooded areas with airborne laser scanner data. ISPRS J Photogramm Remote Sens. 1998;53:193–203.

29. Hodgson ME, Jensen JR, Schmidt L, Schill S, Davis B. An evaluation of LIDAR- and IFSAR-derived digital elevation models in leaf-on conditions with USGS Level 1 and Level 2 DEMs. Remote Sens Environ. 2003;84:295–308.

30. Andersen H-E, Reutebuch SE, McGaughey RJ. A rigorous assessment of tree height measurements obtained using airborne LIDAR and conventional field methods. Can J Remote Sens. 2006;32:355–66.

31. Gonzalez P, Asner GP, Battles JJ, Lefsky MA, Waring KM, Palace M. Forest carbon densities and uncertainties from Lidar, QuickBird, and field measurements in California. Remote Sens Environ. 2010;114:1561–75.

32. Gatziolis D, Andersen H-E. A guide to LIDAR data acquisition and processing for the forests of the Pacific Northwest, Gen. Tech. Rep. PNW-GTR-768. Portland, OR: U.S: Department of Agriculture, Forest Service, Pacific Northwest Research Station; 2008.

33. Säynäjoki R, Maltamo M, Korhonen KT. Forest inventory with sparse resolution Airborne Laser Scanning data – a literature review. Working Papers of the Finnish Forest Research Institute. 2013, 103. 90.

34. Hudak AT, Strand EK, Vierling LA, Byrne JC, Eitel JUH, Martinuzzi S, et al. Quantifying aboveground forest carbon pools and fluxes from repeat LiDAR surveys. Remote Sens Environ. 2012;123:25–40.

35. Jakubowski MK, Guo Q, Kelly M. Tradeoffs between lidar pulse density and forest measurement accuracy. Remote Sens Environ. 2013;130:245–53.

36. Hyyppä H, Yu X, Hyyppä J, Kaartinen H, Kaasalainen S, Honkavaara E, et al. Factors affecting the quality of DTM generation in forested areas. In Proceedings of ISPRS Workshop on Laser Scanning 2005, Vol. XXXVI, 3/W19, 85–90. Netherlands: GITC bv. 12–14 September 2005, Enschede, Netherlands.

37. Liu X. Airborne LiDAR for DEM generation: some critical issues. Prog Phys Geogr. 2008;32:31–49.

38. Lefsky MA. A global forest canopy height map from the moderate resolution imaging spectroradiometer and the geoscience laser altimeter system. Geophys Res Lett. 2010;37:L15401.

39. Feldpausch TR, Lloyd J, Lewis SL, Brienen RJW, Gloor M, Mendoza AM, et al. Tree height integrated into pantropical forest biomass estimates. Biogeosciences. 2012;9:3381–403.

40. Chave J, Réjou-Méchain M, Búrquez A, Chidumayo E, Colgan MS, Delitti WBC, et al. Improved allometric models to estimate the aboveground biomass of tropical trees. Glob Chang Biol. 2014;20:3177–90.

41. Wulder MA, White JC, Nelson RF, Naesset E, Ørka HO, Coops NC, et al. Lidar sampling for large-area forest characterization: a review. Remote Sens Environ. 2012;121:196–209.

42. Zolkos SG, Goetz SJ, Dubayah R. A meta-analysis of terrestrial aboveground biomass estimation using lidar remote sensing. Remote Sens Environ. 2013;128:289–98.

43. Mascaro J, Asner GP, Davies S, Dehgan A, Saatchi S. These are the days of lasers in the jungle. Carbon Bal Manag. 2014;9:7.

44. SMA - Secretaria do Meio Ambiente. Planos de Manejo das Unidades de Conservação: Parque Estadual da Serra do Mar - Núcleo Picinguaba. São Paulo: Plano de Gestao Ambiental - Fase I; 1998.

45. Alves LF, Vieira SA, Scaranello MA, Camargo PB, Santos FAM, Joly CA, et al. Forest structure and live aboveground biomass variation along an elevational gradient of tropical Atlantic moist forest (Brazil). For Ecol Manag. 2010;260:679–91.

46. Joly CA, Assis MA, Bernacci LC, Tamashiro JY, Campos MCR, Gomes JAMA, et al. Floristic and phytosociology in permanent plots of the Atlantic Rainforest along an altitudinal gradient in southeastern Brazil. Biota Neotropica. 2012;12:125–45.

47. Cook BD, Corp LA, Nelson RF, Middleton EM, Morton DC, McCorkel JT, et al. NASA Goddard's LiDAR, Hyperspectral and Thermal (G-LiHT) Airborne Imager. Remote Sens. 2013;5:4045–66.

48. Zhang K, Chen S, Whitman D, Shyu M, Yan J, Zheng C. A progressive morphological filter for removing nonground measurements from airborne LiDAR data. IEEE Trans Geosci Remote Sens. 2003;41:872–82.

49. BCAL LiDAR Tools ver 2.x.x-dev9. Idaho State University, Department of Geosciences, Boise Center Aerospace Laboratory (BCAL), Boise, Idaho. [http://bcal.geology.isu.edu/envitools.shtml]

50. Zandbergen PA. Characterizing the error distribution of lidar elevation data for North Carolina. Int J Remote Sens. 2011;32:409–30.

51. Höhle J, Höhle M. Accuracy assessment of digital elevation models by means of robust statistical methods. ISPRS J Photogramm Remote Sens. 2009;64:398–406.

52. Chen C, Fan Z, Yue T, Dai H. A robust estimator for the accuracy assessment of remote-sensing-derived DEMs. Int J Remote Sens. 2012;33:2482–97.

53. Morsdorf F, Frey O, Meier E, Itten KI, Allgöwer B. Assessment of the influence of flying altitude and scan angle on biophysical vegetation products derived from airborne laser scanning. Int J Remote Sens. 2008;29:1387–406.

## 8

Robustness of model-based high-resolution prediction of forest biomass against different field plot designs

Virpi Junttila[1], Basanta Gautam[2], Bhaskar Singh Karky[3], Almasi Maguya[1,4]* ⓘ, Katri Tegel[2], Tuomo Kauranne[1,2], Katja Gunia[2], Jarno Hämäläinen[2], Petri Latva-Käyrä[2], Ekaterina Nikolaeva[1] and Jussi Peuhkurinen[2]

Abstract

Background: Participatory forest monitoring has been promoted as a means to engage local forest-dependent communities in concrete climate mitigation activities as it brings a sense of ownership to the communities and hence increases the likelihood of success of forest preservation measures. However, sceptics of this approach argue that local community forest members will not easily attain the level of technical proficiency that accurate monitoring needs. Thus it is interesting to establish if local communities can attain such a level of technical proficiency. This paper addresses this issue by assessing the robustness of biomass estimation models based on air-borne laser data using models calibrated with two different field sample designs namely, field data gathered by professional forester teams and field data collected by local communities trained by professional foresters in two study sites in Nepal. The aim is to find if the two field sample data sets can give similar results (LiDAR models) and whether the data can be combined and used together in estimating biomass.

Results: Results show that even though the sampling designs and principles of both field campaigns were different, they produced equivalent regression models based on LiDAR data. This was successful in one of the sites (Gorkha). At the other site (Chitwan), however, major discrepancies remained in model-based estimates that used different field sample data sets. This discrepancy can be attributed to the complex terrain and dense forest in the site which makes it difficult to obtain an accurate digital elevation model (DTM) from LiDAR data, and neither set of data produced satisfactory results.

Conclusions: Field sample data produced by professional foresters and field sample data produced by professionally trained communities can be used together without affecting prediction performance provided that the correlation between LiDAR predictors and biomass estimates is good enough.

Keywords: Above-ground biomass, LiDAR, REDD+, Participatory forest monitoring

Background

Greenhouse gas (GHG) emissions from tropical deforestation and forest degradation contribute about 15–20 % of total annual global GHG emissions, making them the second largest source of greenhouse gases globally [1]. To reduce especially CO_2 (carbon dioxide) emissions from the forestry sector, the United Nations has established a program that would provide payments for the reduction of emissions from deforestation and forest degradation (REDD+). REDD+ is a performance-based policy instrument aimed at reducing anthropogenic emissions of GHG [2, 3].

Nepal is one of the countries participating in REDD+. After successful implementation of a Community Forestry programme, Nepal has taken another leap by piloting innovative REDD+ projects. The International Centre for Integrated Mountain Development (ICIMOD) is one of the first organizations to implement a

*Correspondence: almasi.maguya@gmail.com
[4] Mzumbe University, P.O. Box 1, Mzumbe, Morogoro, Tanzania
Full list of author information is available at the end of the article

community-based REDD+ pilot at micro-watershed level. ICIMOD and its partners, the Federation of Community Forestry Users, Nepal (FECOFUN) and the Asia Network for Sustainable Agriculture and Bioresources (ANSAB) implemented a pilot project from 2009–2013, with support from the Norwegian Agency for Development Cooperation (NORAD) climate and forest Initiative [4]. The major focus of the project was to develop and demonstrate an innovative benefit-sharing mechanism for REDD+ incentives using institutionally and socially inclusive approaches to address the drivers of deforestation and forest degradation and improve forest governance [5] in three micro-watersheds, namely Kayarkhola in Chitwan, Ludikhola in Gorkha and Charanawati in Dolakha districts. The pilot project focused on sequestering carbon through community-based forest management. It is one of the first carbon offset demonstration projects in the world that involves local communities in monitoring the carbon in their forests and providing the necessary training for them to do so. Training on assessing forest carbon pools was provided to the local communities that manage the forest [6]. The trained local communities collected field plot data from 2010 to 2012. The results of this effort are summarized in [7].

In a joint effort, the Forest Resource Assessment Nepal project, Arbonaut Ltd. and ICIMOD carried out a wall-to-wall airborne discrete-return light detection and ranging (LiDAR) data and subsequent field plot collection in two watersheds in Chitwan and Gorkha. The main aim of the work was to estimate accurately and with a high spatial resolution the forest above ground biomass (AGB)/carbon in the watersheds. The field data was collected by professional foresters and technicians.

In measuring biomass for calculating REDD+ compensation the measurements should be conducted in a biennial manner, as has been recently agreed at a UNFCCC meeting in Bonn [8]. As national measurements are required at such a high frequency, traditional sampling-based methods become prohibitively expensive. We therefore propose to use a model-based strategy for biomass prediction, where field plots are only used for model calibration, i.e. model parameter estimation, and biomass prediction models are based on remotely sensed data, such as LiDAR, which is used in this study, or satellite imagery.

In this study, plot-level AGB values estimated from field measurements (labelled as "AGB field estimates" in this article) in given inventory areas are predicted ("predictions") in new validation plot locations ("validation set") using linear model which is calibrated using field measurements obtained from the respective areas ("training set"). The field measurements are collected by two types of forest inventory teams: a professional measurement team ("Prof") and measurement teams trained from the local community members ("Comm"). The local community members were trained to do forest inventory work by professionals from ICIMOD and its project partners that operated independently of the professional measurement team. Inventory work, i.e., field sample plot selection and actual measurement work, is thus performed by two separate groups, which may lead to data sets with different characteristics, even when the measurements are conducted in the same area.

To justify the use of participatory forest monitoring and inventory work of local communities, it is important that there are no significant differences in AGB predictions based on field measurements of both teams. Thus, it is needed to verify if different data sets from the same area lead to similar model based predictions, and if the datasets can be combined and used together in predicting AGB for the areas in the future. In this study, we first analyse the field estimates measured by both teams and validate the correlation between the model auxiliary data, LiDAR predictors, and the corresponding field estimates. We then validate the robustness of high-resolution model-based biomass estimation against different sampling designs from the same area. Models using different sources of training set data but resulting in the same predictions with the same validation data are considered robust against the training set source.

Results

The model-based predictions and corresponding error analyses were performed using data collected from two inventory study sites located in Nepal, namely Gorkha (labelled as "Go") and Chitwan ("Ch"), see Fig. 1 for the location of the sites. The scope in forest inventory by the different inventory teams was different. The professional team measured only closed canopy forest owned by both the state and by the forest communities. The local field teams measured both open and closed canopy forests but only those owned by communities. Only the field plots of similar forest types, plots from closed canopy forest owned by communities, were used in this study. The number of plots measured by the community teams located in closed type forests that are owned by the communities is 151 in study site Gorkha and 151 in study site Chitwan, while the number of plots measured by the professional teams located in closed type forests owned by the communities is 41 in Gorkha and 26 in Chitwan. See Fig. 1 for plot locations and Fig. 2 for the distribution of field estimates.

Examples of the combined distributions of LiDAR predictor values and AGB field estimates of each sample are shown in Fig. 3. The LiDAR predictor—AGB field estimate correlation of both sample in Gorkha, the plots

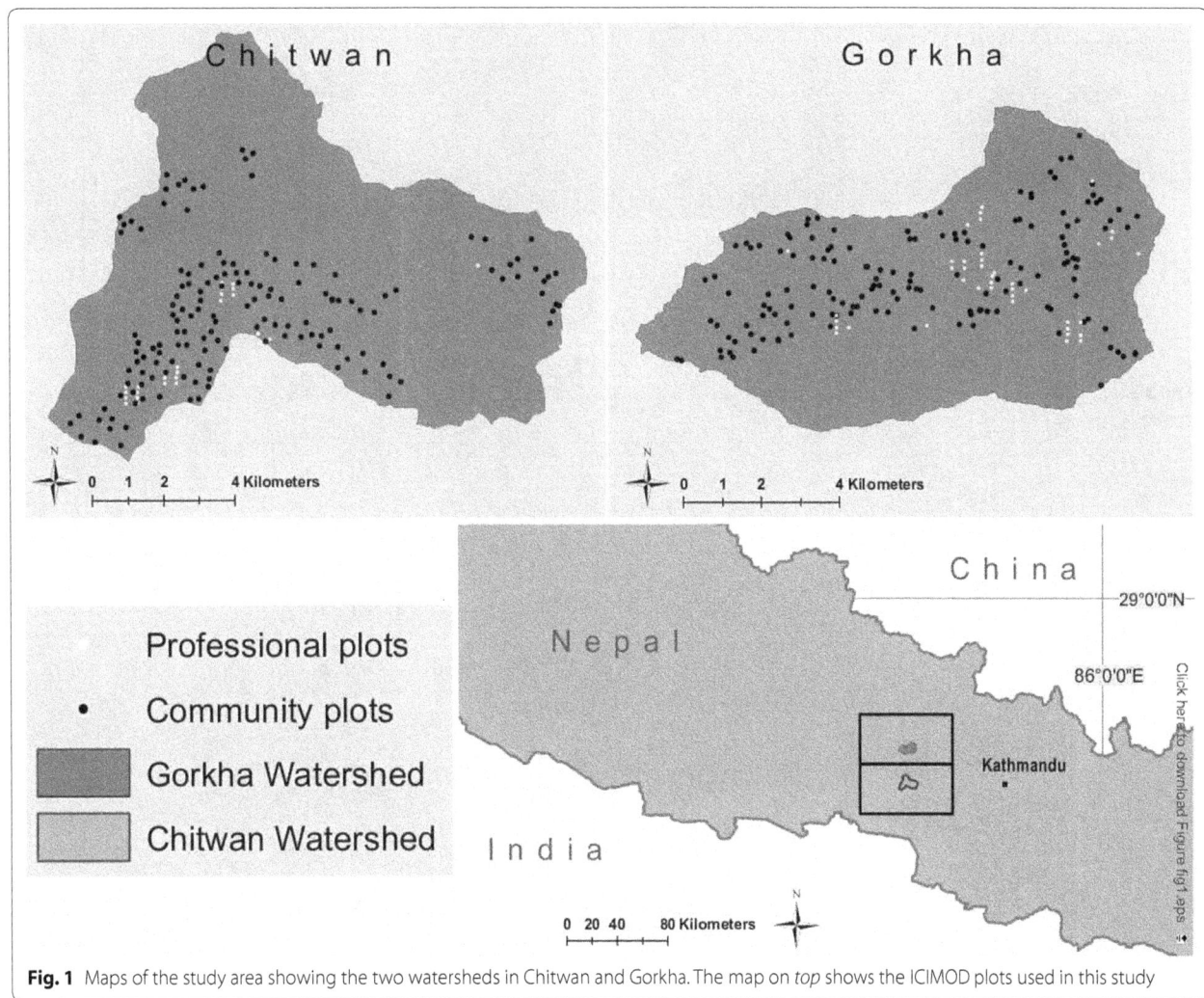

Fig. 1 Maps of the study area showing the two watersheds in Chitwan and Gorkha. The map on *top* shows the ICIMOD plots used in this study

measured by professional teams and the plots measured by the members of local communities, is higher than in Chitwan. The correlation of AGB and the LiDAR example predictor in Gorkha is 0.66 for sample ProfGo, 0.59 for CommGo, and in Chitwan it is 0.52 for ProfCh and 0.45 for CommCh. With visual analysis, the LiDAR predictor—AGB distributions of different inventor teams are similar in both study sites. Similar properties hold for each predictor used in this study.

Results of study site Gorkha

The results for predictions in study site Gorkha are shown in Table 1 and Figs. 4 and 5. In the table, model prediction precision and accuracy are evaluated by relative root mean square error (RMSE %) and relative mean difference (D %). The prediction error distributions in two validation sets (Comm and Prof) resulting from three different model training set data (Comm, Prof and the combined set of these plots, Comm + Prof) are verified. The

differences of the mean values (D %) between the self-validation (validation set and training set are the same) and the cross-validation cases (training set is different than validation set) are tested by a two-sample t test with a 5 % significance level against the case where the validation set and training set are the same. Similarly, the difference of the RMSE % values are tested with a two-sample F test for equal variances with a 5 % significance level. The tests show that distributions of prediction error among different models in the validation set Comm are similar. The mean of the errors are the same as that for training set Comm (t test p values > 0.79 for training sets Prof and Comm + Prof) and also the variance of the error distributions are similar (variance test p values > 0.89). Similar results are shown for the AGB predictions in the plots measured by the professional teams. The statistical tests for the prediction error show no significant difference in error mean or error variance compared to the predictions estimated with training set Prof (t test p values are > 0.84

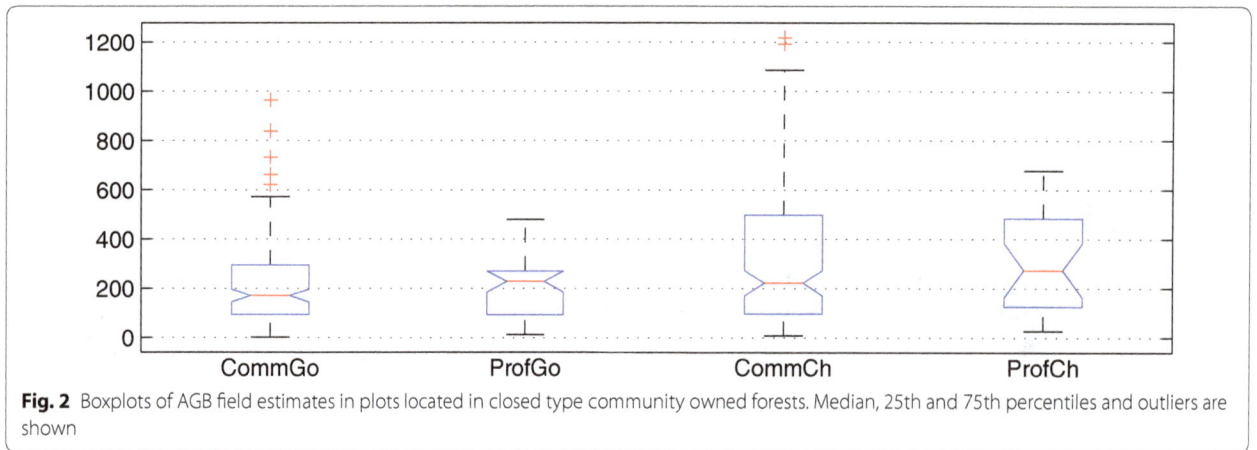

Fig. 2 Boxplots of AGB field estimates in plots located in closed type community owned forests. Median, 25th and 75th percentiles and outliers are shown

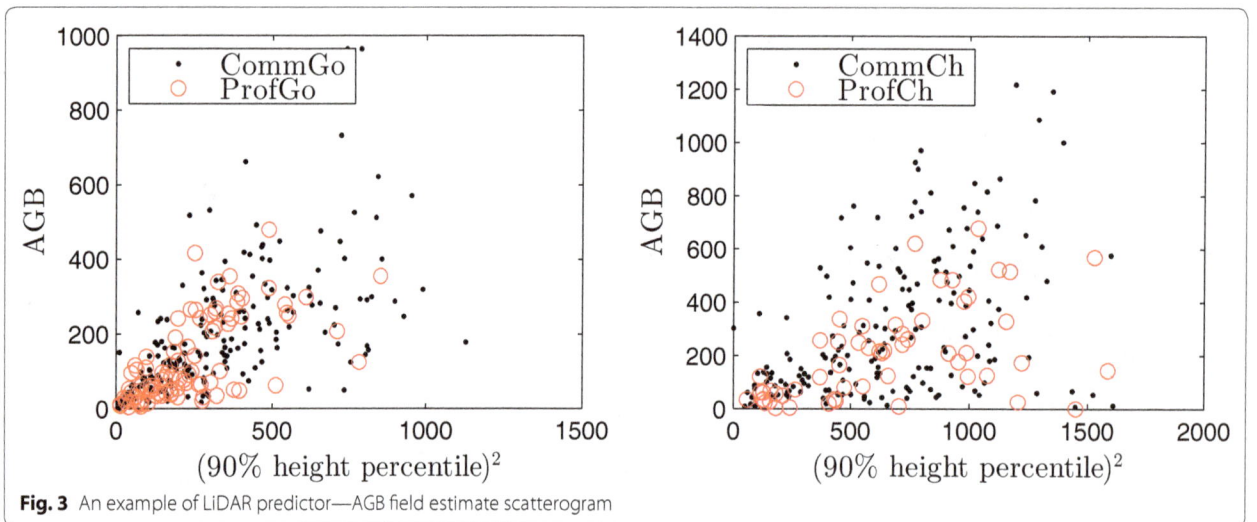

Fig. 3 An example of LiDAR predictor—AGB field estimate scatterogram

Table 1 Results with different combinations of training set–validation set in study site Gorkha

Training set	Error stat.		Test against baseline, p values	
	RMSE %	BIAS %	Variance test	t test for mean
Validation set: Comm				
Comm (base-line)	59.2	0.0		
Prof	59.3	−1.8	0.994	0.794
Comm + Prof	58.6	−0.2	0.888	0.977
Validation set: Prof				
Prof (baseline)	37.5	0.2		
Comm	34.4	1.9	0.586	0.839
Comm + Prof	34.8	1.5	0.644	0.872

and variance test p values are > 0.59 for training sets Comm and Comm + Prof). Figure 4 shows that independent of the training and validation set used, a relatively good model fit is obtained for validation set Comm ($r^2 \geq 0.46$ for all training sets) and for validation set Prof ($r^2 \geq 0.58$ for all training sets).

Also the distributions of the AGB predictions shown in Fig. 5 are close to each other. The lack of ability to predict accurately the largest AGB values in plots measured by the Community teams is similar when using any of the training sets (see the left sub-figure). Up to about 300 Mg/ha, the AGB distributions of all predictions are very close to the distribution AGB field estimates of Community teams, the larger AGB values tend to be underestimated. In case of the AGB field estimates of the

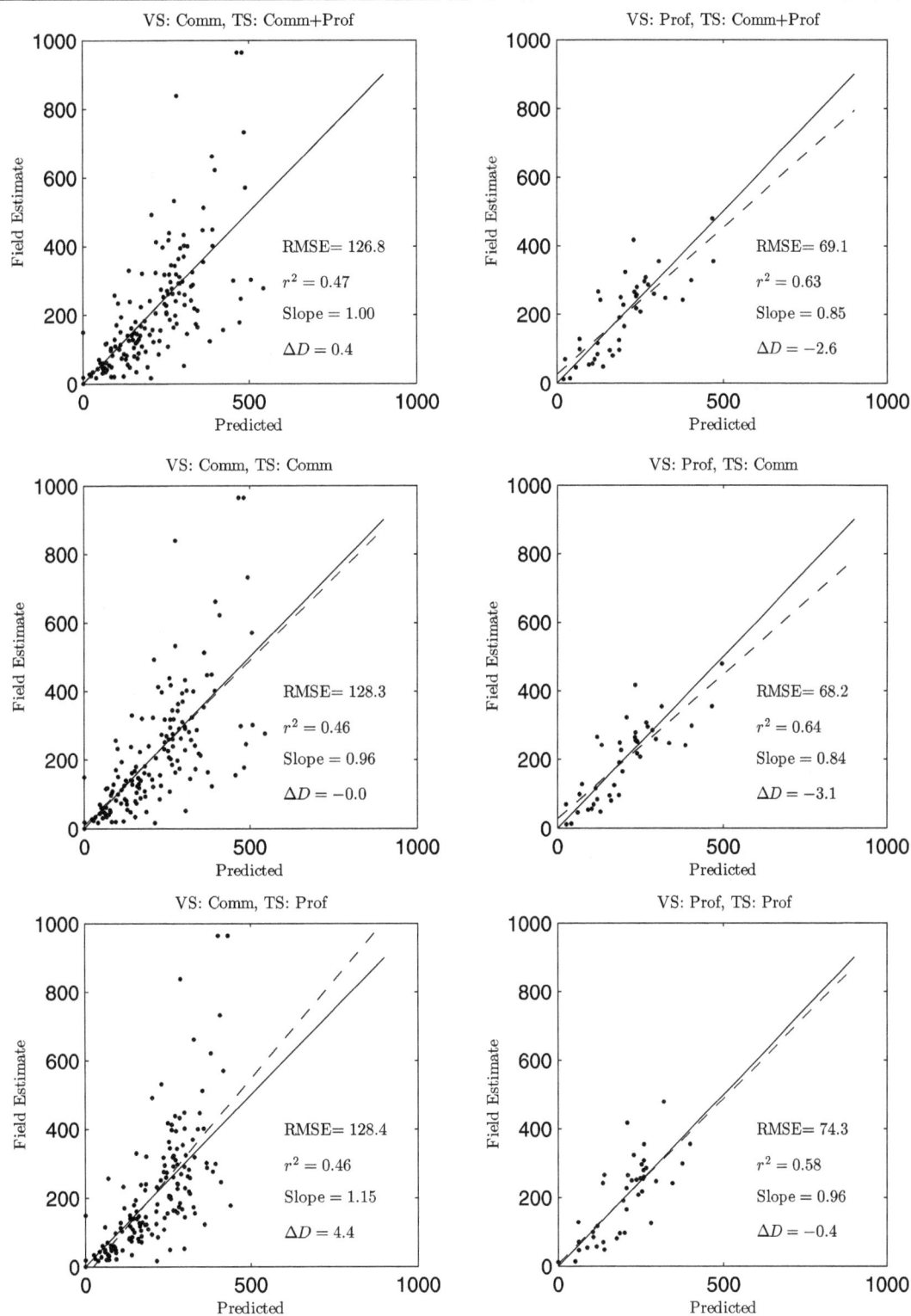

Fig. 4 Scatterogram and prediction error analysis of AGB predictions and field estimates in study site Gorkha. *VS* validation set, *TS* training set

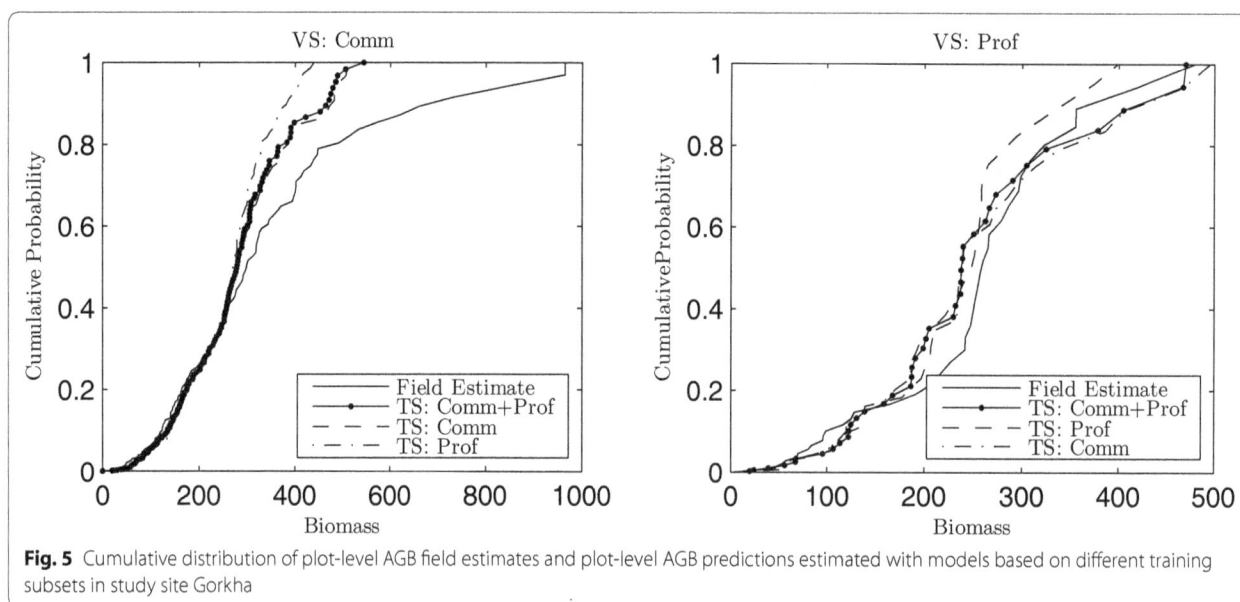

Fig. 5 Cumulative distribution of plot-level AGB field estimates and plot-level AGB predictions estimated with models based on different training subsets in study site Gorkha

Professional teams, predictions estimated by different training sets of either of the teams result in quite correct AGB distributions. No big deviations can be seen up to 400–500 Mg/ha.

Results of study site Chitwan

Results of AGB predictions in study site Chitwan are shown in Table 2 and Figs. 6 and 7. The variance test in Table 2 shows no significant difference between the prediction precision (RMSE %) between the predictions obtained with different training sets for either validation sets (p values > 0.81 for validations set Comm, > 0.43 for Prof). However, the t test for the average values of the average relative differences, D %, shows significant difference (p value $0.008 < 0.05$) for prediction of community

Table 2 Results with different combinations of training set–validation set in study site Chitwan

Training set	Error stat.		Test against baseline, pvalues	
	RMSE %	BIAS %	Variance test	t test for mean
Validation set: Comm				
Comm (baseline)	72.1	−0.5		
Prof	74.2	−22.6	0.806	0.008
Comm + Prof	71.7	−3.5	0.931	0.720
Validation set: Prof				
Prof (baseline)	55.5	−2.5		
Comm	52.9	23.6	0.433	0.079
Comm + Prof	51.3	19.0	0.452	0.146

team plots using model calibrated with field estimates of professional teams. Similarly, predictions for professional team measured plots predicted with model calibrated using the community team plots, the p value is close to 0.05 although not less than it. Thus, there is significant average error in predictions when the training set and validation set are different. Similar doesn't hold when the training set consists also data from the validation set (Training set Comm + Prof).

The one-to-one scatterograms in Fig. 6 shows that the correlation between AGB field estimates and model based predictions is poor in each cross-validation set. Even the predictions estimated using the same training set and validation set (self-validation: validation set Comm estimated with training set Comm and validation set Prof estimated with training set Prof) fail. The RMSE of the predictions is large compared to the variation of the AGB values and the model fit is poor ($r^2 \leq 0.33$) in each case.

In Fig. 7 it can be seen that the models severely underestimate the cumulative probability distribution of AGB for values for AGB greater than 200 Mg/ha in validation set Comm (left sub-figure). This happens regardless of the training set used to train the model, also when the training set is the same as validation set. For validation set Prof, the model over-estimates the cumulative probability distribution of AGB for values of AGB less than 400 Mg/ha regardless of the training set used to train the model (right sub-figure) and for values of AGB greater than 400 Mg/ha the model under-estimates the cumulative probability distribution for each training set.

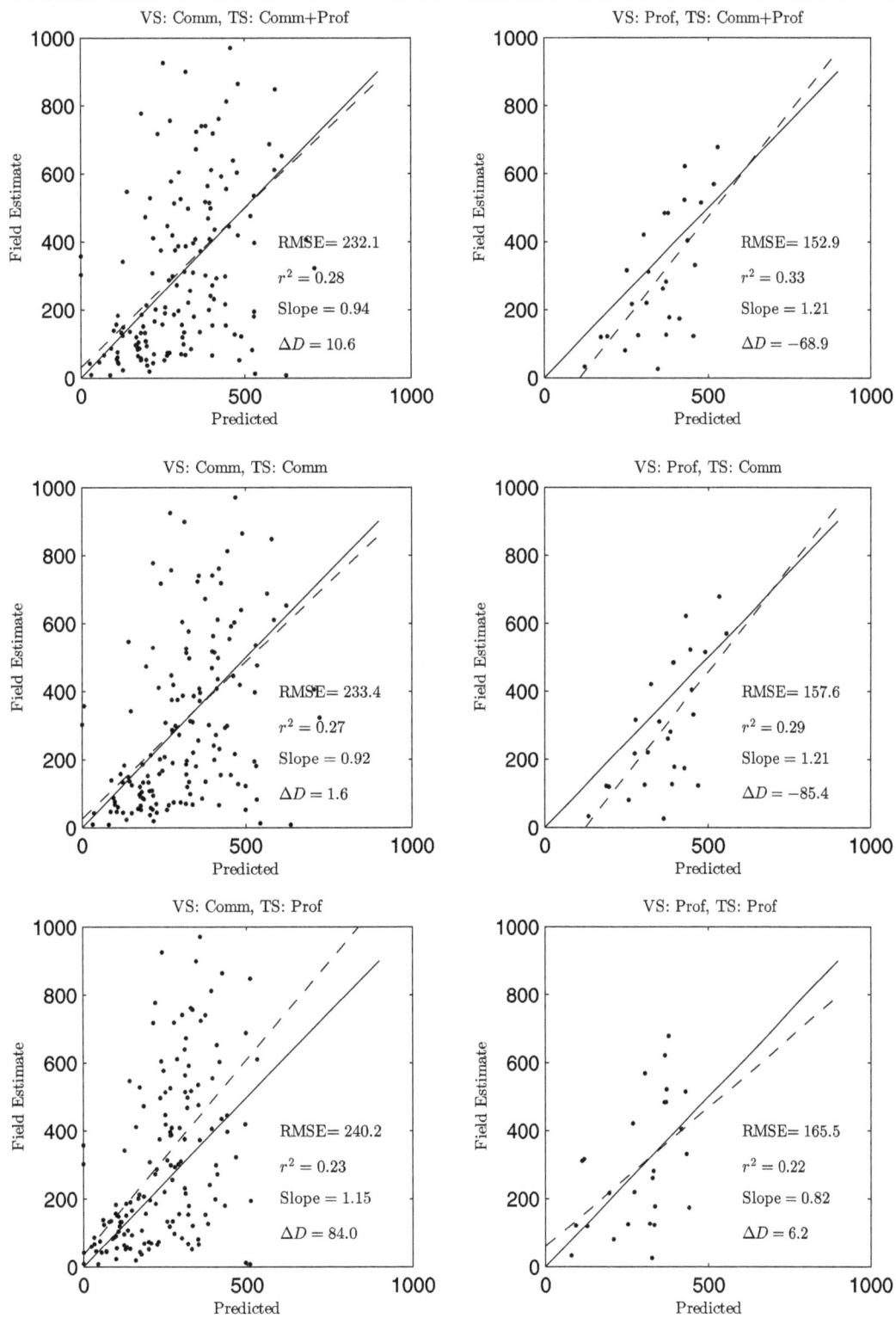

Fig. 6 Scatterogram and prediction error analysis of the AGB predictions ("predicted") and AGB field estimates in study site Chitwan

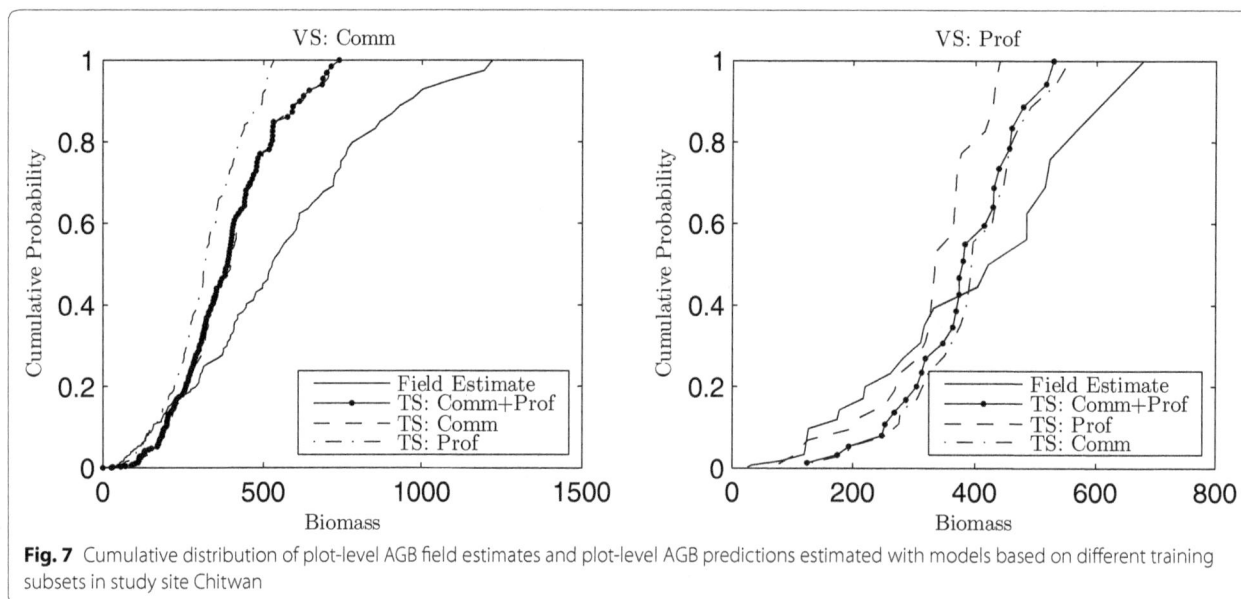

Fig. 7 Cumulative distribution of plot-level AGB field estimates and plot-level AGB predictions estimated with models based on different training subsets in study site Chitwan

Discussion

The results above show that the predictions in study site Gorkha are quite accurate and precise in each case in the cross-validation procedure and no significant difference occurs when different sources of training set data are used. However, there are severe problems in the predictions of AGB in study site Chitwan. The cross-validation procedure shows that there is significant difference in the mean of prediction error when the training set and validation set are different. However, in this study site, the prediction precision overall is weak, the RMSE is large and model fit poor.

In model-based prediction, the correlation between the response, AGB field estimates, and the auxiliary data, LiDAR predictors, define whether the auxiliary data can be used to accurately predict the response over the whole spatial area. In this study, the plot-level prediction map in Fig. 6 and the LiDAR predictor—AGB field estimate correlation in Fig. 3 (right sub-figure) show the basic problem in Chitwan data—the signal (or correlation) between the LiDAR predictors and AGB field estimates is not good enough for precise prediction, and total lack of fit ensues independently of the training set source. Even the predictions of community team plots obtained using community team plots as the training set of the model show no correlation with the field estimates, as seen in the middle left sub-figure in Fig. 6. Even though the error mean is nearly zero, the model fit is poor ($r^2 = 0.27$). Similar results can be seen for the predictions for professional team plots that are obtained using professional team field measurements as the training set (bottom right sub-figure).

With the lack of LiDAR predictor—AGB field estimate correlation, the predictions in study site Chitwan are dominated by the average values of the field estimates in the training sets. In community team measurements the average of the field estimates is larger than in the professional team measurements. The community plot AGB field estimates are thus heavily under-estimated (22.6 % of the average AGB field estimates) when using the training set of professional team measurements as training set, and overestimated in the opposite case.

It is plausible that the lack of fit of model predictions between different training sets in Chitwan is due to different forest populations used in their sampling. The samples generated separately for each population are not necessarily probabilistic samples on the intersection of the populations, i.e. on closed canopy community forests and bias may ensue. However, the prediction analysis of the various models do not support such an interpretation. In both Gorkha and Chitwan model fit is similar, acceptable in Gorkha and very poor in Chitwan, with all combinations of teaching set and validation set, whether these are cross-validated or self-validated. It seems therefore that the reason to the lack-of-fit in Chitwan is caused by some other effect and not inappropriate sampling.

Model-based prediction of biomass uses field plots for model calibration but they can also be used for model validation. When plot collection is conducted by randomly sampling individual plots or plot clusters, such field campaigns can also be used to test model-based predictions for possible lack of fit.

In the current tests between 50 and 200 plots were measured for two areas of roughly 10,000 hectares each.

In heterogeneous forests it is plausible that not every forest type gets an adequate statistical coverage with such a sample. However, even this sampling density far exceeds the cost per area of any foreseeable field sample for nationwide estimates, and is much higher than common sampling densities in national forest inventories. We therefore have to try and use the randomness of sampling as a guarantee against bias at least over sufficiently large forest areas. Once lack of bias has been attained for a model-based estimation method, it becomes much more feasible to provide even high-resolution biomass updates using remote sensing data alone.

Conclusions

The lessons from this study are positive towards using participatory field measurements. The analyses above show that the LiDAR-data based models calibrated with in situ field estimates conducted by professional foresters or by trained community forest members that use different field sample selection methods, different field sample plot sizes and different methods at the field work itself can be used together without degrading model prediction performance, if the correlation between LiDAR predictors and field estimates is good enough. This was evident in study site Gorkha. In Chitwan, the correlation was poor independent of the source of field measurements leading to imprecise predictions. The combined distributions of AGB field estimates with respect to LiDAR predictors were visually assessed to come from the same distribution in both study sites with both field data, Comm and Prof, see Fig. 3. Thus, there is no evidence that the error in Chitwan predictions is caused by the use of different samples, but from the lack of correlation between LiDAR predictors and AGB field estimates. Thus in both study sites, the prediction results were equally good or bad for both participatory and professional field plot measurements.

Previous studies, e.g. [9], concluded that it is possible to utilize data collected with participatory approaches for traditional forest inventories. The authors of [9] studied the feasibility of participatory REDD+ MRV processes in Tanzania. Their study compared the field estimates from the community forest user groups and professional teams and showed that the mean of AGB field estimates differed by no more than 7 % and mostly by 5 %. The variance was higher in the community measurements, therefore indicating that even though the accuracy was as good as professional measurements the precision was weaker.

These types of participatory inventories are limited in the geographic representativeness [10], especially when the aim is the estimation of carbon resources for the remuneration of local communities, since the Community Forests are relatively small compared to the landscape scale.

The present study is the first attempt of its kind to utilize field data collected by community people in LiDAR-based biomass inventories within the REDD+ context. The implementation of participatory methods for the monitoring & measuring, reporting and verification (M&MRV) of forest carbon credits with high accuracy and resolution is a fundamental step for the implementation of REDD+ projects in community forestry level.

The results of this study support the conclusions of a side event on evolving requirements and solutions for REDD+ monitoring, with community focus, at the UN climate change conference in Warsaw in 2013 (COP19) [11]. The side event concluded with the agreement that community monitored data can be scientifically accurate and support also new technology, such as LiDAR. But concerns were raised on whether community monitoring of carbon performance can form the basis for broader financial rewards for REDD+ and whether the data can be integrated into a broader national forest monitoring systems.

This study agrees with the study by [12]. They reported that the overall aboveground biomass estimated by community members differed only slightly from the estimates by the professional foresters. The results of this study therefore show that it is possible to use calibration plots measured by community people in model-based predictions of above-ground biomass. They also show that a model-based analysis can be used to validate the accuracy of field plots by calculating predictions with a model based on different subsets of the plots. If the model predictions thus obtained are compatible and consistent, the field estimates can be regarded as reliable. This approach gives ownership of verified data to different stakeholders which is key to implementing performance based financing mechanism.

The approach described here will hopefully be helpful for unbiased monitoring, reporting, and verification under a result-based payment mechanism in which plot data collected by local communities are integrated with advanced remote sensing-based measurements.

Methods
Study area
The study area consists of two separate sites, Kayarkhola watershed in Chitwan (labeled as Ch) and Ludikhola watershed in Gorkha (Go) located in Nepal, see Fig. 1 for the site locations. The sites are located quite near each other, the distance from northernmost part of Chitwan area is about 20 km to the southernmost part of Gorkha area. The study area in Chitwan is about 12.2 km from west to east, 8.0 km from south to north. The distances in Gorkha are about 10.6 km from east to west, 6.5 km from south to north.

The Kayarkhola watershed is located in Chitwan district, which is a part of the Central Development Region of Nepal. Its total area is 8,002 hectares (ha) and it consists of tropical to sub-tropical forests, covering an altitudinal range of 245–1944 m [5] with 16 Community Forest User Groups (CFUGs) managing this forest. The watershed consists of three different types of forest namely Sal forest, mixed hardwood forest and Riverine forest [13]. The watershed is inhabited by socially and ethnically diverse forest-dependent indigenous communities such as the Chepang and Tamang [14]. These ethnic groups are some of the most marginalized ethnic groups in the country. Chepang and Tamang communities practice shifting cultivation which puts severe pressure on forest resources. The REDD+ pilot project implemented in the area plays a major role to address the issues of forest degradation and deforestation by promoting sustainable forest management practices and linking it with the REDD+ incentive mechanism [13].

The Ludikhola watershed in Gorkha district is located in the southern part of Gorkha. The watershed is located in the Hill region characterized by sub-tropical broad leaved forests, ranging from 318 to 1714 m above sea level. The total area of the watershed is 5750 ha, out of which 4869 ha is forest area, 632 ha is agricultural land and the rest is barren, grassland and natural water bodies. There are 31 CFUGs managing an area of 1888 ha of forests as Community Forests (CFs). The forest in Gorkha represents sub-tropical forests. The watershed was heavily deforested in the past and this has been controlled through sustainable community forest management and conservation through REDD+.

Dominant forest types in the study area are hill sal (*Shorea robusta*) forest, and Schima-Castanopsis forest. Even though Shorea robusta mixed subtropical hill deciduous forest forms the major forest type in Kayarkhola (Chitwan) and Ludikhola (Gorkha), associated species varies between these two watersheds. *Lagerestroemia parviflora, Mallatus phillipinensi* and *Terminelia tomentosa* are dominant associates in Kayarkhola (Chitwan) whereas *Schima wallichii* and *Castanopsis indica* are the most common associates in Ludikhola (Gorkha). According to broader climatalogical categorization of forests, forests in Kayarkhola fall under tropical broadleaved forests and in Ludikhola the forests are of sub-tropical broadleaved forest mostly, with *Shorea robusta* and *Schima wallichi* (sal and chilaune) as principal dominant species.

LiDAR data

Airborne discrete-return LiDAR data was acquired in February/March 2011 from the two watersheds. The two watersheds were scanned in full coverage from 2200

m average height above ground using a local helicopter equipped with a Leica ALS50-II lidar-scanner device. The helicopter flight path was east-west strips at 1 km distance. The scanning parameters are presented in Table 3. The collected LiDAR data were evaluated after each flight, and supporting scans were conducted if data gaps or other problems occurred.

Raw LiDAR data were classified by the vendor into three categories: ground, vegetation and error returns. Further pre-processing included calculation of a digital terrain model (DTM) from the ground returns, removal of overlaps from the raw data, and conversion of height coordinates (zvalues) of the vegetation returns from absolute elevation into distance-to-ground using the DTM [15]. Overlap removal is a procedure to ensure uniform density of points for estimation by the area-based approach (ABA). ABA methods use quantized vertical histograms as the regression variables and it is seen as desirable that their sampling noise, i.e. density of LiDAR points per square meter, is uniformly distributed. From the pre-processed LiDAR data, several LiDAR features were estimated in order to serve as the LiDAR-predictors in the AGB prediction model. The features have been taken from [16] and are an extended and modified version of those published by [17]. They include: (1) different height percentiles for the first-pulse and last-pulse returns, (2) mean height of first-pulse returns above 5 m (high-vegetation returns), (3) standard deviation for first-pulse returns, (4) ratio between first-pulse returns from below 1 m and all first-pulse returns, and (5) ratio between last-pulse returns from below 1 m and all last-pulse returns. These features were estimated from LiDAR points within the plot footprints described below.

Field samples

Field sample plots were selected with two different methods, depending on the inventory team. The professional

Table 3 Specifications for the LiDAR scanning data

Parameter	Value
Average flying altitude above ground level	2200 m
Flying speed	80 knots
Sensor pulse rate	52.9 khz
Sensor scan speed	20.4 lines per second
Nominal outgoing pulse density at ground level	0.8 points per square meter
Scanning field of view (FOW) half angle	20°
Swath width at ground level	1601.47 m
Point spacing on the ground (across-track / along-track)	max. 1.88/2.02 m
Geometric accuracy (horizontal and vertical)	max. 1 m

forest measurement teams and forest measurement teams coming from the local communities collected the field data during the year 2011. The selection criteria for the field plots were different among the two groups of forest measurement teams.

Field plot center coordinates were recorded using Differential GPS (DGPS) with ProMark 3 and MobileMapper CX devices, and corrected in post-processing mode (GNSS Solutions software and MobileMapper Office software). Plots were located with a family of GPS devices where one device is left stationary for all day and it provides differential correction to all other GPS devices used in positioning the plots within a cluster of plots. Subsequent off-line DGPS post-processing was also used and plot geo-location error was estimated to be less than 1 m.

The professionally collected plots were collected as a part of a much larger campaign addressing the REDD+ program in Nepal. This larger program requires sampling from a national, officially accepted forest mask. Since that nationally accepted forest mask does not admit auxiliary information, such as vegetation index, elevation or aspect, clustered random sampling with uniform probability on the area of interest covered with LiDAR was used.

Sampling design and field plot design used by community teams

A stratification was done where forests with more than 70 % of canopy cover were considered as dense strata (i.e. closed canopy) and less than 70 % as sparse strata (i.e. open canopy). In a post-processing step the classification between open and closed canopy was revisited based on LiDAR pulse density so as to obtain a uniform classification of closed canopy for both sets of field plots. Plots deemed not to fall on closed canopy forest were eliminated from the sample by a 70 % canopy cover criterion. The 70 % canopy cover was used as a surrogate variable for selecting as similar a sample of community plots as possible to the professional plot exclusion policy. This 70 % canopy cover was computed as the proportion of vegetation points of all LiDAR points. Since plot sampling was clustered or simple random sampling for professional and community plots, respectively, it was assumed that the plots satisfying the canopy cover criterion reflect different AGB classes in their respective statistical proportions and the estimates are therefore unbiased.

Forest stratification was carried out using high resolution remote sensing imagery (GeoEye image) in ERDAS and ArcGIS software. The random permanent plots that were established during baseline survey were measured for the purpose of monitoring. A total of 365 permanent plots were measured in the field. There were 298 plots in closed canopy forests and 67 plots in open canopy forests.

The size of the plot was fixed to a circle of 8.92 m radius. A sub-plot of 5.64 m radius was established for saplings and a sub-plot with 1 m radius was established for counting regeneration.

Sampling design and field plot design used by professional teams

Before the field campaign, the location of sample plots was designed using a systematic clustered random sampling method. Each cluster contained eight sample plots. Within the clusters, the sample plots were aligned in two parallel columns in North-South direction, with four plots per column. The distance between plots was 300 m, both between columns and rows. The original sampling design generated 16 clusters for the Kayarkhola watershed with a total number of 112 plots while it included 15 clusters for a total of 115 plots for the Ludikhola watershed. The actual number of plots available for the purpose of the study is less than that because some plots were either placed outside the area of study or in non forested areas (water, agricultural and bare soil areas). The total number of plots available for the study was therefore 57 for Kayarkhola and 92 for Ludikhola. The plots were of fixed circular shape with a 12.62 m radius, equivalent to an area of $500\ \text{m}^2$.

The field data were collected in April/May 2011. All the sample plots that were located in forest with at least 10 % canopy cover were measured in the field. The measurements at tree-level included all living trees and shrubs above 5 cm diameter within the plot area.

Above ground biomass measurement estimates

Within each plot, individual tree diameter at breast height measurements for both live and dead trees were taken and used in allometric equations given in [18] to estimate above ground biomass, AGB (stem, branch and foliage biomass). The individual tree AGB field estimates were totaled for each subplot and converted to AGB Mg/ha.

Large AGB field estimate values (over 1500 Mg/ha) are assumed to be outliers, caused by e.g., plot-level AGB estimation errors or measurement errors. At plot level, with relatively small plots, it sometimes occurs that one or a few very thick trees cause the polynomial formula for plot level volume computation to become unstable because of extrapolation. This extrapolation error may cause the volume of a single exceptional tree to be estimated so high that a timber volume of more than 2000 m^3 per hectare is attained, which is not realistic. There is no adequate statistical data available to quantify this phenomenon and revisiting the plots for validation is not feasible either. We therefore resorted to manual removal of probable outliers that are detected as statistical outliers of AGB field estimate distribution and also by visual interpretation from LiDAR predictor—AGB field estimate scatterograms.

In field measurements of the local teams, there were two plots in which the estimated AGB was very high, 2202.6 Mg/ha and 3257.6 Mg/ha, respectively, and could be assumed as outliers. Visual assessment of the LiDAR predictor—AGB field estimate distribution supported this decision. Thus these two plots were discarded as outliers. Otherwise, the estimated AGB values of the plots are treated as the ground truth. After deletion of outliers, a total of 372 field plots measured by local forest measurement teams (182 in the study site Chitwan, 190 in the study site Gorkha), and a total of 149 field plots by the professional measurement teams (57 in Chitwan, 92 in Gorkha) were available for this analysis.

Figure 8 shows the variability of AGB field estimates in each dataset without outliers. Especially data collected by the local forest inventory teams in the study site Chitwan (CommCh) contain a significantly larger range of values than the other subsets. Also, the averages of the ABG field estimates in the field plots measured by local measurement teams were larger than those estimates by the professional inventory teams in both study sites, Gorkha and Chitwan.

Differences between field estimates due to different sampling designs

Different sampling designs affect the distribution of AGB field estimates due to different characteristics of the forest in which the sample plots are located. In particular, the ownership of the forest affects the distribution average and median values, see Table 4 and Fig. 2. The sample collected by local community teams contain fewer plots from privately owned forests compared to samples collected by professional teams. After discarding those plots from the sample of local community teams, i.e., considering only plots within community owned forests, the

average and median values of AGB are larger compared to AGB values obtained by considering all the plots in the professional teams' sample. This effect does not happen in the sample of local community teams. Also, in the case where only plots within closed canopy forests are considered, the average AGB values are slightly larger than in the case where all the measured plots are included.

When only the plots in dense community owned forests are considered, the average AGB values obtained are the largest. The sizes of these subsets are relatively small (only 26–41 field plots). This small sample size may cause problems in the model-based prediction of AGB.

Validation procedure

In this study, a method based on a linear model to predict the AGB values, namely a Bayesian linear model with orthogonal predictors resulting from truncated singular value decomposition is used [19]. This model is designed to give accurate and precise predictions when using a small training set size compared to the number of possibly correlated predictors. It utilizes the singular value decomposition of the normalized predictors, and allows bigger deviation from zero to the regression parameters of the orthonormalized predictors which are known to explain the original predictor variability most, i.e., have biggest singular values. With this method, the effective number of predictors is cut down according to the given data and predictor explanation ratio, and thus it performs better especially with small training sets.

The characteristics and distributions of the AGB field estimates vary among the subsets, e.g., subset CommCh contains data with highest values of AGB, subset ProfGo the lowest values. Variation can be seen also in the LiDAR predictor values among different subsets. If the data can be considered as samples from the same distribution, i.e., there are no significant differences in the forest characteristics nor in the field sample measurement routines, a common model based on all these data should

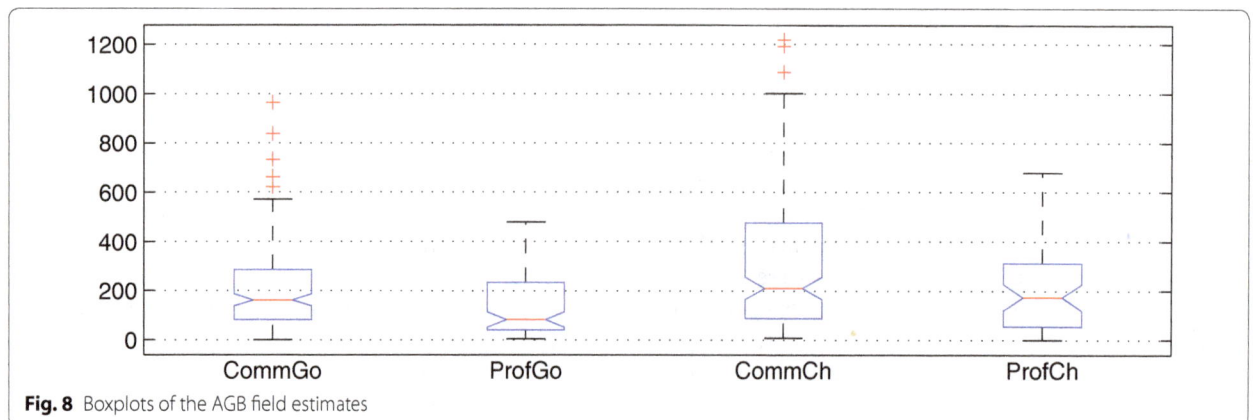

Fig. 8 Boxplots of the AGB field estimates

Table 4 Sizes and average AGB values (Mg/ha) of different measurement team dependent subsets in Chitwan and Gorkha areas

Subset	Comm		Prof	
	N	AGB	N	AGB
Gorkha				
All	190	203.3	92	127.3
Community owned	184	202.9	45	190.4
Closed canopy	157	216.5	82	133.5
Community owned and closed canopy	151	216.5	41	198.3
Chitwan				
All	182	308.4	57	205.8
Community owned	178	313.8	31	290.3
Closed canopy	154	318.8	48	207.0
Community owned and closed canopy	151	323.7	26	298.0

perform well. That is, the prediction model where model parameters are estimated with all the data (in this study Comm + Prof), or with some subset of data (Comm or Prof), should give equally accurate estimates, both in the data belonging to the teaching set subset, or to the other subsets [20]. Thus, to validate whether the data come from the same distribution and if data from one subset can be used to predict the AGB field estimates of another, cross-validation procedure is used. Each subset (Comm and Prof) at time serve as the validation set and the predictions estimated with different subsets (Comm, Prof and Comm + Prof) are validated.

To verify the prediction performance of different data validation subset of size N_v on different models, root mean square error,

$$\text{RMSE} = \sqrt{\frac{\sum_{i=1}^{N_v} \left(\tilde{y}_{v,i} - y_{v,i}\right)^2}{N_v}}, \quad (1)$$

mean difference,

$$D = \frac{\sum_{i=1}^{N_v} \left(\tilde{y}_{v,i} - y_{v,i}\right)}{N_v}, \quad (2)$$

relative RMSE (RMSE % = RMSE × 100 % $/\bar{y}_v$) and relative mean difference (D % = D × 100 % $/\bar{y}_v$) are used. Here $\tilde{y}_{v,i}$ is the the predicted AGB value for plot i, $y_{v,i}$ is the corresponding field estimate ("truth"), and $\bar{y}_v = \sum_{i=1}^{N_v} y_{v,i}/N_v$.

The error statistics are calculated using both leave-one-out cross-validation (LOOCV) procedure and straight cross-validation, depending on the case. In a case, where the training set contains the same plots as the validation set, for example in the full dataset case (Comm + Prof)

and self-validation cases (training and validation sets are the same), LOOCV is used. In LOOCV, one plot of the training set, i, at a time is used as the validation plot, and the rest of the training set is used to estimate the model parameters which are then used to predict the validation plot AGB. With the predicted values $\tilde{y}_{v,i}$, $i = 1, 2, \ldots, N_v$ the error statistics are calculated using formulas (1) and (2). In case where the AGB field estimates of one subset are predicted using a model calibrated with another subset (e.g. validation set Comm plots are predicted using training set Prof), the calibrated model is used as such to predict all the values of the other subset and the error statistics are calculated in a straightforward manner using the given formulas.

Authors' contributions
VJ conducted all lidar modeling work and wrote most of the text. BG lead the professional field teams and served as the liaison between ICIMOD, Arbonaut and LUT teams as well as reviewed the manuscript. BK lead the training of the community forest teams and managed their field collection. AM wrote the final version of the article. KT, KG and PL computed, screened and recomputed several times plot level estimates and LiDAR model features at different stages of the project. JH managed the work of the Arbonaut team. EN classified and compared the field plot designs and identified the field sample designs used. JP supervised the Lidar processing and model building efforts. TK contributed to the mathematical analysis of the results and lead the overall effort at LUT and Arbonaut. All authors read and approved the final manuscript.

Author details
[1] Department of Mathematics and Physics, Lappeenranta University of Technology, P.O.Box 20, 53851 Lappeenranta, Finland. [2] Arbonaut Ltd, Kaislakatu 2, 80130 Joensuu, Finland. [3] International Centre for Integrated Mountain Development (ICIMOD), G.P.O. Box 3226, Khumaltar, Lalitpur, Kathmandu, Nepal. [4] Mzumbe University, P.O. Box 1, Mzumbe, Morogoro, Tanzania.

Acknowledgements
This work has been carried out with the financial and institutional supports from the Governments of Finland and Nepal under the Forest Resource Assessment (FRA) Nepal project and ICIMOD. We would like to thank Finnmap International for successful completion of the airborne laser scanning. Our special thanks go to our field teams and community forest user group members for timely and accurate field data collection.

Competing interests
The authors declare that they have no competing interests.

References
1. IPCC. Climate change 2013: the physical science basis. Contribution of working group 1 to the Fifth Assesment Report: Chapter 6: carbon and other biogeochemical cycles. 2013. Retrieved from http://www.ipcc.ch/report/ar5/wg1/.
2. Angelsen A., Brown S, Loisel C, Peskett L, Streck C, Zarin D. Reducing emissions from deforestation and forest degradation (REDD): an options assessment report. Technical report, Prepared for the Government of Norway by the Meridian Institute. 2009. Retrieved from http://www.REDD-OAR.org.
3. UN FCCC. Guidance for modalities relating to forest reference levels and forest reference emissions levels—a contribution in response to SBSTA's invitation to submit views (see FCCC/SBSTA/2011/L.14, para.4 & annex 1, para.2). 2011. Retrieved from http://unfccc.int/resource/docs/2011/smsn/ngo/332.

4. ICIMOD, ANSAB and FECOFUN. Pilot project: Design and setting up of a governance and payment system for Nepal's Community Forest Management under Reducing Emissions from Deforestation and Forest Degradation. 2011. Retrieved from http://theredddesk.org/countries/initiatives/design-and-setting-governance-and-payment-system-nepal's-community-forest.

5. Karky B, Vadiya R, Karki S, Tulachan B. What is REDD+ additionality in community managed forest for Nepal? J Forest Livelihood. 2013;11(2):37–45.

6. Skutsch MM, Karky BS, Rana EB, Kotru R, Karki S, Joshi L, Pradhan N, Gilani H, Joshi G. Options for Payment Mechanisms Under National REDD+ Programmes. Working Paper 2012/6. Kathmandu: ICIMOD; 2012.

7. Shrestha S, Karky BS, Karki S. Case study report: Redd+ pilot project in community forests in three watersheds of nepal. Forests. 2014;5(10):2425–39.

8. UN FCCC: Bonn climate change conference, the 42nd session of the subsidiary body for scientific and technological advice (SBSTA). 2015. Presentations retrieved from http://www.unfccc.int/meetings/bonnjun2015/session/8855.php.

9. Skutch MM, Van Laake PE, Zahabu E, Karky BS, Phartiyal P. The value and feasibility of community monitoring of biomass under REDD+. 2009. Retrieved from http://www.communitycarbonforestry.org/NewPublications/CIFOR%20paper%20Nov%205%20version.

10. Asner GP, Hughes RF, Varga TA, Knapp DE, Kennedy-Bowdoin T. Environmental and biotic controls over aboveground biomass throughout a tropical rain forest. Ecosystems. 2009;12:261–78.

11. UN FCCC. Warsaw climate change conference, the nineteenth session of the conference of the parties (COP 19). 2013. Presentations retrieved from https://www.seors.unfccc.int/seors/reports/archive.html.

12. Danielsen F, Adrian T, Brofeldt S, van Noordwijk M, Poulsen MK, Rahayu S, Rutishauser E, Theilade I, Widayati A, The An N, Nguyen Bang T, Budiman A, Enghoff M, Jensen AE, Kurniawan Y, Li Q, Mingxu Z, Schmidt-Vogt D, Prixa S, Thoumtone V, Warta Z, Burgess N. Community monitoring for REDD+: international promises and field realities. Ecology Soc. 2013;18(3):41. Retrieved from http://www.dx.doi.org/10.5751/ES-05464-180341.

13. Karna YK. Mapping above ground carbon using worldview satellite image and lidar data in relationship with tree diversity of forests. The Netherlands: Master's thesis, Faculty of Geoinformation Science and Earth Observation, University of Twente; 2012.

14. ICIMOD, ANSAB, FECOFUN. A monitoring report on forest carbon stocks changes in REDD project sites (Ludikhola, Kayarkhola and Charnawati). 2011. http://theredddesk.org/sites/default/files/icimod-ansab-fecofun-redd-carbon-monitoring-report-22-oct-2011.pdf.

15. Gautam B, Peuhkurinen J, Kauranne T, Gunia K, Tegel K, Latva-Käyrä P, Rana P, Eivazi A, Kolesnikov A, Hämäläinen J, Shrestha SM, Gautam SK, Hawkes M, Nocker U, Joshi A, Suihkonen T, Kandel P, Lohani S, Powell G, Dinerstein E, Hall D, Niles J, Joshi A, Nepal S, Manandhar Kandel UY, Joshi C. Estimation of forest carbon using lidar-assisted multi-source programme (LAMP) in nepal. In: International Conference on Advanced Geospatial Technologies for Sustainable Environment and Culture, Pokhara, Nepal; 2013. An event of ISPRS, Technical Commission VI, Education and Outreach, Working Group 6. Retrieved from http://report.arbonaut.com/arbo_site_uploaded_files/pdf/pdf/Estimation of Forest Carbon Using LiDAR-Assisted Multi-source Programme (LAMP) in Nepal.pdf.

16. Junttila V, Kauranne T, Leppänen V. Estimation of forest stand parameters from LiDAR using calibrated plot databases. Forest Science. 2010;56(3):257–70.

17. Næsset E. Predicting forest stand characteristics with airborne scanning laser using a practical two-stage procedure and field data. Remote Sens Environ. 2002;80:88–99.

18. Chave J, Andalo C, Brown S, Cairns MA, Chambers JQ, Eamus D, Folster H, Fromard F, Higuchi N, Kira T, Lescure J-P, Nelson BW, Ogawa H, Puig H, Riera B, Yamakura T. Tree allometry and improved estimation of carbon stocks and balance in tropical forests. Oceologia. 2005;145:87–99.

19. Junttila V, Kauranne T, Finley AO, Bradford JB. Linear models for airborne laser scanning based operational forest inventory with small field sample size and highly correlated LiDAR data. IEEE Trans Geosci Remote Sens. 2015;53(10):1–13.

20. Nikolaeva E. Comparison of model-based above ground biomass predictions from LiDAR. Master's Thesis, Finland: Lappeenranta University of Technology ; 2014.

## 9

Local discrepancies in continental scale biomass maps: a case study over forested and non-forested landscapes in Maryland, USA

Wenli Huang[1]* ⓘ, Anu Swatantran[1], Kristofer Johnson[2], Laura Duncanson[1], Hao Tang[1], Jarlath O'Neil Dunne[3], George Hurtt[1] and Ralph Dubayah[1]

Abstract

Background: Continental-scale aboveground biomass maps are increasingly available, but their estimates vary widely, particularly at high resolution. A comprehensive understanding of map discrepancies is required to improve their effectiveness in carbon accounting and local decision-making. To this end, we compare four continental-scale maps with a recent high-resolution lidar-derived biomass map over Maryland, USA. We conduct detailed comparisons at pixel-, county-, and state-level.

Results: Spatial patterns of biomass are broadly consistent in all maps, but there are large differences at fine scales (RMSD 48.5–92.7 Mg ha^{-1}). Discrepancies reduce with aggregation and the agreement among products improves at the county level. However, continental scale maps exhibit residual negative biases in mean (33.0–54.6 Mg ha^{-1}) and total biomass (3.5–5.8 Tg) when compared to the high-resolution lidar biomass map. Three of the four continental scale maps reach near-perfect agreement at ~4 km and onward but do not converge with the high-resolution biomass map even at county scale. At the State level, these maps underestimate biomass by 30–80 Tg in forested and 40–50 Tg in non-forested areas.

Conclusions: Local discrepancies in continental scale biomass maps are caused by factors including data inputs, modeling approaches, forest/non-forest definitions and time lags. There is a net underestimation over high biomass forests and non-forested areas that could impact carbon accounting at all levels. Local, high-resolution lidar-derived biomass maps provide a valuable bottom-up reference to improve the analysis and interpretation of large-scale maps produced in carbon monitoring systems.

Keywords: Temperate deciduous forest, Lidar, Aboveground biomass, Carbon

Background

Accurate maps of forest aboveground biomass are critical for reducing uncertainties in the carbon cycle and informing carbon management decisions [1–3]. While no method provides direct measurements of biomass over large scales, a combination of remotely sensed data and a well established field inventory is considered suitable for monitoring programs such as REDD+ [4, 5]. Data inputs for biomass estimation have varied widely with tradeoffs between

*Correspondence: wlhuang@umd.edu
[1] Department of Geographical Sciences, University of Maryland, College Park, USA
Full list of author information is available at the end of the article

availability, cost and coverage. Accuracy of estimated biomass has also varied with the sensitivity of data to forest structure, spatial resolution, choice of statistical models, and the accuracy of field training data. Regardless, biomass estimates from different maps seem to agree at very coarse scales [4]. For example, Mitchard et al. [4] found that pan-tropical biomass maps converged at regional scales even though they varied locally. They concluded that uncertainties were largely related to spatial patterns of forest cover change. Langner et al. [5] evaluated pan-tropical biomass maps and successfully combined them into a framework for deriving REDD+ Tier 1 carbon storage estimates. While these findings are encouraging for national and continental scale reporting, there is a need to examine local discrepancies more closely as errors or uncertainty at fine-scales can complicate the use of coarse scale maps in local planning and decision making.

Almost all large-area biomass maps are derived from two-dimensional remote sensing data that have wide coverage but are generally less sensitive to canopy structure, particularly in moderate to high biomass forests (e.g. multispectral and single polarized SAR). Furthermore, they do not currently include fine scale variations in tree cover because of their coarse spatial resolution. Lidar instruments measure three-dimensional canopy structure which improves the accuracy of biomass maps [3] but lidar datasets have limited coverage and are expensive to acquire. An alternative is to use high-resolution lidar derived biomass maps, where available, to evaluate existing coarse scale maps, and make them more compatible for decision-making.

In 2010, NASA initiated the Carbon Monitoring System (CMS) to quantify carbon sources and sinks for an improved understanding of the global carbon cycle [6]. The program combines top-down continental scale approaches with bottom-up local scale approaches. The top-down approach relies on satellite observations to quantify carbon storage and terrestrial fluxes for national reporting. The bottom-up approach focusses on mapping carbon stocks and uncertainties at fine scales. Within the US, continental scale maps use Forest Inventory Analysis (FIA) plot data for model development, and biomass estimates are in turn compared with FIA county or regional averages as a type of validation [7–9]. However, these validations are not based on independent data, and often lack constraints at high spatial resolution. Moreover, field inventories generally do not include trees outside forests [9]. Continental scale maps therefore do not predict biomass outside forested areas and may significantly underestimate carbon balances [10].

A thorough understanding of local-scale discrepancies requires an independently derived high-resolution estimate. Recently, such a map was produced for the state of Maryland as part of CMS [11, 12]. Biomass estimates were derived from lidar data in conjunction with non-FIA field data using machine-learning approaches. The 30 m biomass maps incorporated tree canopy cover at the 1 m resolution, thus including forested and non-forested trees in the process. This local scale effort provides a reference for evaluating existing coarse scale maps.

We present results from a detailed comparison of the biomass map produced over Maryland (hereafter referred to as CMS_RF) with four national scale biomass maps: (A) NBCD2000 [13], (B) Blackard [14], (C) Wilson [15], and (D) Saatchi [16] at the pixel-, county- and state-level. We quantify the degree and spatial patterns of differences

to gain an improved understanding of map discrepancies and their impacts on carbon accounting.

Methods
Study area and field data
Maryland has a land area of ~25,600 km^2 (Fig. 1) and can be divided into 3 major physiographic provinces (or ecoregions) based on species-composition and environmental gradients. These are the Eastern Coastal Plain (hereafter, "Eastern Shore"), the combined Western Coastal Plain and Piedmont (hereafter, "Piedmont") and the combined Blue Ridge, Valley and Central Appalachians (hereafter, "Appalachian"). The wide variability in topography, forest types, and environmental gradients makes it a suitable test-bed for national map comparisons.

We first generated a biomass map using existing lidar data and independent field estimates. Field data were collected in 848 variable and fixed radius plots selected through a stratified sampling of NLCD land cover (evergreen, deciduous, wetlands, mixed and non-forest) and lidar canopy heights (Fig. 1). Tree measurements of diameter at breast height (dbh) and species were recorded in each plot. Allometric estimates of aboveground biomass (Mg ha^{-1}) were calculated for each tree using equations from Jenkins et al. [17] and appropriate blow up factors were applied to estimate biomass density for the variable radius plots. For more details on field data collection, refer to [11, 18]. In addition to these new plots, FIA data were obtained from across the state and used for model validation only [9].

Local scale CMS_RF biomass map
Leaf-off, discrete return lidar data were obtained from the Maryland Department of Natural Resources (DNR) and individual counties. Tree canopy cover and canopy height were mapped at 1 m resolution using a combination of Lidar and high-resolution leaf-on multispectral imagery for every county and seamlessly across the entire state [19, 20]. Lidar canopy height models were masked using high-resolution tree cover to obtain canopy heights over forested and non-forested areas. Lidar metrics such as height percentiles, densities, and canopy cover were calculated within 30 m grid cells corresponding to the NLCD land cover dataset. Field based estimates of biomass were then related to the lidar metrics using Random Forests regression models [21, 22]. Three separate empirical models were developed, one for each physiographic region, and were applied to predict biomass for counties within the region. Predictions over individual counties were merged into a statewide biomass map at 30 m resolution (CMS_RF map). Details of the biomass estimation are available in [18] and [12].

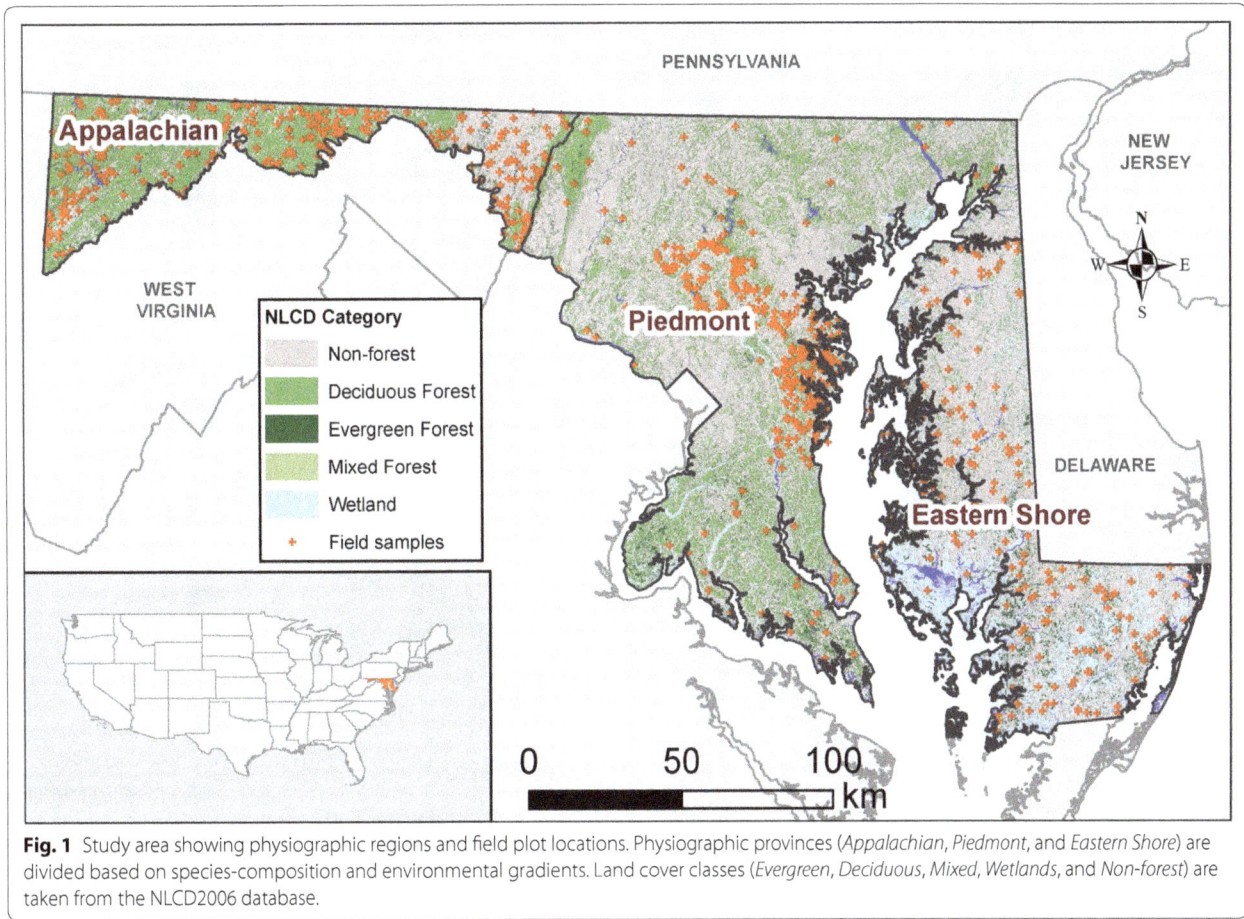

Fig. 1 Study area showing physiographic regions and field plot locations. Physiographic provinces (*Appalachian*, *Piedmont*, and *Eastern Shore*) are divided based on species-composition and environmental gradients. Land cover classes (*Evergreen*, *Deciduous*, *Mixed*, *Wetlands*, and *Non-forest*) are taken from the NLCD2006 database.

Continental scale biomass maps

Four national biomass products (Fig. 2; Table 1) were compared to the CMS_RF map. Each of these maps was derived using medium to coarse resolution satellite imagery. The NBCD2000 was the first 30 m national product developed using InSAR data from the 2000 Shutter Radar Topography Mission (STRM) and Landsat ETM+ data [13, 23]. NBCD2000 provided two versions of biomass: (A) NBCD_FIA map in which tree-level biomass estimates were obtained from tree tables in the FIA database (FIADB); and (B) NBCD_NCE or National Consistent allometric Equations in which biomass estimates were derived from equations developed by Jenkins et al. [17]. We used the NBCD_NCE version for consistency with our field biomass estimates, which were also derived from national allometric equations.

The Blackard map was developed at the 250 m spatial resolution [14] using tree-based regression (i.e., Cubist). It was developed by relating FIA plot data to multi-variable geospatial predictors, including Moderate Resolution Imaging Spectrometer (MODIS) data in 2001, percent tree cover and land cover proportions (from the NLCD

1992 product), topographical variables, and annual climate parameters, etc.

The Wilson map, also developed at the 250 m spatial resolution, was derived from MODIS imagery data from 2002 to 2008 and FIA field plots using a Phenological Gradient Nearest Neighbor (PGNN) imputation approach and canonical correspondence analysis (CCA) models [15, 24]. The Wilson map is a newer and improved version of the Blackard map.

The Saatchi map is a CMS national-scale map derived using a combination of NASA remote sensing data, forest inventory and ancillary data (the same method as [25]). Waveforms from the Geoscience Laser Altimeter System (GLAS) lidar were used to derive Lorey's height, which was then related to FIA biomass. The GLAS shots with predicted biomass estimates were used as ground truth (i.e., biomass plot samples) and related to multiple remote sensing inputs, including MODIS, PALSAR, and Landsat imagery using Maximum Entropy (MaxEnt) models for predicting biomass at the continental scale. An updated version of the Saatchi map (Saatchi et al., personal communication) reported improvements such

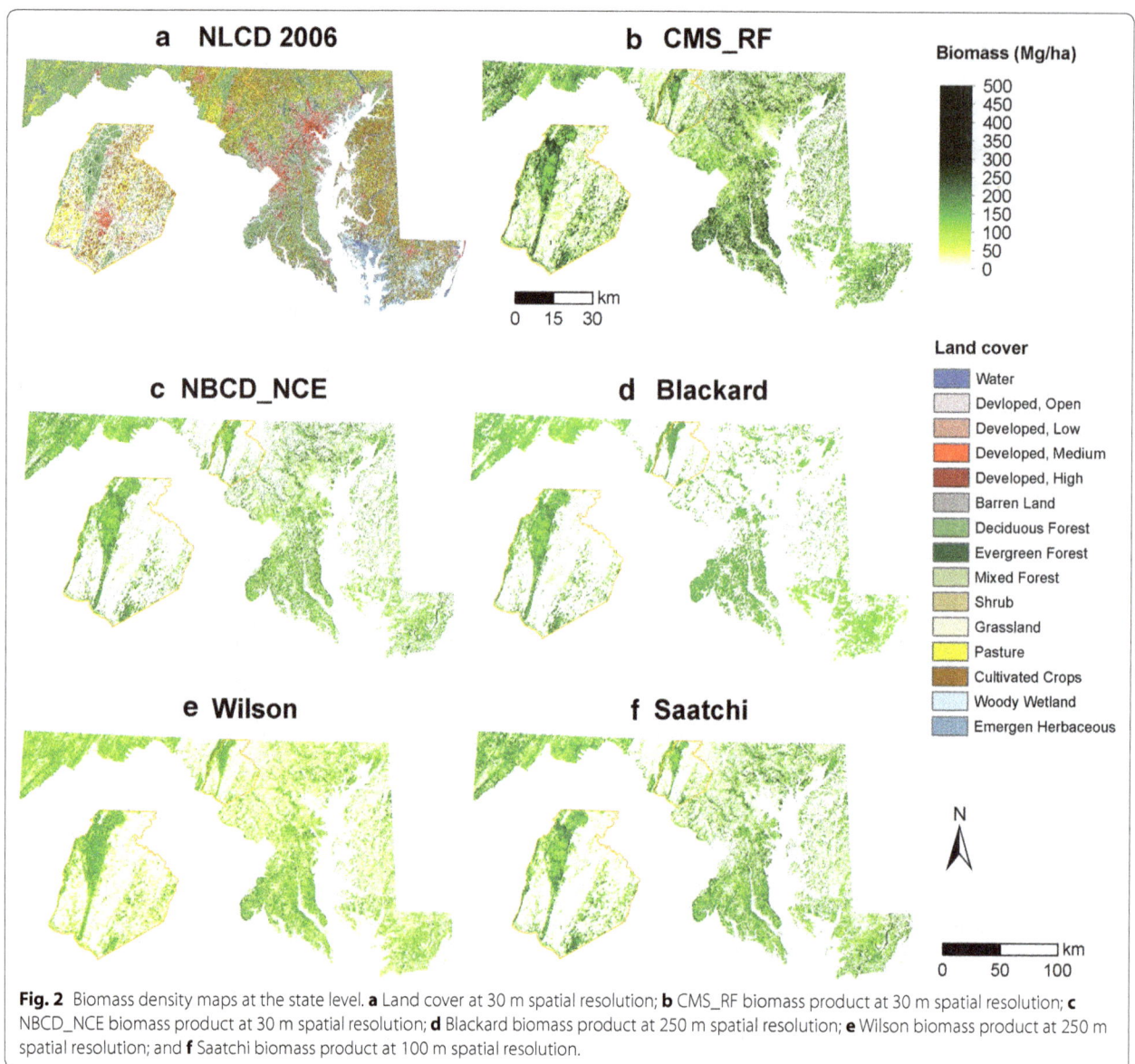

Fig. 2 Biomass density maps at the state level. **a** Land cover at 30 m spatial resolution; **b** CMS_RF biomass product at 30 m spatial resolution; **c** NBCD_NCE biomass product at 30 m spatial resolution; **d** Blackard biomass product at 250 m spatial resolution; **e** Wilson biomass product at 250 m spatial resolution; and **f** Saatchi biomass product at 100 m spatial resolution.

as: (A) reprocessed GLAS data, (B) 15 allometric equations that include three forest types (deciduous, coniferous, and mixed) for 5 regions of the US, and (C) NLCD non-vegetated gaps filled by PALSAR and Landsat data. We present results from the updated version but also include a comparison of the old and new versions in the supplement (Additional file 1: Figure S1 and Additional file 2: Figure S2).

Map comparisons

All maps were warped to a common frame of reference (UTM 18N NAD 83) ensuring minimum distortion to the native projections. Maps were matched to the same extents and pixel sizes. The 30 m biomass density maps

(e.g. CMS_RF and NBCD_NCE) were aggregated to 250 m and coarser resolutions (e.g. 500 m, 1 km, and 4 km). The Wilson, Blackard (originally 250 m), and Saatchi maps (originally ~90 m) were each aggregated to 500 m, 1 km, and 4 km.

A canopy cover mask was used to differentiate between forested and non-forested areas in our comparisons. The mask was created from the NLCD 2006 dataset for consistency with the land cover used in the CMS_RF stratification [18] and [12]. The mask included deciduous forest (41), evergreen forest (42), mixed forest (43), woody wetlands (90) and emergent herbaceous wetlands (95) from the NLCD dataset. NLCD defines forest as more than 20 percent of areas dominated by trees. Therefore, a 20 %

Table 1 Summary of biomass products used in this study

Product	Sensor and year	Field data and year	Resolution	Forest mask	Approach	Uncertainty map	References
CMS_RF	DRL 2004–2012	2011–2014	30 m	NAIP high-res tree canopy cover	Random forest, regression tree	Percentile error (QRF)	[12, 18]
NBCD_NCE	Landsat + SRTM 2000	2000	30 m	NLCD 2001	Random forest, regression tree	Quality voids	[13]
Blackard	MODIS 2001[a]	2005–2009	250 m	NLCD 1992	Cubist, regression tree	Relative error	[14]
Wilson	MODIS 2002–2008	2005–2009	250 m	NLCD 2001 percent tree canopy 25 %	PGNN, kNN		[15]
Saatchi	MODIS + PAL-SAR + Landsat	2005	~250 m v1[b] ~90 m v2[b]	NLCD 2006	MaxEntropy, parametric	Percent error	[16]

[a] Year is national maps in eastern US.

[b] Original maps are in lat/lon, where v1 with 0.00222222 ≅ 250 m and v2 with 0.00083333 deg ≅ 90 m.

DRL Discrete Return Lidar, 1–2 m small footprint lidar aggregate, *NAIP* National Agriculture Imagery Program, *QRF* Quantile Regression Forests, *SRTM* Shuttle Radar Topography Mission, *PALSAR* Phased Array type L-band Synthetic Aperture Radar, *PGNN* Phenological Gradient Nearest Neighbor, *kNN* k-nearest neighbor.

threshold was set while aggregating the mask from 30 to 250 m and other coarse resolutions. Comparisons were made over: forested areas only; non-forested areas only; and over forested and non-forested areas combined.

Statistical indicators such as coefficient of determination (R^2), root mean squared difference (RMSD), RMSD% or CV (coefficient of variation of the RMSD), and mean bias error (MBE) were used to compare the CMS_RF product with the four national maps. The Fuzzy Numerical Index (FNI) is a valuable quantitative descriptor of the spatial similarities and differences between maps and was included in our comparisons, following [26].

$$R^2 = 1 - \frac{\sum_{i=1}^{n}(C_i - M_i)^2}{\sum_{i=1}^{n}(M_i - \overline{M})^2} \quad (1)$$

$$RMSD = \sqrt{\sum_{i=1}^{n}\frac{(C_i - M_i)^2}{n}} \quad (2)$$

$$RMSD\% = \frac{RMSD}{\overline{C}} \times 100 \quad (3)$$

$$MBE = \frac{\sum_{i=1}^{n}(C_i - M_i)}{n} \quad (4)$$

$$FNI = \frac{\sum_{i=1}^{n} 1 - \frac{|C_i - M_i|}{\max(C_i, M_i)}}{n} \quad (5)$$

M_i is the value of national map; C_i is the CMS_RF predicted value; i is the sample index; $\overline{C}$ and $\overline{M}$ are the means of CMS_RF and national map respectively; and n is the sample size.

Results

Spatial patterns of biomass were consistent with land cover and physiographic gradients in visual comparisons. Within forested areas, all maps showed distinct dendritic patterns corresponding to riparian zones that had higher biomass than surrounding areas. Similar spatial patterns of biomass were also noted along ridges, valleys and forested patches with high structural variability.

Although spatial patterns were similar, biomass densities and levels of detail varied considerably (Fig. 2). The CMS_RF biomass map provided greater detail over urban/suburban landscapes (Fig. 3, e.g. trees along roadsides, hedges and backyards) when compared visually with high-resolution [1 m] land cover map and high-resolution imagery (Google Earth). The other maps predicted little or no biomass in non-forested areas. Differences over heterogeneous areas were particularly large (Fig. 3). Results ranged between 36,600 and 119,679 Mg, showing wide local-scale differences.

FNI provides a spatial representation of similarities and differences when calculated at a pixel-level. However, it does not capture the positive and negative deviations with respect to the CMS_RF map. We therefore calculated a mean FNI value for each map comparison with values ranging from 0 (perfect dissimilarity) to 1 (perfect similarity). A combination of map differences and FNI index values provided additional spatial and quantitative understanding of map discrepancies (Fig. 4; Table 2). Differences between maps were prominent in the Piedmont region, over counties in southern Maryland and along the Appalachians in the West. The Saatchi map was most similar to the CMS_RF map (FNI = 0.53) while the Blackard Map (FNI = 0.26) was the most dissimilar. The Wilson map had almost an equal proportion of similar

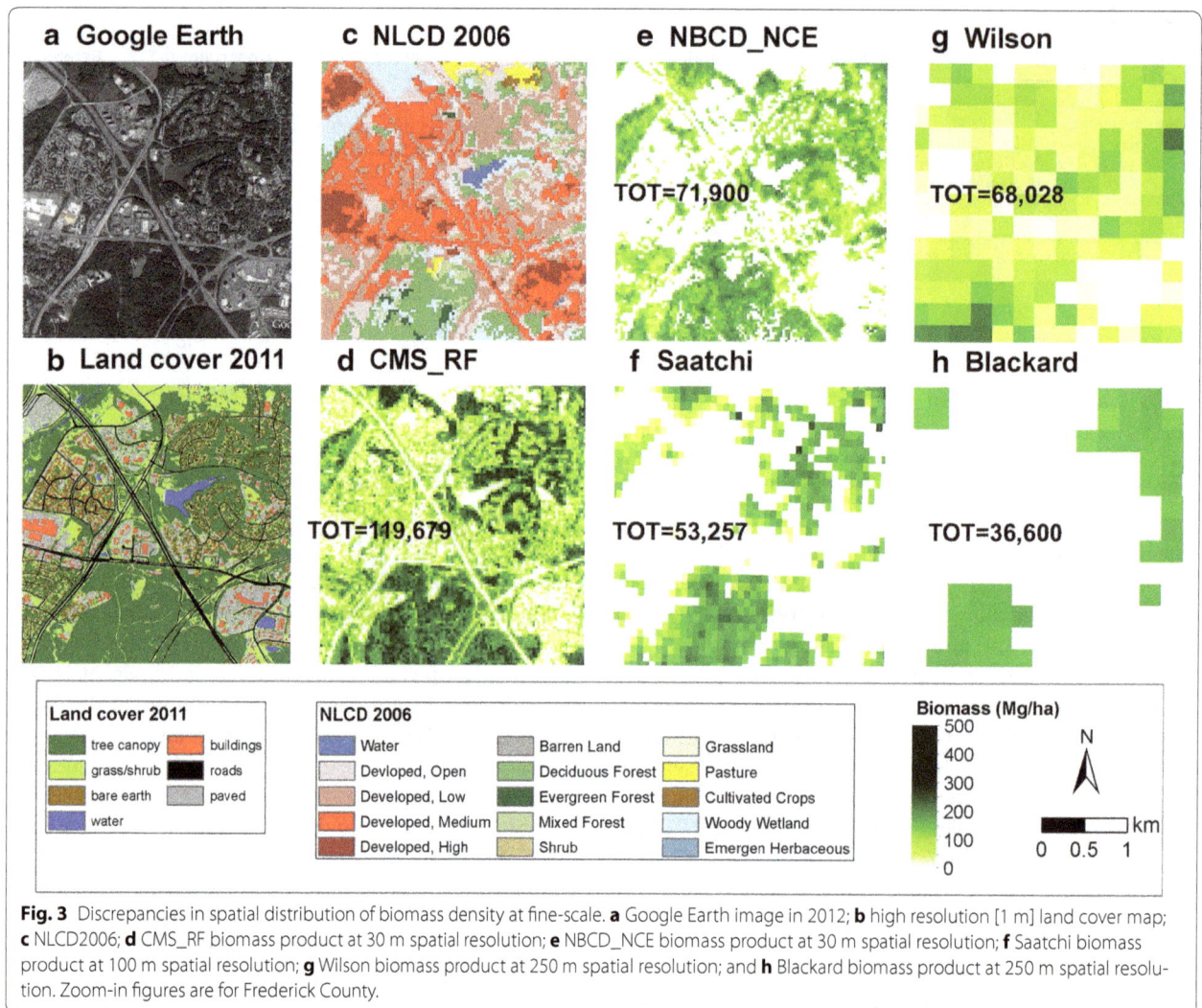

Fig. 3 Discrepancies in spatial distribution of biomass density at fine-scale. **a** Google Earth image in 2012; **b** high resolution [1 m] land cover map; **c** NLCD2006; **d** CMS_RF biomass product at 30 m spatial resolution; **e** NBCD_NCE biomass product at 30 m spatial resolution; **f** Saatchi biomass product at 100 m spatial resolution; **g** Wilson biomass product at 250 m spatial resolution; and **h** Blackard biomass product at 250 m spatial resolution. Zoom-in figures are for Frederick County.

and dissimilar pixels (FNI = 0.49) while the NBCD map was slightly lower with an FNI of 0.48.

Comparisons at the pixel level

(i)   High-resolution comparisons [30 m]

Pixel level comparisons between the NBCD_NCE and CMS_RF biomass products showed wide scatter with a large number of zero biomass predictions from the NBCD map (Fig. 5). Most areas that did not have biomass values on the NBCD map had predictions in the CMS_RF map. The NBCD biomass values were biased lower than the 1:1 line with an overall RMSD of 75.0 Mg ha^{-1}. Biomass distributions from CMS_RF and NBCD_NCE maps showed large differences over total and non-forested regions (Fig. 6). The NBCD_NCE distribution was bimodal with modes shifted toward the left (or lower biomass ranges). The CMS_RF map had higher and more widely distributed values over non-forested regions. The NBCD_NCE dataset did not predict biomass outside forests but the non-forest histograms had some high biomass values. This was because an older NLCD (2000) forest/non-forest mask was used to generate the NBCD_NCE map. The time lag between the maps and the difference in forest/non-forest masks complicated the comparisons but did not affect the overall trend in the forested and non-forested scatter plots (Fig. 5).

(ii)   Comparisons with field data

We compared predictions from the CMS_RF and NBCD_NCE maps with biomass estimates from FIA data (average of four sub-plots)

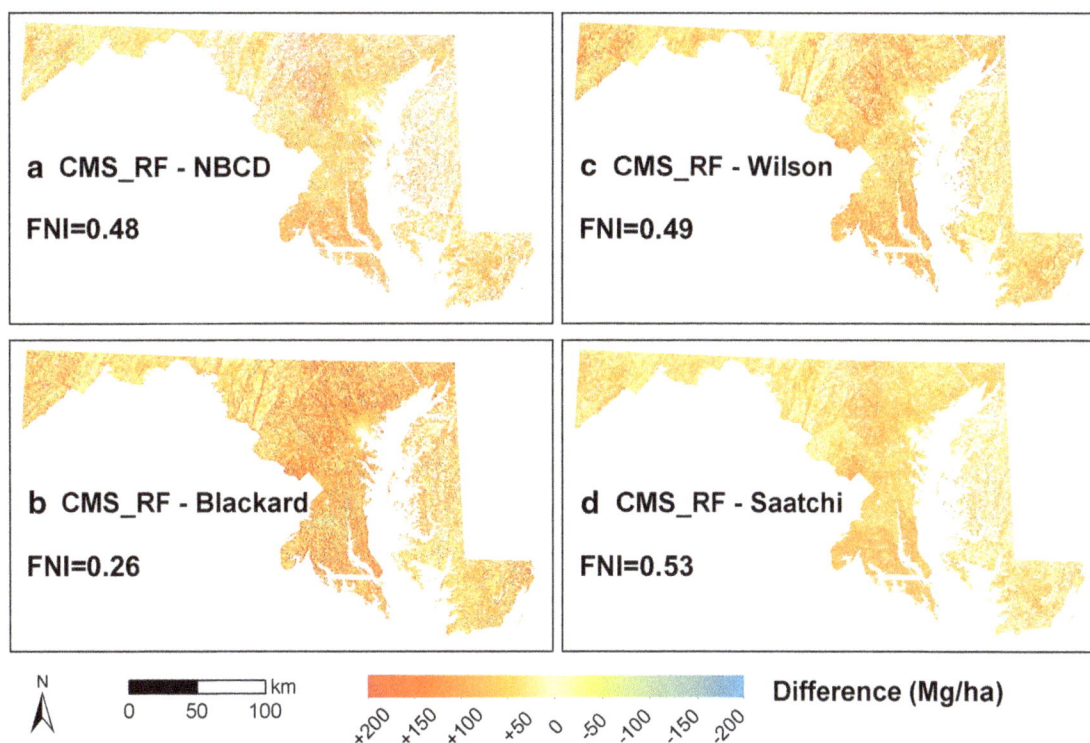

Fig. 4 Difference maps of biomass density. **a** CMS_RF-NBCD_NCE at 30 m spatial resolution; **b** CMS_RF-Blackard at 250 m spatial resolution; **c** CMS_RF-Wilson at 250 m spatial resolution; and **d** CMS_RF-Saatchi at 100 m spatial resolution. Areas in *red* have lower values and areas in *blue* have higher values than the CMS_RF map. Fuzzy Numerical Index (*FNI*) quantifies overall similarity between the national biomass maps and the CMS_RF map, ranging from 0 (fully distinct) to 1 (fully identical).

Table 2 Mean Fuzzy Numeric Index

Name	All	Forest	Non-forest
NBCD_NCE	0.48	0.62	0.22
Blackard	0.26	0.38	0.04
Wilson	0.49	0.52	0.41
Saatchi	0.53	0.69	0.25

Values calculated from maps at 250 m resolution.

(Fig. 7a) and our variable radius field plots (Fig. 7b, c). The Random Forests model used to generate the CMS_RF map explained ~50 % variability in biomass from variable radius field plots ($R^2 = 0.49$, RMSE = 89.3 Mg ha^{-1}, n = 848). A cross-validation of the CMS_RF map with plot level FIA data showed higher agreement, partly due to higher sample number ($R^2 = 0.69$, RMSE = 58.2 Mg ha^{-1}, n = 1,055). On the other hand, a cross validation of the NBCD_NCE map with variable radius estimates resulted in substantially weaker relationships ($R^2 = 0.14$, RMSE = 125.1 Mg ha^{-1}, n = 433).

(iii) Comparisons at the pixel level [250 m]

Large disagreements were observed in the scatter plots and associated errors at the 250 m resolution (Fig. 8). Overall RMSD values ranged between 48.5 and 92.7 Mg ha^{-1}. The RMSD values ranged between 55.0 and 90.0 Mg ha^{-1} over forested regions, and between 33.9 and 103.9 Mg ha^{-1} over non-forested regions. The Saatchi and NBCD maps agreed more closely with the CMS_RF map with fewer zero biomass values after spatial aggregation. The updated version of Saatchi map agreed closely with the NBCD and CMS_RF map, while the original version showed a large difference (Additional file 1: Figure S1 & Additional file 2: Figure S2). The Blackard map was the least correlated with the CMS_RF map while the Wilson map had a large scatter around the 1:1 line. Histograms of biomass in intervals of 10 Mg ha^{-1} were generated and analyzed over the entire range (0–400 Mg ha^{-1}) (Fig. 9). There was little agreement among the maps across the entire range of predicted values. The only similarities were between the NBCD_NCE and the Saatchi map

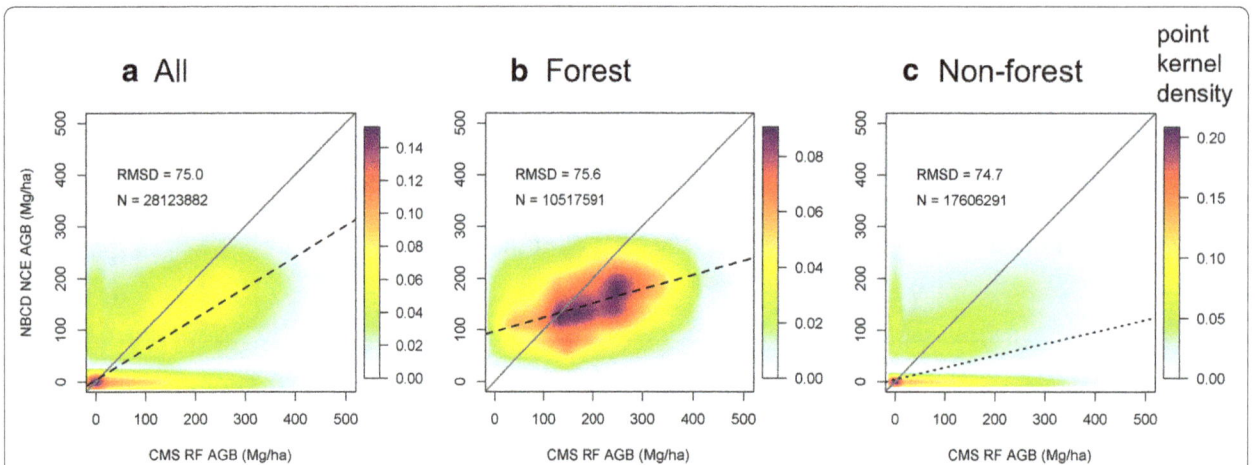

Fig. 5 Scatter plots of NBCD_NCE biomass product versus CMS_RF biomass product (30 m) at state-level. **a** All; **b** forest; and **c** non-forest. The *x* axis in each plot represents biomass values from CMS_RF, and the *y* axis represents the biomass values from NBCD_NCE product. *N* is the number of pixel used in calculation of RMSD. *Black dashed line* is the fitted regression lines, *gray solid line* is the 1:1 line, and *light blue to dark red colors* represent point kernal density. Forest and non-forest categories are derived from NLCD2006 dataset at 30 m spatial resolution.

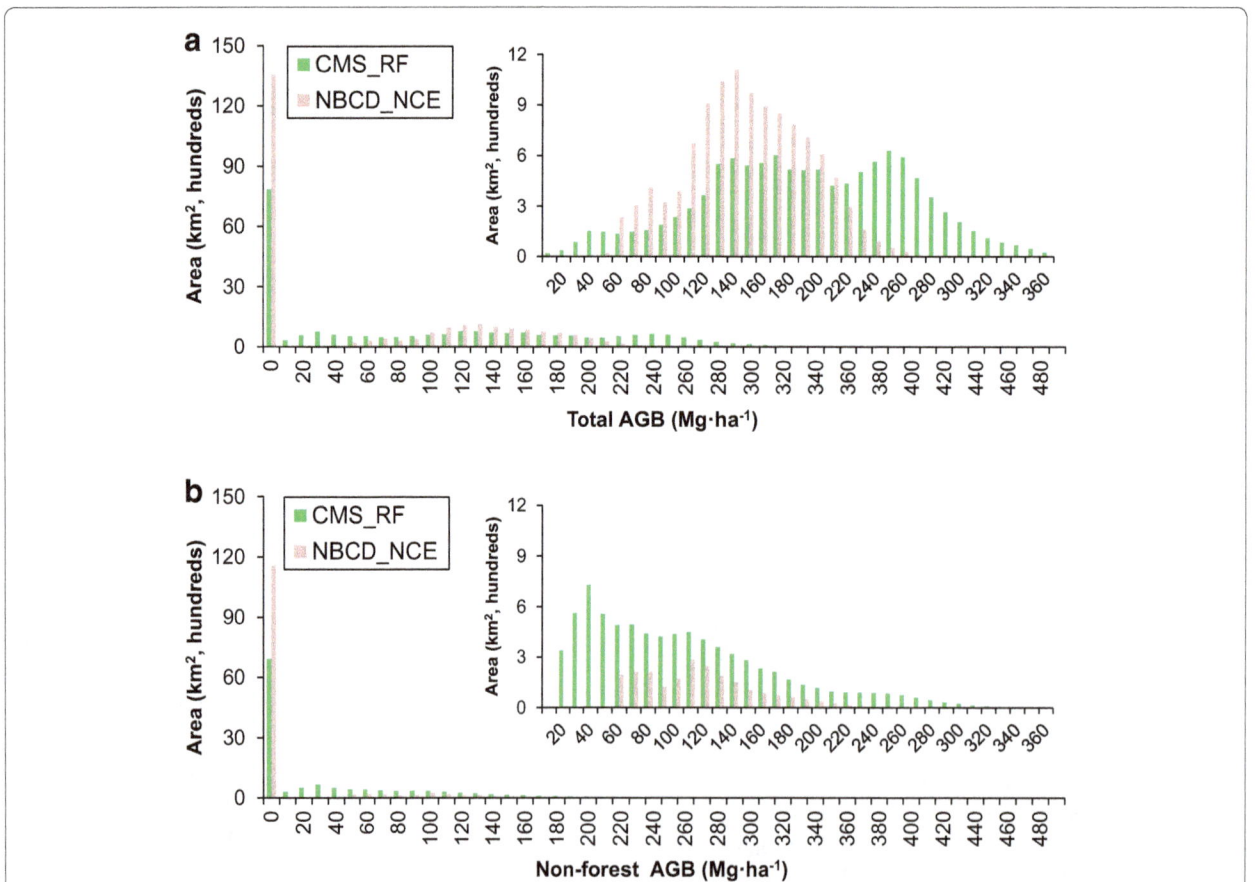

Fig. 6 *Histograms* showing the biomass distribution of CMS_RF and NBCD_NCE products over the state of Maryland at 30 m resolution in 10 Mg ha^{-1} bins. **a** All and **b** non-forest. Note that zero values are ignored in the *inset* plots. Non-forest category is derived from NLCD2006 dataset.

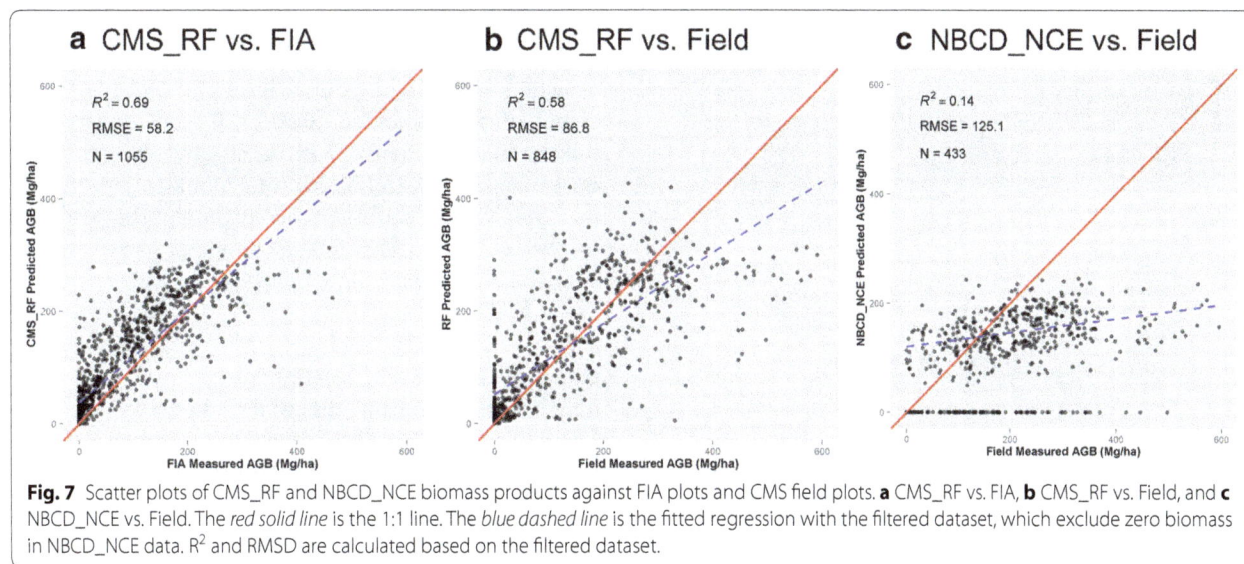

Fig. 7 Scatter plots of CMS_RF and NBCD_NCE biomass products against FIA plots and CMS field plots. **a** CMS_RF vs. FIA, **b** CMS_RF vs. Field, and **c** NBCD_NCE vs. Field. The *red solid line* is the 1:1 line. The *blue dashed line* is the fitted regression with the filtered dataset, which exclude zero biomass in NBCD_NCE data. R^2 and RMSD are calculated based on the filtered dataset.

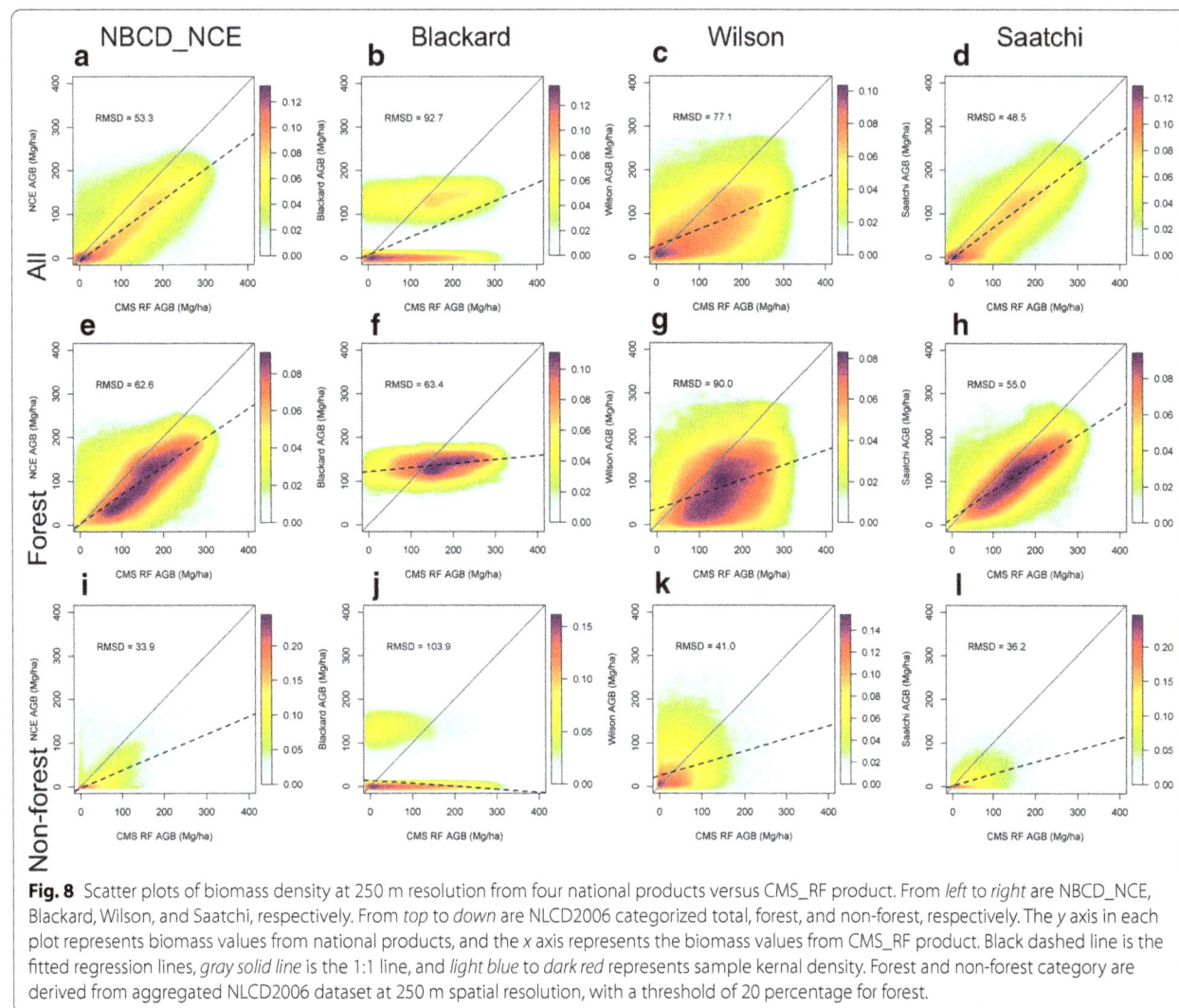

Fig. 8 Scatter plots of biomass density at 250 m resolution from four national products versus CMS_RF product. From *left* to *right* are NBCD_NCE, Blackard, Wilson, and Saatchi, respectively. From *top* to *down* are NLCD2006 categorized total, forest, and non-forest, respectively. The *y* axis in each plot represents biomass values from national products, and the *x* axis represents the biomass values from CMS_RF product. Black dashed line is the fitted regression lines, *gray solid line* is the 1:1 line, and *light blue* to *dark red* represents sample kernal density. Forest and non-forest category are derived from aggregated NLCD2006 dataset at 250 m spatial resolution, with a threshold of 20 percentage for forest.

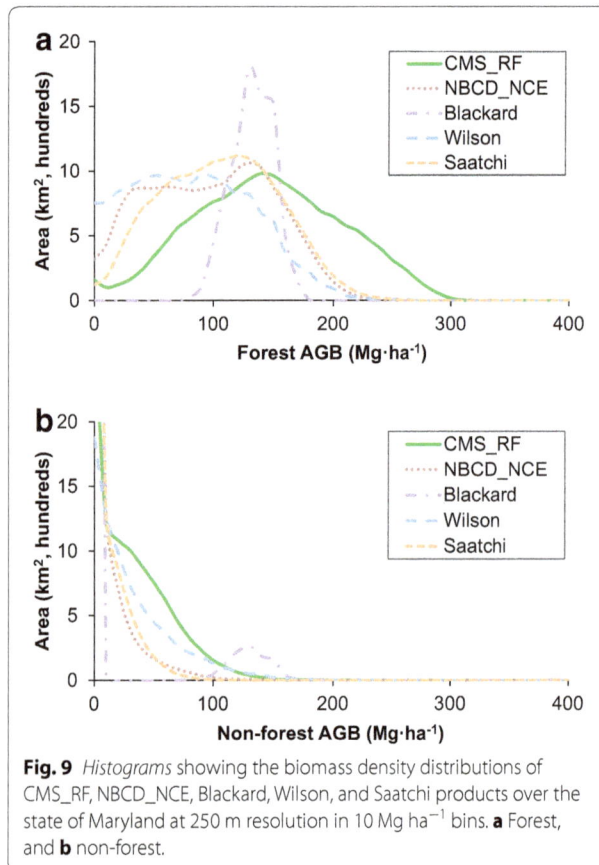

Fig. 9 *Histograms* showing the biomass density distributions of CMS_RF, NBCD_NCE, Blackard, Wilson, and Saatchi products over the state of Maryland at 250 m resolution in 10 Mg ha^{-1} bins. **a** Forest, and **b** non-forest.

above 125 Mg ha^{-1}. The distribution of biomass in different ranges was also vastly different. Biomass values in the Blackard map were predominantly between 100 and 150 Mg ha^{-1} while all other datasets had values less than 250 Mg ha^{-1}. Only the CMS_RF maps had predictions in ranges greater than 250 Mg ha^{-1}.

Comparisons at the county level

At the county level, the four maps showed improved correlation with the CMS_RF map in both mean (Fig. 10) and total biomass (Fig. 11). Among the three physiographic regions, the counties in Appalachian region were closer to 1:1 line in all four products. Counties in Piedmont region had more evenly distributed biomass values in all products except the Blackard map. Counties in Eastern Shore region were more clustered, ranging between 40.1 and 79.2 Mg ha^{-1} for mean, and 4.6 and 7.8 Tg for total biomass respectively. Despite the improved correlation, the MBE was high in all four products, ranging between −33.0 and −54.6 Mg ha^{-1} for mean, and −3.5 and −5.8 Tg for total biomass respectively.

County totals from the continental scale maps and the CMS_RF map were also compared with FIA totals (Fig. 12). For this comparison, we used the gap-filled

Jenkins estimate from FIA data as it includes non-forested biomass [9]. Continental scale maps were strongly correlated with FIA at county level and had high coefficients of determination (0.63–0.80), but consistently underestimated biomass with a negative bias, ranging between −3.4 and −1.1 Tg for total biomass (Fig. 12a–d). The CMS_RF map showed good agreement too but had a positive bias and overestimated biomass, particularly in counties that had many low biomass areas such as in the Piedmont (Fig. 12e).

Comparisons at the state level

There were significant differences between the biomass totals at the state level (Fig. 13). The national maps estimated state totals between 126.0 and 170.6 Tg and seemed to converge but were much lower when compared to the CMS_RF map. A detailed breakdown of mean and total biomass from all the maps is provided in Tables 3 and 4. The CMS_RF had higher mean (Tables 3, 4) and total biomass values (Fig. 13) over both forested and non-forested regions. The CMS_RF map also had higher total biomass than what is traditionally reported by FIA (164 Tg, 2008–2012 collection period) (Additional file 3: Table S1, [27]). However, we note that FIA does not measure trees in areas defined as "non-forest" and the allometric approach used by FIA to calculate tree biomass is known to give lower estimates in this region [9]. Adjusting for these nuances in the FIA data achieved better agreement with CMS_RF, although the FIA estimate was still lower by 43 Tg (Additional file 3: Table S1, [28]).

Lastly, we examined the coefficient of determination and corresponding errors as a function of resolution to detect trends and convergence between the maps (Fig. 14). The R^2 values for both total and mean biomass increased with decreasing resolution, gradually moving closer to 0.90. Correspondingly, RMSD values decreased gradually stabilizing at ~35 Mg ha^{-1}. The NBCD_NCE, Saatchi, and Wilson maps converged with near perfect agreement at around 4 km and onward. The Blackard map showed similar trends but less convergence with other products. Despite the improved agreement, the maps did not converge with the CMS_RF map at any scale considered in this study.

Discussion

Spatial patterns of similarities and differences were consistent with land cover and physiographic gradients. Geographically, the greatest spatial discrepancies were in the Piedmont region. This is not unexpected, given the urban development and suburban sprawl in the region. Coarse scale maps did not capture the heterogeneity of urban-suburban landscapes as finely as the CMS_RF map, hence the difference (Fig. 3). Distinct spatial patterns of

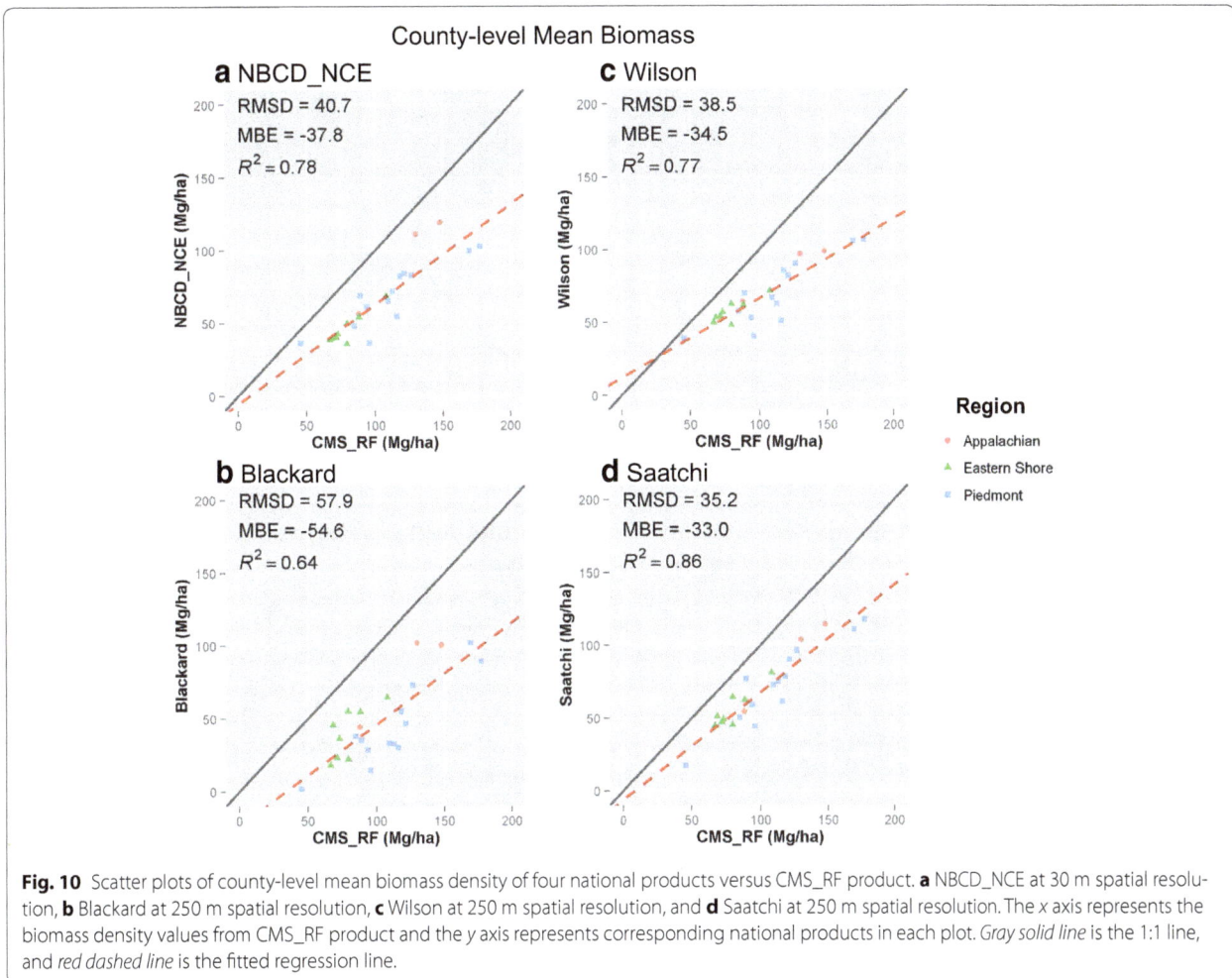

Fig. 10 Scatter plots of county-level mean biomass density of four national products versus CMS_RF product. **a** NBCD_NCE at 30 m spatial resolution, **b** Blackard at 250 m spatial resolution, **c** Wilson at 250 m spatial resolution, and **d** Saatchi at 250 m spatial resolution. The x axis represents the biomass density values from CMS_RF product and the y axis represents corresponding national products in each plot. *Gray solid line* is the 1:1 line, and *red dashed line* is the fitted regression line.

differences were also observed in Western Maryland and Southern Piedmont. These areas corresponded to dense forests where estimates from all the continental scale maps were lower. The Eastern shore had fewer discrepancies, probably because of sparse tree cover and lower biomass. However, unusually high values were noted in the national maps over several low-lying areas. This could be because of the mixed reflectance of water and vegetation over wetlands that is not easily separated in coarse resolution imagery [29].

We expected the 30 m NBCD_NCE map to be most similar to the CMS_RF map because it closely matched the spatial patterns in the CMS_RF map and had finer details than the other maps. However, the enhanced Saatchi map agreed more closely (Figs. 8, 4), despite having a coarse resolution ($\sim$90 m) and fewer predictions beyond 250 Mg ha^{-1}. This was probably because the Saatchi map had more predictions in the 50–100 Mg ha^{-1} range than the NBCD_NCE map. The NBCD_NCE map had many pixels with very low biomass

values (Fig. 9) which reduced its overall agreement with the CMS_RF map. Another surprising digression was the 250 m Wilson map that had a higher overall similarity index (FNI) than the NBCD_NCE map and the best agreement (Table 2) with the CMS_RF map over non-forested regions. A closer examination revealed that the Wilson map had better predictions in non-forested areas than any other map because it did not include a forest/non-forest mask and was developed using different models for areas greater than and less than 50 % NLCD forest cover. Thus, the agreement of continental scale maps with high-resolution estimates is not necessarily a function of spatial resolution but depends more on modeling approaches, time-lags and forest/non-forest definitions.

Choice of statistical/modelling approach was less critical in the CMS_RF estimation [18] but affected biomass predictions in other maps. The Blackard and Wilson maps used similar inputs yet had entirely different spatial distributions and histograms (Fig. 9) because of the difference in the regression models (Table 1). Similarly,

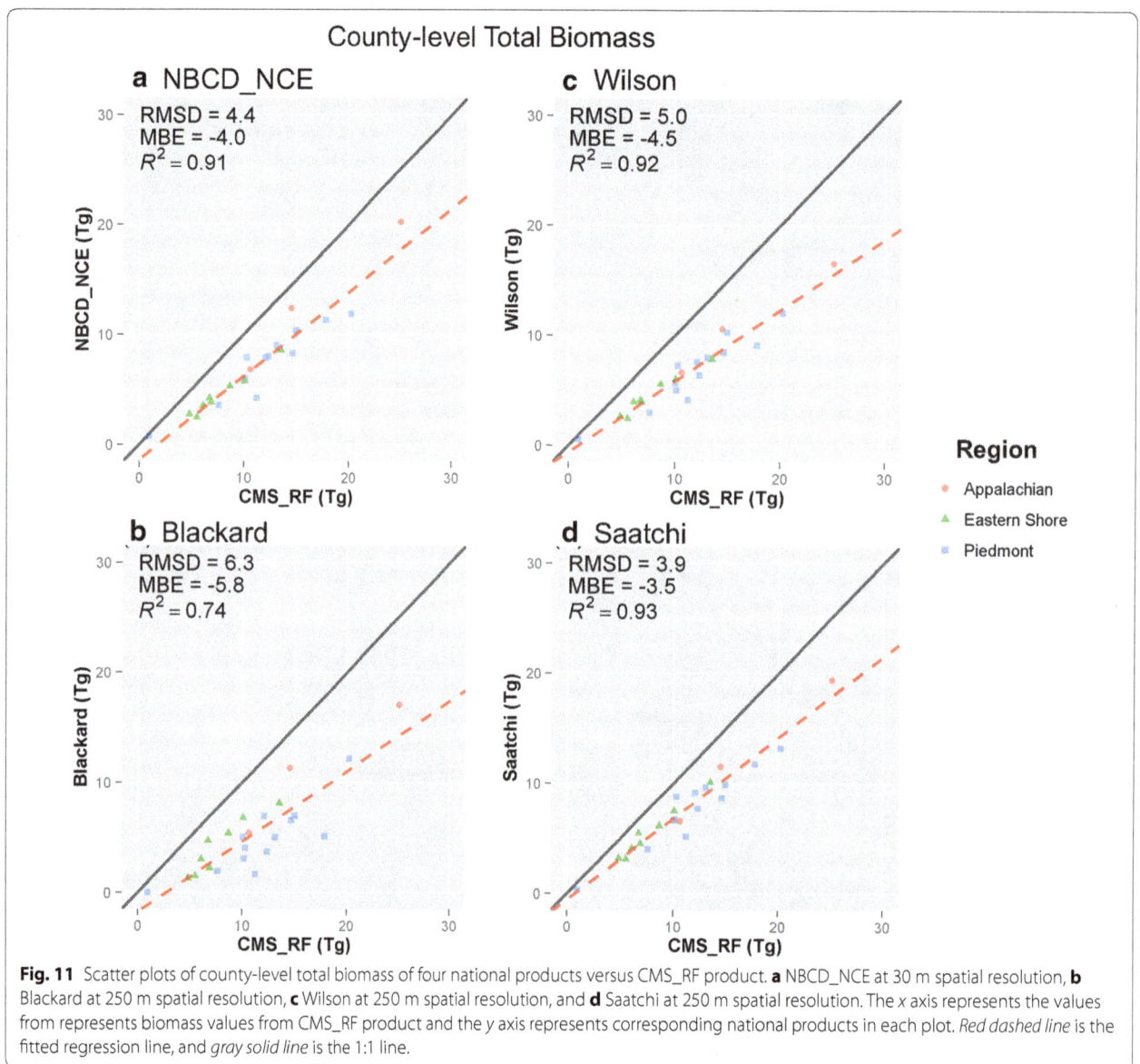

Fig. 11 Scatter plots of county-level total biomass of four national products versus CMS_RF product. **a** NBCD_NCE at 30 m spatial resolution, **b** Blackard at 250 m spatial resolution, **c** Wilson at 250 m spatial resolution, and **d** Saatchi at 250 m spatial resolution. The *x* axis represents the values from represents biomass values from CMS_RF product and the *y* axis represents corresponding national products in each plot. *Red dashed line* is the fitted regression line, and *gray solid line* is the 1:1 line.

we noted a strong influence of the MaxEnt model in the form of stratified predictions (Additional file 1: Figure S1) from the original Saatchi map. Such discrepancies are not easily detected in a broad comparison but are evident in a pixel-by-pixel comparison, as demonstrated in this study.

Continental scale maps (except the Wilson map) did not predict values outside forested areas because of limited FIA field plots for model development. This reduced their total biomass estimates and increased pixel-level discrepancies with the CMS_RF map. While we acknowledge that a fair comparison cannot be made over non-forested regions, we quantified the effect of excluding non-forest biomass on county and state level totals. Our results indicate that the underestimation is non-trivial, particularly in heterogeneous landscapes such as our

study area. We provide further corroboration to findings of [10] and support the need for including biomass outside forests in carbon reporting.

Some apparent non-forested biomass values crept into the national map totals (Fig. 13) because of time lags between maps and inconsistencies in forest/non-forest masks from the NLCD product. We noticed high values (greater than 100 Mg ha^{-1}) in non-forest histograms from some national maps (Fig. 9). These could be artifacts of forested areas that were converted since the production of the maps or differences in NLCD classifications over time. Some non-forested biomass was a result of edge effects in the coarse scale maps. Discrepancies could also be attributed to canopy cover thresholds used for comparisons (e.g. 20 % in this study). Larger

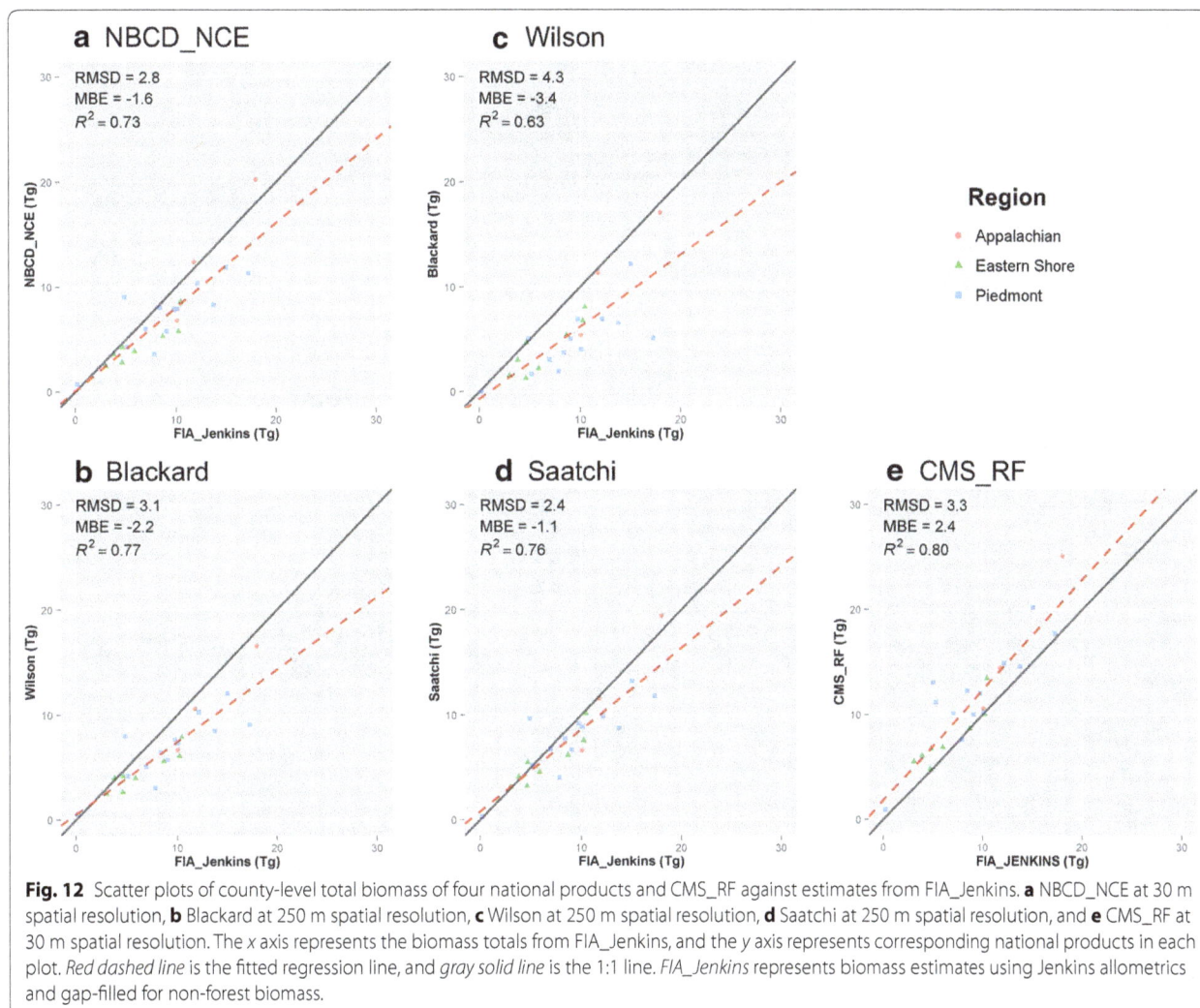

Fig. 12 Scatter plots of county-level total biomass of four national products and CMS_RF against estimates from FIA_Jenkins. **a** NBCD_NCE at 30 m spatial resolution, **b** Blackard at 250 m spatial resolution, **c** Wilson at 250 m spatial resolution, **d** Saatchi at 250 m spatial resolution, and **e** CMS_RF at 30 m spatial resolution. The x axis represents the biomass totals from FIA_Jenkins, and the y axis represents corresponding national products in each plot. *Red dashed line* is the fitted regression line, and *gray solid line* is the 1:1 line. *FIA_Jenkins* represents biomass estimates using Jenkins allometrics and gap-filled for non-forest biomass.

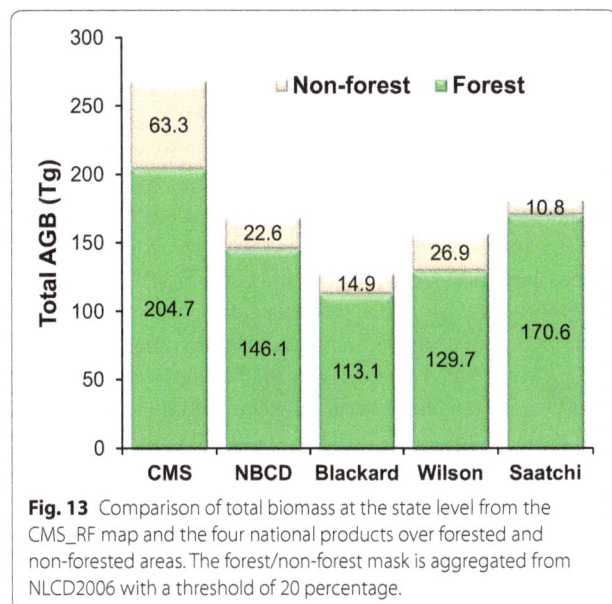

Fig. 13 Comparison of total biomass at the state level from the CMS_RF map and the four national products over forested and non-forested areas. The forest/non-forest mask is aggregated from NLCD2006 with a threshold of 20 percentage.

Table 3 Mean and total biomass for CMS_RF and NBCD_NCE products at 30 m resolution by forest and non-forest class

Type	CMS_RF		NBCD_NCE		Area compared[a]
	Mean (Mg ha^{-1})	Total (Tg)	Mean (Mg ha^{-1})	Total (Tg)	(km^2)
Forest	175.8	204.7	125.5	146.1	11,642
Non-forest	46.3	63.3	16.6	22.6	13,670
All	105.9	268.0	66.7	168.8	25,312
	Mg ha^{-1}	Tg	Mg ha^{-1}	Tg	km^2

[a] Summarized from CMS_RF and NBCD_NCE products at 30 m resolution.

thresholds can lead to lower non-forest biomass and vice versa. Some of these inconsistencies can be reduced by including sub-pixel estimates of tree cover [30] instead of a forest/non-forest mask in future continental scale mapping projects similar to the Wilson map [15]. This may

Table 4 Mean and total biomass for three national products at 250 m resolution by forest and non-forest class

Type	Blackard		Wilson		Saatchi		Area compared
	Mean (Mg ha^{-1})	Total (Tg)	Mean (Mg ha^{-1})	Total (Tg)	Mean (Mg ha^{-1})	Total (Tg)	(km^2)
Forest	97.2	113.1	111.4	129.7	146.6	170.6	11,642
Non-forest	10.9	14.9	19.7	26.9	7.9	10.8	13,670
All	50.6	128.0	61.9	156.6	71.7	181.5	25,312
	Mg ha^{-1}	Tg	Mg ha^{-1}	Tg	Mg ha^{-1}	Tg	km^2

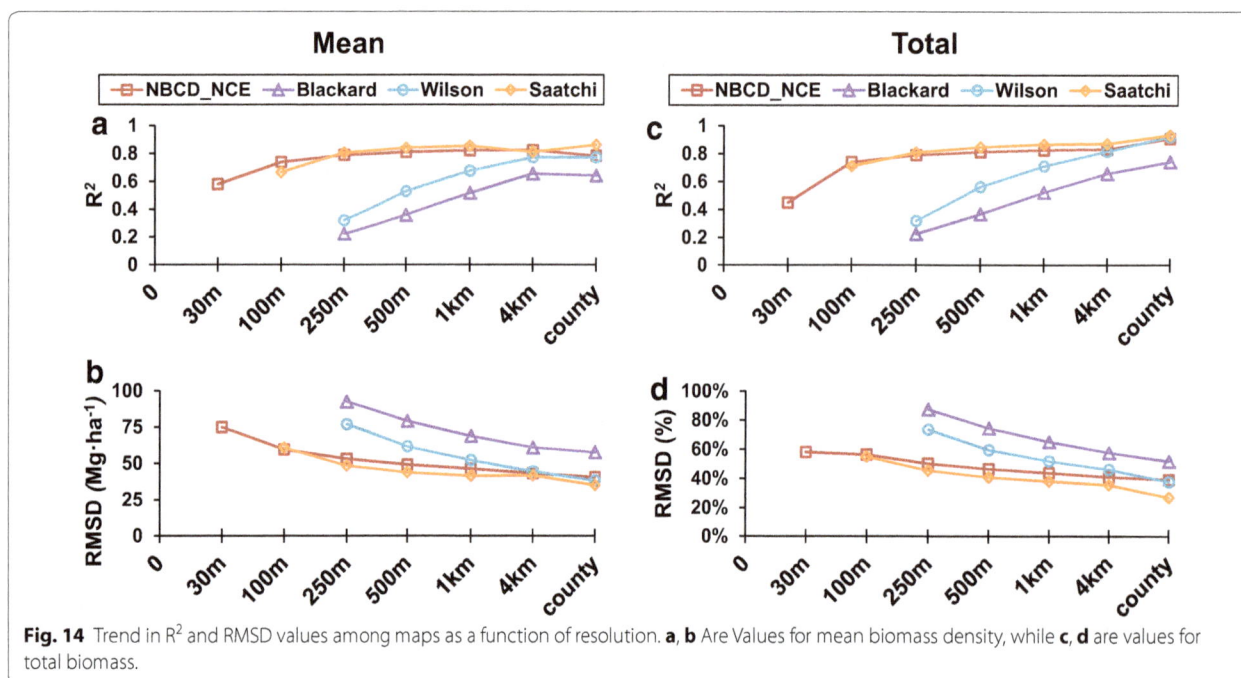

Fig. 14 Trend in R^2 and RMSD values among maps as a function of resolution. **a**, **b** Are Values for mean biomass density, while **c**, **d** are values for total biomass.

greatly improve the agreement among maps, particularly in the 0–250 Mg ha^{-1} range (Fig. 8).

Another important difference between the continental scale maps and CMS_RF map was in high biomass forests. Continental scale maps had few predictions greater than 250 Mg ha^{-1}. This was because they were developed using passive multispectral/radar data that were not sensitive enough to canopy structure in medium to high biomass ranges [31]. We expected some improvement in the enhanced Saatchi map as it included space-borne lidar data but did not observe any. This was probably because lidar data were used for model calibration rather than prediction. Biomass predictions were therefore influenced more by the 2D remote sensing data than the lidar inputs. One way of improving estimates beyond the 250 Mg ha^{-1} range is by including lidar measurements with higher resolution such as those from GEDI (expected launch in 2018) [32] or ICESAT-2 (expected

launch in 2017) as predictor variables individually or through fusion with other datasets.

Some discrepancy in total biomass values between the different maps can be attributed to the differences in allometric models applied to the field dataset used to develop the maps. For example, re-calculating tree biomass from field data in Maryland with Jenkins equations [17] instead of the Component Ratio Method (CRM) [33] that is currently used by FIA, increases the total biomass by 11 %. Therefore, it is possible that the difference between the CMS_RF map and the maps derived from field data that applied the CRM (i.e., Blackard, Wilson, and Saatchi), could be somewhat lower than calculated in Table 4.

In general, the CMS_RF map had higher values than all the other maps because the discrete-return airborne lidar were effective in predicting biomass beyond 250 Mg ha^{-1} and the high-resolution tree cover mask ensured estimates for virtually all trees in the State. We noted some

Local discrepancies in continental scale biomass maps: a case study over forested and non-forested landscapes...

129

overestimation, particularly in the low biomass ranges, when we compared the CMS_RF map to FIA county totals (Fig. 12). This could be attributed to the Random Forests model that predicted higher biomass in areas with very low canopy height and cover or limitations with FIA estimates. More research is needed to understand these differences but the cross-validation of the CMS_RF map with FIA data at plot level was strong (Fig. 7) indicating the overall robustness of the CMS_RF map and its suitability as a reference for map comparisons.

Interestingly, all maps showed better agreement at the county scale despite large discrepancies at finer scales. One reason for this could be that all maps captured some of the variability in biomass as a function of canopy structure, land cover type and physiographic gradient, irrespective of inputs and modeling approaches. This was evident from similarities in spatial patterns and FNI values. All maps (except the Blackard map) had greater than 45 % similarity with the CMS_RF map at the pixel level, contributing to the agreement. Secondly, all continental scale maps were developed using statistical regressions with FIA data which is meant to provide an unbiased state-wide estimate. A regression model, by nature, estimates the mean of the predictor data. This is also applicable to the Random Forest model used to generate the CMS_RF map. Since all maps were more or less accurate in predicting mean biomass, there was a fundamental agreement despite the variability at fine scales. Outliers reduced on spatial aggregation and the agreement between maps increased, as observed from the decreasing RMSD and increasing goodness of fit (Fig. 14).

Continental scale maps showed increasing agreement at coarse scales and converged between 4 km and 10 km (county-scale). This is similar to trends observed in Mitchard et al. [4] and Avitabile et al. [26]. However, the agreement is misleading, as these maps do not converge with the CMS_RF map at any scale considered in this study (Fig. 14). The mismatch was because of a relatively constant negative bias in all the continental scale estimates when compared to the CMS_RF and was as high as 30 % (Figs. 10, 11) at county scale. The negative bias was primarily because of the underestimation in high biomass ranges and lack of predictions in non-forested areas. Since this difference does not diminish with coarsening resolution, we argue that local-scale discrepancies may affect carbon reporting at all levels and should not be ignored.

Conclusion

A detailed validation with high-resolution estimates can be valuable in identifying discrepancies and making continental-scale maps truly applicable to carbon accounting applications. We demonstrated one example over

temperate forested and non-forested landscapes in Maryland. More studies across different biomes are required to confirm these findings. Armed with a comprehensive understanding from such validations, we can improve and integrate multi-source datasets to inform carbon monitoring efforts.

Additional files

Additional file 1: Figure S1. Scatter plots of biomass at 250 m resolution from old Saatchi (v1) and new Saatchi (v2) maps versus CMS_RF product.

Additional file 2: Figure S2. Histograms showing the distribution of forest biomass from old Saatchi (v1), new Saatchi (v2), and CMS_RF maps.

Additional file 3: Table S1. Total Maryland biomass, in Tg, for CMS_RF at 30 m resolution and FIA calculations (2008–2012 cycle) by forest and non-forest classification. FIA forest and non-forest definitions do not follow NLCD landcover classes, rather by on the ground plot conditions. Non-forest biomass in the FIA dataset was calculated by methods described in [28].

Authors' contributions
The study was designed by AS, WH and RD. WH performed the analysis and produced the figures using data layers and tables provided by KJ, LD, and HT. WH and AS wrote the manuscript and were co-first authors. All co-authors provided valuable feedback, interpretation of results and manuscript revisions. All authors read and approved the final manuscript.

Author details
[1] Department of Geographical Sciences, University of Maryland, College Park, USA. [2] USDA Forest Service, Northern Research Station, Newtown Square, PA, USA. [3] Rubenstein School of the Environment and Natural Resources, University of Vermont, Burlington, USA.

Acknowledgements
This work was funded by the NASA's CMS project (NNX10AT74G—PI, Ralph Dubayah). We are thankful to Sassan Saatchi for providing us the old and new versions of the CMS continental scale maps. We are thankful to Katelyn Dolan for her valuable inputs and comments that made the manuscript stronger. The datasets used in this study are available on line from: (a) CMS_RF [18] is available at http://carbonmonitoring.umd.edu/data.html, (b) NBCD [13] is available at http://www.whrc.org/mapping/nbcd/, (c) Blackard et al. [14] is available at http://webmap.ornl.gov/biomass/biomass.html, (d) Wilson et al. [15] at http://www.fs.usda.gov/rds/archive/Product/RDS-2013-0004, (e) Saatchi et al. [16] is available at http://carbon.jpl.nasa.gov/data/dataMain.cfm.

Compliance with ethical guidelines

Competing interests
The authors declare that they have no competing interests.

References
1. Houghton R, Lawrence K, Hackler J, Brown S (2001) The spatial distribution of forest biomass in the Brazilian Amazon: a comparison of estimates. Glob Chang Biol 7(7):731–746
2. Lu D (2006) The potential and challenge of remote sensing-based biomass estimation. Int J Remote Sens 27(7):1297–1328
3. Goetz S, Dubayah R (2011) Advances in remote sensing technology and implications for measuring and monitoring forest carbon stocks and change. Carbon Manag 2(3):231–244

4.   Mitchard E, Saatchi S, Baccini A, Asner G, Goetz S, Harris N et al (2013) Uncertainty in the spatial distribution of tropical forest biomass: a comparison of pan-tropical maps. Carbon Balance Manag 8(1):10

5.   Langner A, Achard F, Grassi G (2014) Can recent pan-tropical biomass maps be used to derive alternative Tier 1 values for reporting REDD+ activities under UNFCCC? Environ Res Lett 9(12). doi:10.1088/1748-9326/9/12/124008

6.   Hurtt G, Wickland D, Jucks K, Bowman K, Brown M, Duren R et al (2014) NASA Carbon Monitoring System: prototype monitoring, reporting, and verification, 1–37

7.   Zhang X, Kondragunta S (2006) Estimating forest biomass in the USA using generalized allometric models and MODIS land products. Geophys Res Lett 33(9). doi:10.1029/2006gl025879

8.   Cartus O, Santoro M, Kellndorfer J (2012) Mapping forest aboveground biomass in the Northeastern United States with ALOS PALSAR dual-polarization L-band. Remote Sens Environ 124:466–478. doi:10.1016/j.rse.2012.05.029

9.   Johnson K, Birdsey R, Finley A, Swantaran A, Dubayah R, Wayson C et al (2014) Integrating forest inventory and analysis data into a LIDAR-based carbon monitoring system. Carbon Balance Manag 9(1):3

10. Jenkins J, Riemann R (2003) What does nonforest land contribute to the global C balance? In: Proceedings of the 3rd annual forest inventory and analysis symposium. U.S. Department of Agriculture, Forest Service, North Central Station

11. Dubayah R, Swatantran A, Johnson K, Hurtt G, Zhao M, Finley A (2014) High resolution carbon estimation using remote sensing and ecosystem modeling in NASA's carbon modeling system. ForestSAT2014 open conference system

12. Swatantran A, Huang W, Duncanson L, Johnson K, Dunne JON, Hurtt G et al. High-resolution aboveground biomass mapping for carbon monitoring in Maryland (manuscript in preparation)

13. Kellndorfer J, Walker W, LaPoint E, Bishop J, Cormier T, Fiske G et al (2012) NACP aboveground biomass and carbon baseline data, V. 2 (NBCD 2000), U.S.A., 2000. Data set. Oak Ridge, Tennessee, U.S.A.: ORNL DAAC

14. Blackard J, Finco M, Helmer E, Holden G, Hoppus M, Jacobs D et al (2008) Mapping U.S. forest biomass using nationwide forest inventory data and moderate resolution information. Remote Sens Environ 112(4):1658–1677. doi:10.1016/j.rse.2007.08.021

15. Wilson BT, Woodall CW, Griffith DM (2013) Imputing forest carbon stock estimates from inventory plots to a nationally continuous coverage. Carbon Balance Manag 8(1):1. doi:10.1186/1750-0680-8-1

16. Saatchi S, Yifan Y, Fore A, Nuemann M, Chapman B, Nguyen et al (2005) CMS biomass pilot project: US Forest Biomass Maps

17. Jenkins JC, Chojnacky DC, Heath LS, Birdsey RA (2003) National-scale biomass estimators for United States tree species. For Sci 49(1):12–35

18. Dubayah R (2012) County-scale carbon estimation in NASA's carbon monitoring system. Biomass Carbon Storage

19. O'Neil-Dunne JPM, MacFaden SW, Royar AR, Pelletier KC (2013) An object-based system for LiDAR data fusion and feature extraction. Geocarto Int 28(3):227–242. doi:10.1080/10106049.2012.689015

20. O'Neil-Dunne J, MacFaden S, Royar A, Reis M, Dubayah R, Swatantran A (2014) An object-based approach to statewide land cover mapping. Proceedings of ASPRS 2014 annual conference, 23–28 March 2014, Louisville, KY, USA

21. Breiman L (2001) Random forests. Mach Learn 45(1):5–32

22. Cutler DR, Edwards TC Jr, Beard KH, Cutler A, Hess KT, Gibson J et al (2007) Random forests for classification in ecology. Ecology 88(11):2783–2792. doi:10.2307/27651436

23. Kellndorfer JM, Walker WS, LaPoint E, Kirsch K, Bishop J, Fiske G (2010) Statistical fusion of lidar, InSAR, and optical remote sensing data for forest stand height characterization: a regional-scale method based on LVIS, SRTM, Landsat ETM+, and ancillary data sets. J Geophys Res 115:G00E8. doi:10.1029/2009jg000997

24. Wilson BT, Lister AJ, Riemann RI (2012) A nearest-neighbor imputation approach to mapping tree species over large areas using forest inventory plots and moderate resolution raster data. For Ecol Manag 271:182–198. doi:10.1016/j.foreco.2012.02.002

25. Saatchi SS, Harris NL, Brown S, Lefsky M, Mitchard ETA, Salas W et al (2011) Benchmark map of forest carbon stocks in tropical regions across three continents. Proc Natl Acad Sci 108(24):9899–9904. doi:10.1073/pnas.1019576108

26. Avitabile V, Herold M, Henry M, Schmullius C (2011) Mapping biomass with remote sensing: a comparison of methods for the case study of Uganda. Carbon Balance Manag 6(1):7

27. Miles P (2014) Forest Inventory EVALIDator web-application version 1.5.1.06. In: http://apps.fs.fed.us/Evalidator/evalidator.jsp. U.S. Department of Agriculture, Forest Service, Northern Research Station, St. Paul, MN

28. Johnson K, Birdsey RA, Cole J, Swantaran A, O'Neil-Dunne J, Dubayah R et al. Integrating LiDAR and forest inventories to fill the trees outside forests data gap. Environ Monit Assess (in review)

29. Adam E, Mutanga O, Rugege D (2010) Multispectral and hyperspectral remote sensing for identification and mapping of wetland vegetation: a review. Wetl Ecol Manag 18(3):281–296

30. Sexton JO, Song X-P, Feng M, Noojipady P, Anand A, Huang C et al (2013) Global, 30-m resolution continuous fields of tree cover: landsat-based rescaling of MODIS vegetation continuous fields with lidar-based estimates of error. Int J Digit Earth 6(5):427–448

31. Lu D, Chen Q, Wang G, Liu L, Li G, Moran E (2014) A survey of remote sensing-based aboveground biomass estimation methods in forest ecosystems. Int J Digit Earth 1–64. doi:10.1080/17538947.2014.990526

32. Dubayah R, Goetz S, Blair JB, Luthcke S, Healey S, Hansen M et al (2014) The Global Ecosystem Dynamics Investigation (GEDI) Lidar. ForestSAT2014 open conference system

33. Heath L, Hansen M, Smith J, Miles P, Smith W (2008) Investigation into calculating tree biomass and carbon in the FIADB using a biomass expansion factor approach. In: McWilliams W, Moisen G, Czaplewski R (eds) Proceedings of Forest Inventory and Analysis Symposium 2008. USDA Forest Service, Rocky Mountain Research Station, Fort Collins, Colorado, USA; Park City, Utah

## 10

Effects of field plot size on prediction accuracy of aboveground biomass in airborne laser scanning-assisted inventories in tropical rain forests of Tanzania

Ernest William Mauya[1*], Endre Hofstad Hansen[1], Terje Gobakken[1], Ole Martin Bollandsås[1], Rogers Ernest Malimbwi[2] and Erik Næsset[1]

Abstract

Background: Airborne laser scanning (ALS) has recently emerged as a promising tool to acquire auxiliary information for improving aboveground biomass (AGB) estimation in sample-based forest inventories. Under design-based and model-assisted inferential frameworks, the estimation relies on a model that relates the auxiliary ALS metrics to AGB estimated on ground plots. The size of the field plots has been identified as one source of model uncertainty because of the so-called boundary effects which increases with decreasing plot size. Recent research in tropical forests has aimed to quantify the boundary effects on model prediction accuracy, but evidence of the consequences for the final AGB estimates is lacking. In this study we analyzed the effect of field plot size on model prediction accuracy and its implication when used in a model-assisted inferential framework.

Results: The results showed that the prediction accuracy of the model improved as the plot size increased. The adjusted R^2 increased from 0.35 to 0.74 while the relative root mean square error decreased from 63.6 to 29.2%. Indicators of boundary effects were identified and confirmed to have significant effects on the model residuals. Variance estimates of model-assisted mean AGB relative to corresponding variance estimates of pure field-based AGB, decreased with increasing plot size in the range from 200 to 3000 m^2. The variance ratio of field-based estimates relative to model-assisted variance ranged from 1.7 to 7.7.

Conclusions: This study showed that the relative improvement in precision of AGB estimation when increasing field-plot size, was greater for an ALS-assisted inventory compared to that of a pure field-based inventory.

Keywords: Airborne laser scanning; Model-assisted estimation; Plot size; Aboveground biomass

Background

Tropical forests play an important role in the global carbon cycle as they store about 40% of the global terrestrial carbon, and absorb larger amounts of CO_2 from the atmosphere than any other vegetation type [1]. Despite their potential, tropical forests continue to be exploited at alarming rates, by being converted into secondary forest and many other forms of land use. In an effort to conserve tropical forests, the United Nations Framework Convention on Climate Change (UNFCCC) has developed the mechanism called Reducing Emissions from Deforestation and Forest Degradation in tropical countries (REDD+). There is high interest in seeing such initiatives to take form, but a key limitation for successful implementation of REDD+ is reliable methods for quantifying forest aboveground biomass (AGB) [2,3]. Such methods are important because payments for carbon offsets under REDD+ are based on estimates of carbon stock and stock changes over time. Moreover, AGB information is also useful for understanding the contribution of the tropical forests to the global carbon cycle and ecosystem processes [4].

* Correspondence: ernest.mauya@nmbu.no
[1]Department of Ecology and Natural Resource Management, Norwegian University of Life Sciences, P.O. Box 5003, Oslo, NO 1432, Ås, Norway
Full list of author information is available at the end of the article

Airborne laser scanning (ALS) has emerged as one of the most promising remote sensing technologies to support AGB forest inventories in boreal-, temperate-, and tropical forests [5]. A particular strength of ALS for forest applications is its ability to accurately characterize the three-dimensional (3D) structure of the forest canopy [6]. Such information is more useful for forest inventories than the information from other remote sensing techniques see e.g. [7]. Height and density metrics derived from the ALS data has been reported to be highly correlated with AGB see e.g. [8,9]. Furthermore, ALS has shown to be superior to other remote sensing data sources because the relationship between AGB and the remotely sensed information has a much higher saturation level for ALS compared to other types remote sensing. Because of this, ALS is a highly appropriate choice of technique in high-biomass forests. Based on its potential, ALS has recently been recommended for Monitoring, Reporting and Verification (MRV) systems under REDD+ initiatives [10].

Estimation of AGB using ALS is often carried out according to the area-based approach (ABA) [11]. In ABA, empirical models between various metrics derived from the ALS data and AGB values obtained in geo-referenced field sample plots are fitted. The area of interest is then tessellated into grid cells [12] with the same size as the plots [13,14] and the developed models are used to provide cell-wise predictions of AGB. Finally, estimates for the particular area of interest (forest stand, forest property, village, district, or nation) are provided by summing the individual cell predictions. For some estimation approaches, adjustment of model prediction bias [15] is also carried out.

As indicated above, the modeled relationship between ALS metrics and ground-based values is of fundamental importance for the outcome of the ALS-assisted estimation. The use of field plot data for model development requires co–registration of field plot location with the ALS data [16,17]. In an ALS-assisted inventory, the point cloud is extracted only within the plot perimeter. However, in field measurements trees are treated as being inside plots if the center point of the stem is inside the plot. This is a challenge in ALS-assisted forest inventory, since the crowns of trees just outside the plot border partly extend into the plot area which means that the ALS data will be affected by trees that are not registered in field. Conversely, also trees just inside the plot extend their crowns beyond the plot boundary. This means that there may be mismatch between the data captured in field and from the air.

In order to reduce these boundary effects, it has been suggested in a number of studies to use larger plots in ALS-assisted forest inventory see e.g. [18,19]. This is because, as plot size increases, the perimeter to area ratio decreases and thus the plots include a lower proportion of boundary-related elements. Similarly, the relative and negative influence of a given plot positioning error is reduced because the relative overlap between the field- and ALS-data becomes larger as plot size increases. Reduction in model errors are also expected by increasing plot size due to so-called spatial averaging of the errors [20], because both the field observations and the ALS data capture more of the spatial variation as they increase in size. Thus, as plot sizes increase, the variances of field-based and ALS-assisted estimates are expected to be reduced, which means that fewer plots are needed to reach a certain precision of an AGB estimate. However, large plots also have disadvantages by being more complicated to measure, which may affect the time consumption for collecting field measurements [21], This makes it challenging to select the "optimal" plot size that balances the tradeoff between plot size, sample size (number of plots), on-plot costs, traveling costs and precision of ALS-assisted AGB estimates in different forest types.

As indicated above, plot size has a profound effect on the precision of ALS-assisted AGB estimates for several reasons. Likewise, the plot size has an impact on the precision of pure field-based estimates for reasons mentioned above; larger plots capture more of the variability in the area of interest and thus precision will tend to improve as long as the sample size is kept constant. A key question is therefore if larger plots will favor ALS-assisted estimation precision to the same extent as it favors field-based estimation precision. Different responses to plot size should have a direct impact on how tropical ALS-assisted field sample surveys should be designed as their designs currently are "optimized" for pure field-based estimation.

Forest sample surveys are often designed according to design-based (probability-based) principles. Simple random sampling is one of these principles, and analytical and so-called design-unbiased estimators and corresponding variance estimators exist for a great number of such designs. When auxiliary data such as those acquired by ALS are at hand for the entire area of interest, or at least with partial coverage of the area of interest, use of these data can greatly improve the precision over a pure field-based estimate assuming the same design. The inferential framework applied under probability sampling when a model is used to predict AGB using the ALS data is known as design-based model-assisted (MA) estimation. In the MA framework, the model is used to predict AGB for grid cells and then AGB is summed over all grid cells as indicated in the ABA, but in addition to that, the model predictions for the ground samples are used to provide an estimate of bias in the model predictions, which corrects the pure model-based estimate. Several studies see e.g. [22-24] have indicated the potential of MA estimation in reducing the variance of AGB estimates in boreal forests, but apart from some indications provided by [23], neither of them has analyzed how the variance of the estimates is affected by changes in field plot sizes. In tropical forests where the current study was conducted, there is even less knowledge regarding performance of MA estimation using ALS with varying plot sizes. Several tropical studies have examined the effects of plot size on model prediction accuracy See e.g. [25-27], but none of them have assessed the effects on the precision of AGB estimates and compared such precision estimates with corresponding precision of field-based AGB estimates using the same sampling design, which is of fundamental importance for designing future sample surveys serving multiple purposes and estimation approaches.

The objectives of this study were to (1) examine the effects of field plot size on AGB regression model quality, (2) assess plot boundary effect and its impact on model quality based on the field data, and (3) quantify the precision of ALS-assisted estimates of AGB relative to field-based estimates of AGB assuming the same design for different plot sizes. The study was conducted in tropical rain forest in Tanzania with high AGB densities, which was expected to represent a particular challenge in terms of large boundary effects.

Results

Effects of field plot size on ALS AGB predictions

To assess the effect of plot size on ALS assisted forest inventory, we first fitted the regression models for each of the plot sizes. The independent variables selected varied between the models developed for the different plot sizes (Table 1). The number of variables varied between two and three. For all models, the parameter estimates were significantly different from zero ($p < 0.05$) and the VIF values were <10, indicating acceptable levels of multicolinearity. The variability explained by separate models (i.e. adjusted R^2) improved as the plot size increased, with few exceptions (Figure 1a). The adjusted R^2 ranged from 0.35 for the plot size of 200 m^2 to 0.74 for the plot size of 3000 m^2. The RMSE% values for LOOCV decreased non-linearly with increasing plot size, from 63.8 to 29.2% (Figure 1b). The MPE% values (Figure 1b) and the pattern of under predictions for plots with high AGB were relatively lower for larger plots compared smaller (Figure 2). However, it should be noted that the number of the larger plots was relatively small.

Boundary effects

Boundary effects were studied by analyzing how the relative residual errors of the models were affected by the ground reference AGB of the trees in an outer buffer zone for different field plot sizes. Our results showed

Table 1 Selected ALS metrics for different plot sizes

Plot size (m^2)	n	Selected variables[a]		
200	30	D_0.F	D_1.L	log.H_{80}.F
300	30	D_1.L	log.H_{90}.F	log.D_0.L
400	30	H_{80}.F	D_1.L	
500	30	H_{70}.F	D_1.L	
600	30	H_{70}.F	D_1.L	
700	30	H_{90}.F	D_1.L	
800	30	H_{90}.F	D_1.L	
900	30	H_{90}.L	D_1.L	
1000	30	Hsd.L	D_1.L	log.D_0.F
1100	30	D_3.F	D_2.L	log.H_{10}.F
1200	30	H_{mean}.F	D_1.L	
1300	30	H_{70}.F	D_1.L	
1400	30	D_3.F	D_2.L	log.H_{10}.F
1500	30	H_{70}.F	D_1.L	
1600	30	H_{60}.F	D_1.L	
1700	30	H_{60}.F	D_1.L	
1800	30	H_{60}.F	D_1.L	
1900	30	H_{60}.F	D_1.L	
2000	25	H_{60}.F	D_1.L	
2100	25	H_{60}.F	D_1.L	
2200	24	H_{60}.F	D_1.L	
2300	24	H_{60}.F	D_1.L	
2400	24	H_{70}.F	D_1.F	
2500	22	H_{70}.F	D_1.F	
2600	22	H_{70}.F	D_1.F	
2700	22	H_{70}.F	D_1.F	
2800	22	H_{70}.F	D_1.F	
2900	22	H_{70}.F	D_1.F	
3000	22	H_{60}.F	D_1.F	

[a]D_0.F, D_1.F and D_3.F = Canopy densities corresponding to the proportion of first echoes above fraction #0 (2 m), #1 and #3 (see text). [a]D_0.L, D_1.L and D_2.L = Canopy densities corresponding to the proportion of last echoes above fraction #0 (2 m), #1 and #2 (see text). H_{10}.F, H_{60}.F, H_{70}.F, H_{80}.F and H_{90}.F = ALS height percentiles of the canopy height for the first echo. Hsd.L = Standard deviation of the canopy height of the first echoes. H_{mean}.F = Arithmetic mean of the first echo ALS canopy height.

that SAGB$_{buffer}$ and MAGB$_{buffer}$ contributed to explaining the variation in the relative residual errors (Table 2). Relating the absolute value of the relative residual with plot size using simple linear regression model indicated that there was a highly significant effect of plot size ($p < 0.0001$). Furthermore, the parameter estimate for plot size was negative showing that the relative residual is larger in absolute terms for small plots compared to larger plots (Table 3).

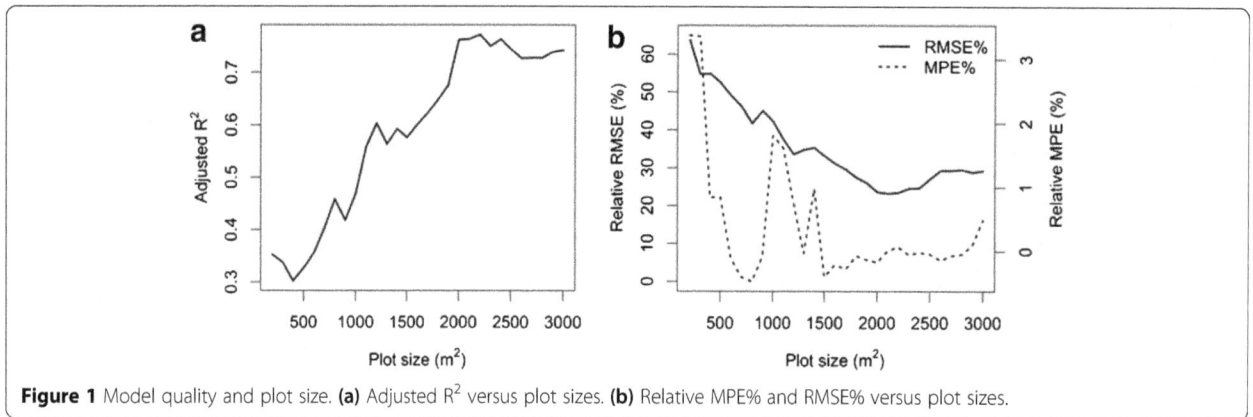

Figure 1 Model quality and plot size. **(a)** Adjusted R^2 versus plot sizes. **(b)** Relative MPE% and RMSE% versus plot sizes.

Efficiency of ALS-assisted AGB estimation

The SE estimates for the field-based AGB estimates were larger than the corresponding model-assisted SE estimates (Figure 3). For the plot sizes that allowed consistent analysis for all 30 sizes, i.e. from 200 to 1900 m^2, the field-based SE estimates decreased from 58.0 Mg ha^{-1} to 28.7 Mg ha^{-1}, while the model-assisted SE estimates decreased from 44.3 Mg ha^{-1} to 15.5 Mg ha^{-1}. Relative to the mean of field reference AGB for the plot size from 200 to 1900 m^2, the field –based SE estimates decreased from 14.1% to 8.2% , while for the model-assisted estimates decreased from 10.8% to 4.4%. Similarly, for the

larger plots (up to 3000 m^2) for which 22 observations were available for consistent analysis, the SE estimates for model-assisted were relatively much smaller compared to the field-based inventory. In both cases the SE was higher for smaller plots compared to the larger plots. Generally, the effectiveness of the ALS-assisted estimates was more improved as the plot size increased compared to the field-based estimates. This indicates that larger plots are relatively more favorable for ALS-assisted estimation than for pure field-based estimation. The RE values were >1 with a maximum value of 3.4 (Figure 4) for the plot sizes ranging from 200–1900 m^2

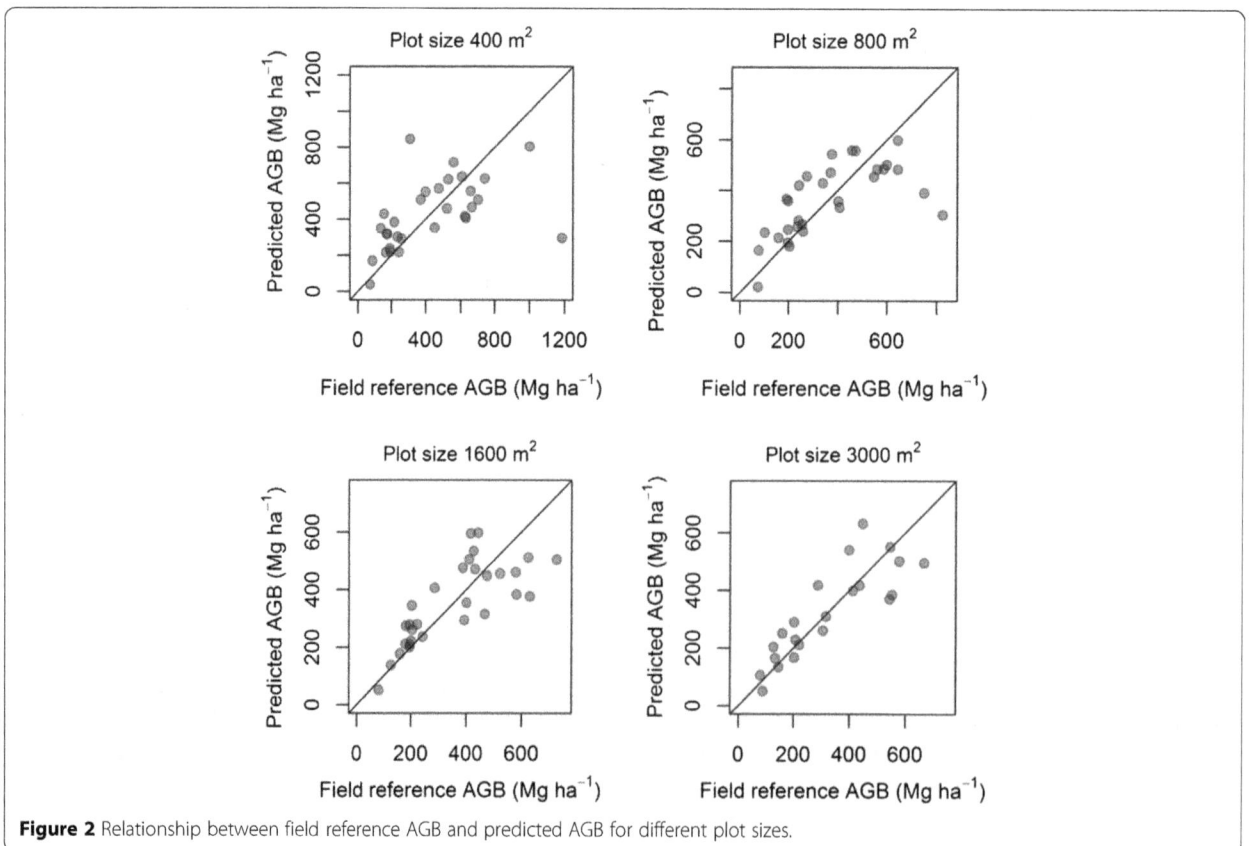

Figure 2 Relationship between field reference AGB and predicted AGB for different plot sizes.

Table 2 Coefficient estimates for models explaining residual errors of AGB using information extracted from buffer zones

Models[1]	Model parameter	3 m buffer			6 m buffer		
		Parameter estimate	p-value	AIC	Parameter estimate	p-value	AIC
Model 1	Intercept	−0.0159	0.8022	297	−0.1835	0.0206	654
	[2]SAGB$_{buffer}$	0.0838	<0.0001		0.1892	<0.0001	
Model 2	Intercept	−0.0321	0.5826	266	0.0501	0.4674	663
	[2]MAGB$_{buffer}$	0.4865	<0.0001		0.7015	<0.0001	

[1]Models = Two models; Model 1 uses SAGB$_{buffer}$ as fixed effects with plot identity as random effect. Model 2 uses MAGB$_{buffer}$ with plot identity as random effect (see text).
[2]SAGB$_{buffer}$ = Ratio of either sum of AGB at the buffer to the ground reference AGB per hectare, MAGB$_{buffer}$ = ratio of Maximum AGB at the buffer to the ground reference AGB per hectare (see text).

for which we have a complete dataset of 30 plots. For the other set with plot size up to 3000 m^2 the maximum RE value was 7.7. It should be noted that the peak in relative efficiency for the smallest dataset (22 plots) in Figure 4 was caused by considerable change in the observed AGB for a single plot when increasing the plot size beyond 2000 m^2. The increasing AGB was due to a large tree that was included in the plot measurements once the plot radius exceeded 25 m. This illustrates that in a small dataset the results can be sensitive to individual observations and even to the presence of individual trees.

Discussion

The findings of this study demonstrated the importance of choosing appropriate field plot sizes in ALS-assisted forest inventories in tropical forests. This is particularly important given that field campaigns are expensive and time consuming, and linking field measurements with remotely sensed data in the most effective manner would benefit both REDD+ implementations, together with all other studies related to forest carbon cycle. The current study extends previous research conducted in tropical forests, by having a dataset with a wide range of plot sizes. Furthermore, most of the previous studies have used rectangular plots.

See e.g. [18,26], whereas in this case circular plots have been used. Circular plots are more convenient for remote sensing studies compared to square or rectangular plots because only a single coordinate together with a plot radius are needed to match the two data sources geographically [19,28,29]. Circular plots are also within certain sizes easier to establish in the field because they have one dimension (i.e. radius) that defines the plot

boundary. The use of circular plots minimizes the plot boundary effects because of a smaller circumference to area ratio than all other plot shapes. However, the visibility from the plot center to the perimeter on a circular plot is increasingly hampered as the plots get larger, which increase per tree measurement time for the border trees. An increase of the area of a rectangular plot would not necessarily mean increased marginal cost (cost of including one more tree) if the width of the plot is kept constant and inclusion of trees are made with reference to the long side. However, rectangular plots are in general more difficult to establish. For example, in rugged terrain it can be difficult to keep the sides parallel.

Our findings demonstrated empirically the positive effects of increasing plot sizes on improved predictive power of the AGB models. The model fit (adjusted R^2) of the regression models was improved as plot size increased. Reduced circumference to area ratio, spatial averaging, and less effect of positioning errors are probably the main reasons. The fit of our models are in line with previous ALS-based studies in both tropical forests and temperate forests. For example, [30] reported R^2 of 0.78 in the tropical rainforest of Hawaii islands while [31] reported R^2 of 0.64 in a tropical rainforest of West Africa. Furthermore, results from the cross-validation showed smaller RMSE% and MPE% (Figure1b) for larger plots compared to smaller plots. Similar trends have been reported and discussed by other authors in both temperate and tropical forests see e.g. [32].

Plot boundary effects have been discussed in previous studies see e.g. [16,33] as one among the sources of model error in ALS-assisted inventories, particularly when relying on small plots. We demostrated this in two steps; first by relating relative residuals to the sum of AGB per hectare for all trees in the buffer (SAGB$_{buffer}$) and the maximum AGB per hectare for the largest tree in the buffer (MAGB$_{buffer}$) where we noted that their importance were depending on the size of the buffer. The buffer conditions as expressed both by (MAGB$_{buffer}$) and (SAGB$_{buffer}$), seemed to have more impact on the

Table 3 Parameter estimates for the model relating relative residual in absolute form and plot sizes

Coefficients	Parameter estimates	p-value
Intercept	0.5060	<0.0001
Plot size	−0.0002	<0.0001

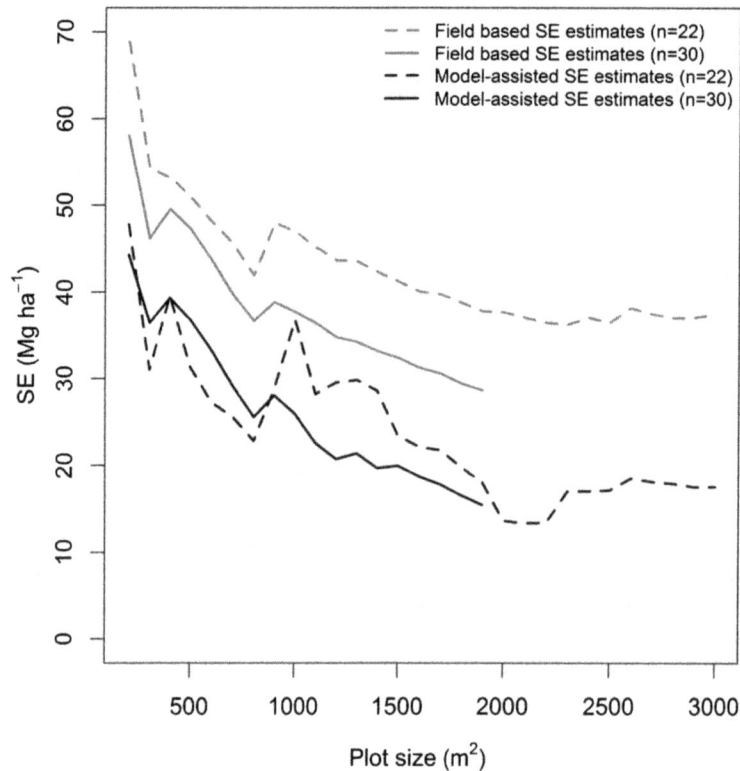

Figure 3 Field-based and model-assisted SE estimates for different plot sizes covered in two sample datasets (i.e. 200 to 1900 and 200 to 3000 m²).

residual error with decreasing distance to the plot judged by the AIC values (Table 2), which is logical. Furthermore, when comparing the two variables, SAGB$_{buffer}$ seemed to lose less explanatory power by going from 3 meter to 6 m buffer than MAGB$_{buffer}$. This result was also expected because the represetation of the whole buffer by SAGB$_{buffer}$ is less prone to be changed by the increase in size compared to MAGB$_{buffer}$ which is calculated from a single tree. Furthermore, the decrease in ALS model residuals (Table 3) with increasing plot sizes is a clear indication that smaller plots are more prone to boundary effects compared to larger plots.

Contribution of ALS data in improving precision of AGB estimates was also demonstrated within varying ranges of plot sizes. The RE values were > 1, indicating that ALS-assisted estimation is more efficient compared to pure field-based estimation. To achieve similar precision of a pure field-based estimate relying on simple random sampling, would mean to increase the sample size for the field-based inventory by a factor equivalent to the value of RE, which would have a substantial effect on field inventory costs. In general, the gain in relative efficiency was more pronounced as plot size increased, suggesting that larger plots are more favorable when ALS-data are used to assist in the estimation. Even though we did not undertake any analysis of cost-efficiency, the trend

would be toward larger and fewer plots as one introduces ALS to support in the estimation. Even this finding can be attributed to the effects discussed above, namely reduced boundary effects and co-registration errors.

Despite the potential of improving the efficiency of ALS-assisted inventories by use of larger plots, choice of an "optimal" plot size must be seen in a broader context by considering a number of factors including; sample sizes, on-plot costs, traveling costs and overall field inventory design. Several authors see e.g. [20,23,30] have indicated that selection of the plot size also will depend on forest types, available resources and the needed precision. Based on our findings, there is larger potential of gaining efficiency of using ALS data in this type of forest when the field plot size is larger than 1200 m². Finally, even though our study was limited to the tropical rainforests of Tanzania, the major findings are of interest and efforts should be taken to upscale to other tropical forests by considering more factors that would lead to selection of "optimal" plot size.

Conclusions

To conclude, our study has demonstrated that field plot size effect the prediction accuracy of ALS-assisted AGB estimation in the tropical forests. Generally, there was substantial improvement in prediction accuracy from

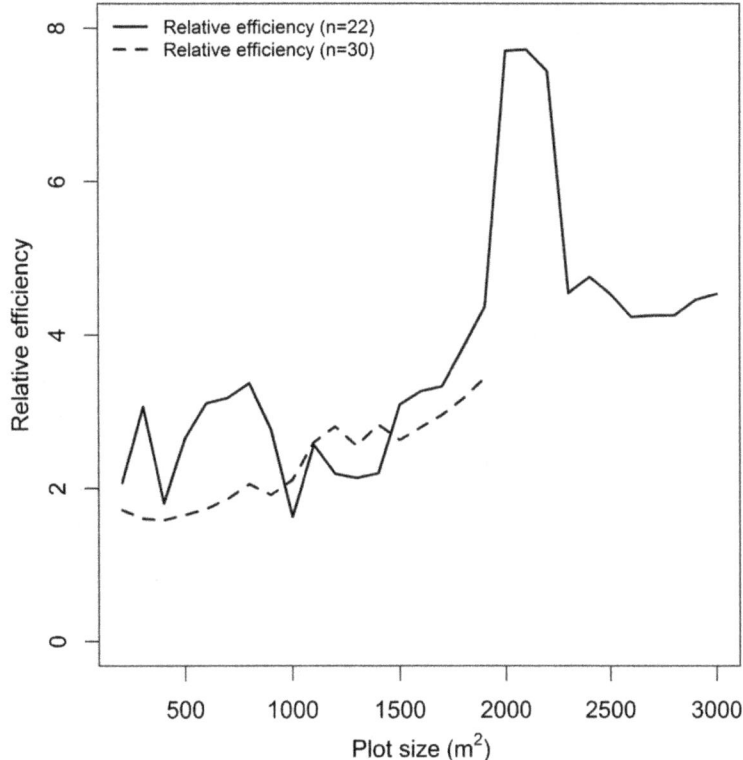

Figure 4 Relative efficiency for different plot sizes.

larger plots compared to smaller plots. Indicators of boundary effects were also identified and confirmed to have significant effects on the model quality. From a purely technical point of view, our results suggested that it is relatively more favorable to increase the plot size when ALS is used to enhance the estimates. This study showed that there is a relative improvement in precision of ALS-assisted AGB estimation, compared to pure field-based estimation up to around 3000 m^2 in this type of forest. However, the maximum plot size of 3000 m^2 in the current study leaves an open question as to whether there are any additional gains in relative precision beyond this size. Future studies should be conducted to quantify the contribution of ALS to improve estimation precision for even larger plots as the basis for design of future inventories in tropical rainforests. Similar studies should also be conducted in other types of tropical forests.

Methods
Site description
The study was conducted in Amani nature reserve (ANR), which is situated in the southern part of the East Usambara Mountains in northern Tanzania (Figure 5). It was gazetted in 1997 with a protected area of 8,380 ha. ANR lies between 5°14' - 5° 04' S and 38° 30' - 38°40' E, with an altitudinal range of 190 to 1130 m above sea level [34]. Rainfall is heavy at higher altitudes and in the

southeast of the mountain, with an average of 1900 mm annually. The dry seasons are from June to August and January to March, but rainfall is frequent throughout the year. The mean annual temperature is 20.6°C [35].

Data collection
Sampling design
An initial probability sample of 173 field plots with an average size of 900 m^2 were established across ANR according to a systematic design (450 m × 900 m distance between plots) in 1999–2000 by a non-governmental conservation and development organization, Frontier Tanzania [34] (Figure 5). The plots were revisited and re-measured in 2008–2012. In order to analyse plot size effects on AGB estimates, a small sub-sample of 30 large plots was established. Measurements on the 30 plots were acquired in a separate campaign after completion of measurements of the large sample. Due to high travel costs and long walking distances in the very steep and rough terrain, establishing a probability sample of 30 large plots across the entire study area was cost-prohibitive. Instead we developed a sampling strategy by which we took advantage of the a priori knowledge of the distribution of AGB in the large probability sample and selected purposefully three sub-regions within the study area in which the initial plots were revisited. There is a strong altitude-dependent AGB gradient in

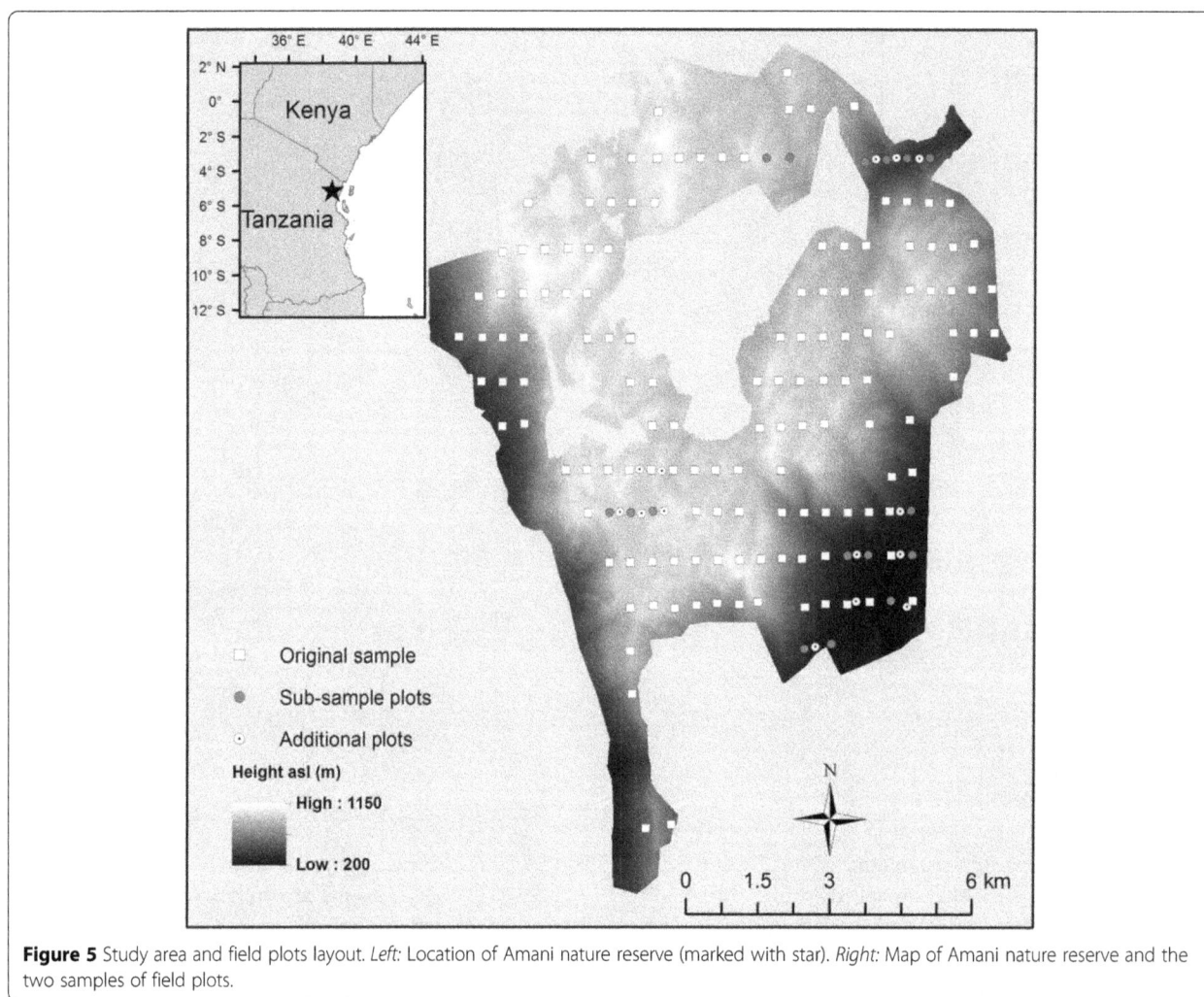

Figure 5 Study area and field plots layout. *Left:* Location of Amani nature reserve (marked with star). *Right:* Map of Amani nature reserve and the two samples of field plots.

the study area. It was therefore important to capture the altitude gradient in each of the three sub-regions in order to resemble the AGB distribution in the initial probability sample.

In the sampled sub-regions, we first selected 16 of the plots in the initial probability sample for measurement. We also established 14 new and additional plots along the grid-lines of the probability sample and located them exactly mid-way between two existing plots. Thus, the distance between our plots was 225 m rather than 450 m.

Although the resulting sample of 30 large plots was not selected according to probabilistic principles, it closely resembled essential properties of the large probability sample. First of all the AGB distributions of the two samples were similar (Figure 6). The mean AGB of the 30 plots with an area of 900 m^2 was 366.0 Mg ha^{-1} (Table 4, Figure 6), while it was 461.9 Mg ha^{-1} for the large probability sample (Figure 6). The AGB range was 69.4-908.3 Mg ha^{-1} (standard deviation of 216.3 Mg ha^{-1}) while it was 43.2-1147.1 Mg ha^{-1} (standard deviation of 214.7 Mg ha^{-1}) for the large sample. Furthermore, the 30

plots covered an elevation range of 200 to 1000 m above sea level (Figure 7a) so that both the lowland forests (<800 m above sea level) and the sub mountain forests (>800 m above sea level) were represented. The 30 plots also covered a wide range of tree sizes (Figure 7b).

Field data

Field data were collected during November 2012, about six mounts after completion of the field work on the large probability sample. On each of the 30 plots, we registered all trees within a radius limited by the maximum distance measuring range of a Vertex hypsometer [36], which was used to measure the horizontal distance from the plot centre to each tree. The maximum measuring range of the hypsometer varied among the plots due to differences in terrain ruggedness and forest density. The radius distribution among the 30 plots was as follows; 31 m (22 plots), 28 m (2 plots), 26 m (1 plot) and 25 m (5 plots). For each tree with diameter at breast height (*dbh*) larger than 5 cm, scientific name, local name, distance to plot centre and *dbh* was registered. A

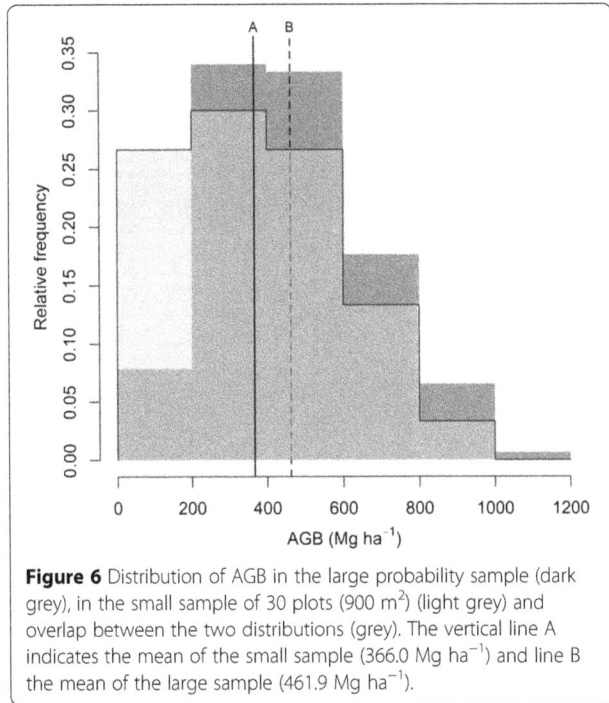

Figure 6 Distribution of AGB in the large probability sample (dark grey), in the small sample of 30 plots (900 m^2) (light grey) and overlap between the two distributions (grey). The vertical line A indicates the mean of the small sample (366.0 Mg ha^{-1}) and line B the mean of the large sample (461.9 Mg ha^{-1}).

diameter tape, rather than a calliper, was used to gauge diameters since tree trunks in this forest type tend to be both oval and large in size. The distance was measured from plot center to the front of each tree, and half of the tree diameter was added to get the total horizontal distance. The distance measures enabled us to generate any plot size within the limit of the maximum radius. For this study, we decided to select radii between 7.98 m (200 m^2) and 30.90 m (3000 m^2) (Table 4) for further analysis. Three trees (largest, medium and smallest in terms of diameter) per plot were measured for height (h) using a Vertex hypsometer.

Precise field coordinates were determined in the centre of each plot by means of differential Global Navigation Satellite Systems (dGNSS). Topcon Legacy 40 channels dual frequency receivers, observing both pseudo-range and carrier phase of the Global Positioning System (GPS) and the Global Navigation Satellite System (GLONASS) were used as rover and base station. The post-processing reports from Pinnacle version 1.0 software [37] indicated an average error of 19 cm for the planimetric coordinates. The error was computed as two times the standard deviations of the corrected single observations reported from Pinnacle output [38].

Field estimates of AGB

For each plot AGB was estimated by using the local allometric AGB model developed by [39] with both *dbh* and *h* as predictor variables (Eq. 2). Using models with both *dbh* and *h* is reported to moderate the effect of large *dbh*-values on AGB estimates as compared to models

with *dbh* only [40-42]. Before calculating AGB, a height model (Eq. 1), was developed using the observations of tree height and corresponding diameters from each plot. A number of model forms for diameter–height relationship [43-48] were tested using non-linear mixed effect approach. Best model fit, judged by the Akaike information criterion (AIC), was obtained using the model form by [46]

$$h = 1.3 + 45.5103\left[exp\left(-2.7163 * exp\left(-0.0354 * dbh\right)\right)\right]$$
(1)

This model was used to predict height for trees without height measurements. AGB was calculated for individual trees within each plot according to [39] i.e.,

$$AGB = 0.4020 * (dbh)^{1.4365}(h)^{0.8613}$$
(2)

and then summed to obtain total AGB for the respective plot. The AGB values were finally scaled to per ha values for the different plot sizes (Table 4). The calculated AGB values are henceforth denoted field reference AGB.

Laser scanner data

ALS data were collected during the period from 19 January to 18 February 2012 using a Leica ALS70 sensor (Leica Geosystems AG, Switzerland) carried by a Cessna 404 fixed-wing aircraft. Mean flying altitude was 800 m above ground covering the entire area of ANR (i.e. wall to wall) at a ground speed of 75 m s^{-1}. The scanning rate was 58.6 Hz and the instrument operated at a pulse repetition frequency of 339 kHz with a resulting average pulse density of 10.6 points m^{-2}.

Processing of the ALS data started with classification of each ALS echo as ground or vegetation using the progressive irregular triangular network densification method [49] implemented in the TerraScan software [50]. A Triangular Irregular Network (TIN) was created using the ALS echoes classified as ground echoes. The heights above the ground surface were calculated for all echoes by subtracting the respective TIN heights from the height values of all echoes recorded. Up to five echoes were registered per pulse and we used the three echo categories classified as "single", "first of many", and "last of many". The "single" and "first of many" echoes were pooled into one dataset denoted as "first" echoes, and correspondingly, the "single" and "last of many" echoes were pooled into a dataset denoted as "last" echoes.

Several variables were extracted from the ALS data for each of the field plot sizes as described by [51]. For each plot size, height distributions of both first and last echoes were first created. A height threshold of 2.0 m was applied in order to remove the effect of low vegetation and echoes from ground features falsely classified as vegetation. Then, heights at nine percentiles (10th,

Table 4 Summary of field data

Plot size (m²)	Number of plots (n)	Mean (Mg ha⁻¹)	Standard deviation (Mg ha⁻¹)	Minimum (Mg ha⁻¹)	Maximum (Mg ha⁻¹)
200	30	411.4	323.2	53.3	1179.5
300	30	401.0	257.3	48.2	816.0
400	30	424.5	275.8	72.6	1185.2
500	30	413.8	263.4	77.8	1148.0
600	30	395.3	243.9	87.9	1066.6
700	30	371.8	221.5	75.4	931.7
800	30	363.1	204.4	74.3	824.3
900	30	366.0	216.3	69.4	908.3
1000	30	367.1	210.1	62.4	859.7
1100	30	365.6	203.0	66.4	839.5
1200	30	365.0	193.7	78.4	797.6
1300	30	361.0	190.5	82.1	757.9
1400	30	352.3	184.7	87.3	707.0
1500	30	354.2	180.4	85.5	757.8
1600	30	353.2	174.1	82.2	725.5
1700	30	355.0	170.2	95.6	702.6
1800	30	355.9	163.9	91.5	696.5
1900	30	351.1	159.6	90.7	703.3
2000	25	352.2	170.8	89.6	669.3
2100	25	350.4	168.0	85.5	646.2
2200	24	344.7	169.3	89.3	631.1
2300	24	343.0	167.8	88.5	639.8
2400	24	344.2	171.3	87.9	677.7
2500	22	332.1	175.0	84.4	661.5
2600	22	334.1	183.1	91.8	669.9
2700	22	328.0	179.8	88.6	674.7
2800	22	322.7	177.7	85.4	665.9
2900	22	323.5	177.9	82.5	655.6
3000	22	321.0	179.7	79.7	666.7

Number of plots for the different plot sizes together with mean field reference AGB values with corresponding standard deviation, minimum, and maximum.

20th, ..., 90th) of both the first- and last echo distributions were computed to represent canopy height and labeled $H_{10}.F$, $H_{20}.F$, ..., $H_{90}.F$ (first echoes) and $H_{10}.L$, $H_{20}.L$, ..., $H_{90}.L$ (last echoes), respectively. Measures of canopy density were also derived for first and last echoes of each plot size. The range between the lowest ALS canopy height (>2 m) and the 95th percentile height was divided into 10 vertical fractions of equal height. Canopy densities were then computed as the proportion of ALS echoes above each fraction to total number of first echoes and labeled $D_0.F$ (>2 m), $D_1.F$, ..., $D_9.F$. Density variables for the last echo distribution were calculated the same way (relative to total number of last echoes) and labeled $D_0.L$, $D_1.L$, ..., $D_9.L$. Furthermore, for both first and last echo height distributions on each plot, the maximum height ($H_{max}.F$ and $H_{max}.L$), mean

values ($H_{mean}.F$ and $H_{mean}.L$), standard deviation ($H_{sd}.F$ and $H_{sd}.L$), coefficient of variation ($H_{cv}.F$ and $H_{cv}.L$), and skewness ($H_{skewness}.F$ and $H_{skewness}.L$) were computed.

Data analyses
Model development
Multiple linear regression analysis with ordinary least square regression (OLS) was used to develop the statistical models relating the field reference AGB and the predictor variables from the ALS data. To ensure that our modelling approaches met the basic assumptions of OLS, the response variable was transformed to logarithmic scale [11,52], while for the predictors both log transformed and non-transformed variables were used. Separate models with log transformed response and combination of log transformed and non-transformed

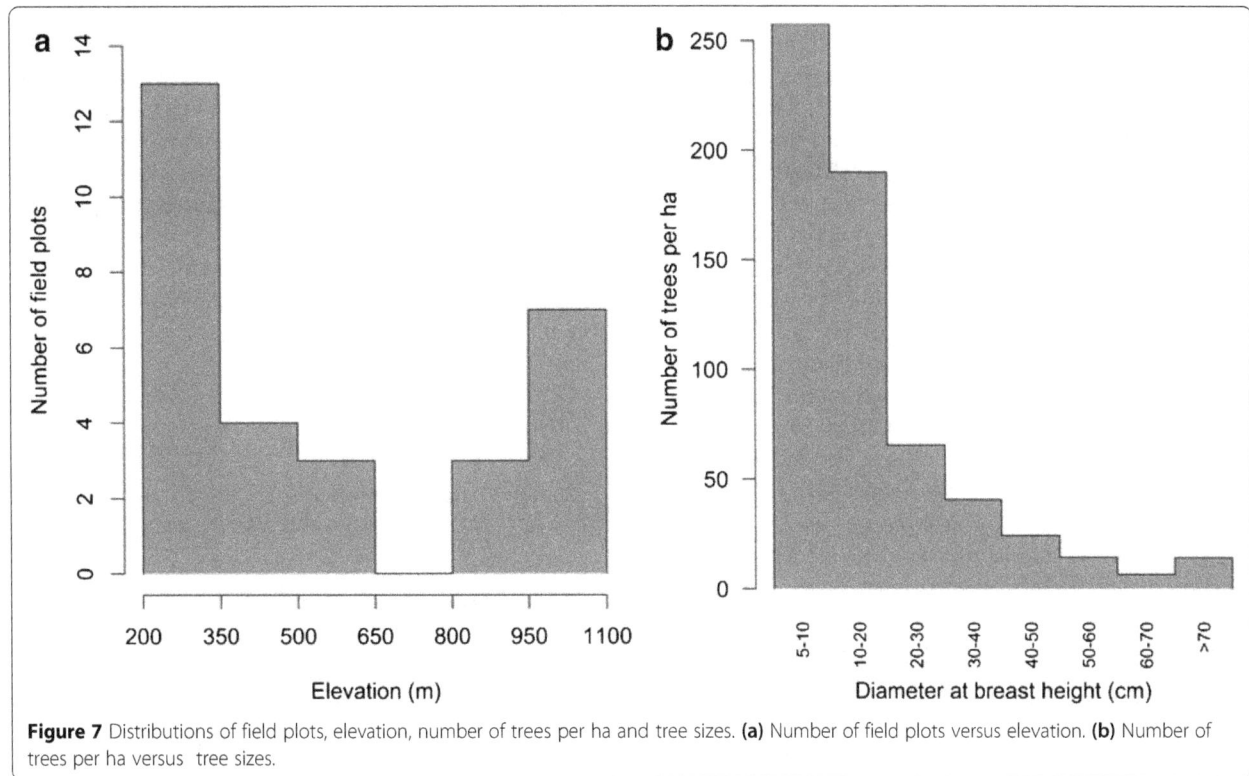

Figure 7 Distributions of field plots, elevation, number of trees per ha and tree sizes. **(a)** Number of field plots versus elevation. **(b)** Number of trees per ha versus tree sizes.

predictor variables were fitted for each of the plot sizes. We decided to fit separate models (unique variable combinations) for each of the plot sizes, because we wanted the model for each plot size to be the "best" and not be constrained by forcing specific variables into the model.

Variable selection was conducted by using reg-subset in the leaps package in R [53]. The selection of the variables was limited to the best combinations of three or fewer variables in order to avoid multicollinearity among candidate predictors. The preferred models were chosen based on the Bayesian information criterion (BIC) [54]. Adjusted R^2 was also used for assessing the model fit while multicollinearity was assessed by computing the variance inflation factors (VIF). The VIF values were determined for the individual β parameters. VIF values greater than 10 were regarded as an indication of multicollinearity problems [55].

Log-transformation of the response variable introduces a bias when back-transforming to the arithmetic scale. The model for AGB was therefore adjusted for logarithmic bias according to [56] by adding half of the model mean square error to the constant term before transformation to arithmetic scale.

Model validation and accuracy assessment

In order to assess the performance of the models for each plot size, leave-one-out cross–validation (LOOCV)

was performed. One field plot at a time was excluded from the dataset, and the model was fitted based on n-1 plots to predict the AGB of the left out plot. Here, n denotes the number of field plots, where i = 1,..., n. Relative root mean square error (RMSE %) and the mean prediction error (MPE%) were used as the measures of reliability and calculated according to

$$\text{RMSE\%} = \frac{\sqrt{\sum_{i=1}^{n}(y_i-\widehat{y}_i)^2/n}}{\bar{y}} \times 100 \qquad (3)$$

$$\text{MPE\%} = \frac{\sum_{i=1}^{n}(y_i-\widehat{y}_i)/n}{\bar{y}} \times 100 \qquad (4)$$

Where y_i and $\widehat{y}_i$ denote field reference AGB and predicted AGB for plot i, respectively, and ȳ denotes mean field reference AGB for all plots. RMSE% is a good measure of how accurately the model predicts the response and is the most important criterion for fit if the main purpose of the model is prediction [57].

Analysis of boundary effects

To analyze the boundary effects we studied how the residual errors of the models were related to the field reference AGB of the trees in an outer buffer zone for different field plot sizes. To archive this, we extracted field reference AGB values for 3 m and 6 m buffers outside the field plots for the plot sizes of 200–1500 m² and

200–1100 m^2, respectively. We selected the trees with *dbh* > 10 cm and computed AGB per hectare for the largest tree in the buffer and the total AGB per hectare for all trees in the buffer. To obtain the model residual error, we first subtracted the ground reference AGB from the predicted AGB. Then we calculated the ratio between the residuals and the total field reference AGB for the respective plot (i.e., relative residual). Similar ratios between (1) sum of AGB per hectare for all trees in the buffer (SAGB$_{buffer}$) and the field reference AGB for the plot and (2) the maximum AGB per hectare for the largest tree in the buffer (MAGB$_{buffer}$,) and the field reference AGB for the plot were also computed. Two empirical models explaining the variation in the relative residual values using either SAGB$_{buffer}$ or MAGB$_{buffer}$ as explanatory variables were developed. Linear mixed effects (LME) regression using *nlme* add-on package [58] in R was used for model fitting. LME models are linear regression models in which parameters are the sum of the fixed and random effects. In this case the fixed effects were either SAGB$_{buffer}$ or MAGB$_{buffer}$ while plot identity was treated as the random effect. We assumed that each plot will have different random error structures and that the distribution of AGB within these plots is not independent of one another. To test the effect of plot sizes on relative residual, we also fitted the linear regression model which relates relative residuals in absolute form and plot sizes. Absolute value was used because we were interested in the magnitude of the residual regardless of its sign.

Efficiency of ALS-assisted AGB estimation

ALS-assisted estimation of AGB within the design-based and model-assisted inferential framework can greatly improve the precision compared to pure field-based estimation. The purpose of this analysis was to quantify the gain in estimated precision of using ALS data relative to a pure field-based estimate for increasing plot sizes.

A basic requirement for validity of design-based inference is the availability of a probability sample [59]. As stated above, the current sample of 30 plots was obtained as a subsample of a probability sample, but the sub-sampling was not conducted according to strict probabilistic principles. However, the sub-sample was selected to resemble important properties of the large probability sample as closely as practically feasible. Thus, a comparison of variances using the current data and assuming a probabilistic design will most likely introduce a bias in the estimators of unknown magnitude. Likewise, when a systematic sample is obtained, it is common to adopt design-based estimators assuming e.g. simple random sampling (SRS) although it is well-known that SRS variance estimators usually are positively biased under systematic sampling. The magnitude

of the bias is always unknown for a particular sample because bias is a property of an estimator and not a particular sample. The current analysis was conducted under the assumption that the sample at hand would give a meaningful quantification of the effect of plot size on relative variance estimates. Thus, in the current study we adopted design-based variance estimators assuming simple random sampling and complete cover of ALS data.

Assuming SRS, the variance estimator for the field-based AGB estimate ignoring corrections for finite population is [60].

$$\widehat{V}_{field} = \frac{\sum_{i=1}^{n}(y_i - \bar{y})^2}{n(n-1)} \tag{5}$$

For model-assisted estimation, the variance estimator of the so-called generalized regression estimator is [60].

$$\widehat{V}_{ALS} = \frac{\sum_{i=1}^{n}(\hat{e}_i - \bar{e})^2}{n(n-1)} \tag{6}$$

where $\hat{e}_i = y_i - \widehat{y}_i$ is the model prediction residual for plot i and $\bar{e} = \frac{\sum_{i=1}^{n}\hat{e}_i}{n}$ is the mean residual for all plots. Standard error (SE) was computed as the square root of the variance estimates. Finally, the relative efficiency (RE) of ALS-assisted inventory relative to field-based inventory was calculated for different plot sizes as the ratio of the two variance estimates, i.e.,

$$RE = \widehat{V}_{field} \Big/ \widehat{V}_{ALS} \tag{7}$$

Values of RE greater than 1.0 indicates higher efficiency of ALS-assisted estimates than field-based estimates for a given plot size. To achieve consistency in the analysis across different plot sizes, the dataset was divided into two major groups. The first group subject to analysis comprised all the 30 plots and allowed consistent analysis of plot size ranging from 200–1900 m^2. The second group allowing analysis from 200 to 3000 m^2 consisted of 22 of the plots.

Abbreviations

ALS: Airborne laser scanning; ABA: Area-based approach; AGB: Aboveground biomass; AIC: Akaike information criterion; ANR: Amani nature reserve; BIC: Bayesian information criterion; dGNSS: differential Global Navigation Satellite Systems; GLONASS: Global Navigation Satellite System; GPS: Global Positioning System; LME: Linear mixed effects; LOOCV: leave-one-out cross–validation; MA: Model-assisted; MAGB$_{buffer}$: Maximum AGB per hectare for the largest tree in the buffer; MPE: Mean prediction error; MRV: Monitoring, Reporting and Verification; OLS: Ordinary least square regression; REDD+: Reducing Emissions from Deforestation and Forest Degradation in tropical countries; RMSE: Root mean square error; SAGB$_{buffer}$: Sum of AGB per hectare for all trees in the buffer; SAR: Synthetic aperture RADAR; SE: Standard error; SRS: Simple random sampling; TIN: Triangular Irregular Network; UNFCCC: United Nations Framework Convention on Climate Change; RE: Relative efficiency.

Competing interests
The authors declare that they have no competing interests.

Author's contributions
All the authors have made substantial contribution towards successful completion of this manuscript. Authors; EWM and OMB have been involved in designing the study, drafting the manuscript, data analysis and write up. EHH has been involved in data analysis and quality control of both raw data and results. EN and TG have been responsible for designing the ALS acquisition and they were involved in revising the manuscript. REM was involved in critical discussion on the field inventory design and all logistics related to the field data acquisition. All authors read and approved the final manuscript.

Author's information
EWM and EHH are PhD students in forest inventory at Norwegian university of Life Sciences (NBMU).They are both associated with the forest mensuration group in the university. OMB is the researcher in the same group specialized on the application of ALS in forestry. EN and TG are senior scientists and professors in ALS and forest sampling at NMBU. Both EN and TG, are resource persons for the forest mensuration group at NMBU. REM is professor in forest inventory and mensuration at Sokoine university of Agriculture, Tanzania.

Acknowledgements
The financial support for this research was provided by Government of Norway through the two projects entitled "Climate Change Impacts, Adaptation and Mitigation (CCIAM) in Tanzania" and "Enhancing the Measuring, Reporting and Verification (MRV) of forests in Tanzania through the application of advanced remote sensing techniques". We are highly acknowledging our field team in Tanzania, and Terratec Norway, for collecting and processing of the ALS data. We are also grateful to the administration of ANR for all support, and especially for provision of office space for establishment of the GPS base station.

Author details
[1]Department of Ecology and Natural Resource Management, Norwegian University of Life Sciences, P.O. Box 5003, Oslo, NO 1432, Ås, Norway.
[2]Department of Forest Mensuration and Management, Sokoine University of Agriculture, P.O. Box 3013, MorogoroTanzania.

References
1. Lewis SL, Lopez-Gonzalez G, Sonké B, Affum-Baffoe K, Baker TR, Ojo LO, et al. Increasing carbon storage in intact African tropical forests. Nature. 2009;457:1003–6.
2. Joseph S, Herold M, Sunderlin WD, Verchot LV. REDD+ readiness: early insights on monitoring, reporting and verification systems of project developers. Environ Res Lett. 2013;8:034038.
3. Herold M, Skutsch M: Monitoring, reporting and verification for national REDD plus programmes: two proposals. Environ Res Lett. 2011;6:014002.
4. Keith H, Mackey BG, Lindenmayer DB. Re-evaluation of forest biomass carbon stocks and lessons from the world's most carbon-dense forests. Proc Natl Acad Sci. 2009;106:11635–40.
5. Hyyppä J, Hyyppä H, Leckie D, Gougeon F, Yu X, Maltamo M. Review of methods of small-footprint airborne laser scanning for extracting forest inventory data in boreal forests. Int J Remote Sens. 2008;29:1339–66.
6. Vauhkonen J, Maltamo M, McRoberts RE, Næsset E: Introduction to Forestry Applications of Airborne Laser Scanning. In: Maltamo M, Næsset E, Vauhkonen J, editors. Forestry applications of airborne laser scanning – concepts and case studies. Dordrecht, Netherlands: Springer; 2014. p. 1–16.
7. Coops NC, Wulder MA, Culvenor DS, St-Onge B. Comparison of forest attributes extracted from fine spatial resolution multispectral and lidar data. Can J Remote Sens. 2004;30:855–66.
8. Hansen EH, Gobakken T, Bollandsås OM, Zahabu E, Næsset E. Modeling Aboveground Biomass in Dense Tropical Submontane Rainforest Using Airborne Laser Scanner Data. Remote Sens. 2015;7:788–807.
9. Ioki K, Tsuyuki S, Hirata Y, Phua M-H, Wong WVC, Ling Z-Y, et al. Estimating above-ground biomass of tropical rainforest of different degradation levels in Northern Borneo using airborne LiDAR. For Ecol Manage. 2014;328:335–41.
10. Gautam B, Peuhkurinen J, Kauranne T, Gunia K, Tegel K, Latva-Käyrä P, et al. Estimation of Forest Carbon Using LiDAR-Assisted Multi-Source Programme (LAMP) in Nepal. In: Proceedings of the International Conference on Advanced Geospatial Technologies for Sustainable Environment and Culture, Pokhara, Nepal. 2013. p. 12–3.
11. Næsset E. Predicting forest stand characteristics with airborne scanning laser using a practical two-stage procedure and field data. Remote Sens Environ. 2002;80:88–99.
12. Næsset E. Estimating timber volume of forest stands using airborne laser scanner data. Remote Sens Environ. 1997;61:246–53.
13. Næsset E, Bjerknes K-O. Estimating tree heights and number of stems in young forest stands using airborne laser scanner data. Remote Sens Environ. 2001;78:328–40.
14. Næsset E: Area-Based Inventory in Norway–From Innovation to an Operational Reality. In: Maltamo M, Næsset E, Vauhkonen J, editors. Forestry applications of airborne laser scanning – concepts and case studies. Dordrecht, Netherlands: Springer; 2014. p. 215–240.
15. McRoberts RE, Cohen WB, Naesset E, Stehman SV, Tomppo EO. Using remotely sensed data to construct and assess forest attribute maps and related spatial products. Scand J For Res. 2010;25:340–67.
16. Frazer GW, Magnussen S, Wulder MA, Niemann KO. Simulated impact of sample plot size and co-registration error on the accuracy and uncertainty of LiDAR-derived estimates of forest stand biomass. Remote Sens Environ. 2011;115:636–49.
17. Gobakken T, Næsset E. Assessing effects of positioning errors and sample plot size on biophysical stand properties derived from airborne laser scanner data. Can J Forest Res. 2009;39:1036–52.
18. Mascaro J, Detto M, Asner GP, Muller-Landau HC. Evaluating uncertainty in mapping forest carbon with airborne LiDAR. Remote Sens Environ. 2011;115:3770–4.
19. Næsset E, Bollandsås OM, Gobakken T, Gregoire TG, Ståhl G. Model-assisted estimation of change in forest biomass over an 11 year period in a sample survey supported by airborne LiDAR: A case study with post-stratification to provide "activity data". Remote Sens Environ. 2013;128:299–314.
20. Zolkos S, Goetz S, Dubayah R. A meta-analysis of terrestrial aboveground biomass estimation using lidar remote sensing. Remote Sens Environ. 2013;128:289–98.
21. Asner GP, Mascaro J, Muller-Landau HC, Vieilledent G, Vaudry R, Rasamoelina M, et al. A universal airborne LiDAR approach for tropical forest carbon mapping. Oecologia. 2012;168:1147–60.
22. Gregoire TG, Ståhl G, Næsset E, Gobakken T, Nelson R, Holm S. Model-assisted estimation of biomass in a LiDAR sample survey in Hedmark County, Norway This article is one of a selection of papers from Extending Forest Inventory and Monitoring over Space and Time. Can J Forest Res. 2010;41:83–95.
23. Næsset E, Gobakken T, Solberg S, Gregoire TG, Nelson R, Ståhl G, et al. Model-assisted regional forest biomass estimation using LiDAR and InSAR as auxiliary data: A case study from a boreal forest area. Remote Sens Environ. 2011;115:3599–614.
24. Ene LT, Næsset E, Gobakken T, Gregoire TG, Ståhl G, Nelson R. Assessing the accuracy of regional LiDAR-based biomass estimation using a simulation approach. Remote Sens Environ. 2012;123:579–92.
25. Asner GP, Clark JK, Mascaro J, Vaudry R, Chadwick KD, Vieilledent G, et al. Human and environmental controls over aboveground carbon storage in Madagascar. Carbon balance and management. 2012;7:2.
26. Asner GP, Mascaro J. Mapping tropical forest carbon: Calibrating plot estimates to a simple LiDAR metric. Remote Sens Environ. 2014;140:614–24.
27. Mascaro J, Asner GP, Dent DH, DeWalt SJ, Denslow JS. Scale-dependence of aboveground carbon accumulation in secondary forests of Panama: A test of the intermediate peak hypothesis. For Ecol Manage. 2012;276:62–70.
28. Adams T, Brack C, Farrier T, Pont D, Brownlie R. So you want to use LiDAR?-a guide on how to use LiDAR in forestry. N Z J For. 2011;55:19–23.
29. White JC, Wulder MA, Varhola A, Vastaranta M, Coops NC, Cook BD, et al. A best practices guide for generating forest inventory attributes from airborne laser scanning data using an area-based approach. For Chron. 2013;89:722–3.
30. Asner GP. Tropical forest carbon assessment: integrating satellite and airborne mapping approaches. Environ Res Lett. 2009;4:034009.

31. Chen Q, Vaglio Laurin G, Battles JJ, Saah D. Integration of airborne lidar and vegetation types derived from aerial photography for mapping aboveground live biomass. Remote Sens Environ. 2012;121:108–17.

32. Gobakken T, Næsset E. Assessing effects of laser point density, ground sampling intensity, and field sample plot size on biophysical stand properties derived from airborne laser scanner data. Can J Forest Res. 2008;38:1095–109.

33. Wulder MA, White JC, Nelson RF, Næsset E, Ørka HO, Coops NC, et al. Lidar sampling for large-area forest characterization: A review. Remote Sens Environ. 2012;121:196–209.

34. Doody K, Howell K, Fanning E. Amani Nature Reserve-A biodiversity survey. East Usambara Conservation Area Management Programme, Technical Paper 52. In: Ministry of Natural Resources and Tourism Tanzania and Frontier-Tanzania. Tanga. 2001.

35. Hamilton AC, Bensted-Smith R: Forest conservation in the East Usambara mountains, Tanzania. Gland, Switzerland: IUCN; 1989.

36. Haglöf A. Users guide Vertex III and Transponder T3. Långsele, Sweden: Haglöf Sweden, AB; 2002.

37. Anon: Pinnacle User's Manual; Javad Positioning Systems. In: CA. Edited by Jose S. USA; 1999.

38. Naesset E. Effects of differential single-and dual-frequency GPS and GLONASS observations on point accuracy under forest canopies. Photogramm Eng Remote Sens. 2001;67:1021–6.

39. Masota A: Tree allometric models for predicting above- and belowground biomass of tropical rainforests in Tanzania. in press.

40. Feldpausch T, Banin L, Phillips O, Baker T, Lewis S, Quesada C, et al. Height-diameter allometry of tropical forest trees. Biogeosciences. 2011;8:1081–106.

41. Banin L, Feldpausch TR, Phillips OL, Baker TR, Lloyd J, Affum-Baffoe K, et al. What controls tropical forest architecture? Testing environmental, structural and floristic drivers. Glob Ecol Biogeogr. 2012;21:1179–90.

42. Mugasha WA, Bollandsås OM, Eid T. Relationships between diameter and height of trees in natural tropical forest in Tanzania, Southern Forests. J For Sci. 2013;75:221–37.

43. Nilsson U, Agestam E, Ekö P-M, Elfving B, Fahlvik N, Johansson U, et al. Thinning of Scots pine and Norway spruce monocultures in Sweden. 2010.

44. Ratkowsky DA, Giles DE. Handbook of nonlinear regression models. New York: Marcel Dekker; 1990.

45. Richards F. A flexible growth function for empirical use. J Exp Bot. 1959;10:290–301.

46. Winsor CP. The Gompertz curve as a growth curve. Proc Natl Acad Sci U S A. 1932;18:1.

47. Wykoff WR, Crookston NL, Stage AR. User's guide to the stand prognosis model. In: US Department of Agriculture, Forest Service, Intermountain Forest and Range Experiment Station. 1982.

48. Yang RC, Kozak A, Smith JHG. The potential of Weibull-type functions as flexible growth curves. Can J Forest Res. 1978;8:424–31.

49. Axelsson P. Processing of laser scanner data—algorithms and applications. ISPRS J Photogramm Remote Sens. 1999;54:138–47.

50. Axelsson P. DEM generation from laser scanner data using adaptive TIN models. Int Arch Photo Remote Sensing. 2000;33:111–8.

51. Næsset E. Practical large-scale forest stand inventory using a small-footprint airborne scanning laser. Scand J For Res. 2004;19:164–79.

52. Hudak AT, Crookston NL, Evans JS, Falkowski MJ, Smith AM, Gessler PE, et al. Regression modeling and mapping of coniferous forest basal area and tree density from discrete-return lidar and multispectral satellite data. Can J Remote Sens. 2006;32:126–38.

53. Team RC: R: a language and environment for statistical computing. 2013. R Foundation for Statistical Computing, Vienna, Austria. In.: ISBN 3-900051-07-0; 2013.

54. Schwarz G. Estimating the dimension of a model. Ann Stat. 1978;6(2):461–4.

55. Fox J, Weisberg S: An R companion to applied regression. United Kingdom: Sage; 2011.

56. Goldberger AS: The interpretation and estimation of Cobb-Douglas functions. Econometrica: J Econc Soci. 1968:464–472

57. Yoo S, Im J, Wagner JE. Variable selection for hedonic model using machine learning approaches: A case study in Onondaga County, NY. Landsc Urban Plan. 2012;107:293–306.

58. Pinheiro J, Bates D, DebRoy SS, Sarkar D: D., and the R Development Core Team 2013. nlme: Linear and Nonlinear Mixed Effects Models. R package version:3.1-103.

59. McRoberts RE, Næsset E, Gobakken T. Inference for lidar-assisted estimation of forest growing stock volume. Remote Sens Environ. 2013;128:268–75.

60. Sarndal C-E, Swensson B, Wretman J: Model assisted survey sampling. New York: Springer-Verlag; 1992.

## 11

Forest biomass change estimated from height change in interferometric SAR height models

Svein Solberg[1*], Erik Næsset[2], Terje Gobakken[2] and Ole-Martin Bollandsås[2]

Abstract

Background: There is a need for new satellite remote sensing methods for monitoring tropical forest carbon stocks. Advanced RADAR instruments on board satellites can contribute with novel methods. RADARs can see through clouds, and furthermore, by applying stereo RADAR imaging we can measure forest height and its changes. Such height changes are related to carbon stock changes in the biomass. We here apply data from the current Tandem-X satellite mission, where two RADAR equipped satellites go in close formation providing stereo imaging. We combine that with similar data acquired with one of the space shuttles in the year 2000, i.e. the so-called SRTM mission. We derive height information from a RADAR image pair using a method called interferometry.

Results: We demonstrate an approach for REDD based on interferometry data from a boreal forest in Norway. We fitted a model to the data where above-ground biomass in the forest increases with 15 t/ha for every m increase of the height of the RADAR echo. When the RADAR echo is at the ground the estimated biomass is zero, and when it is 20 m above the ground the estimated above-ground biomass is 300 t/ha. Using this model we obtained fairly accurate estimates of biomass changes from 2000 to 2011. For 200 m² plots we obtained an accuracy of 65 t/ha, which corresponds to 50% of the mean above-ground biomass value. We also demonstrate that this method can be applied without having accurate terrain heights and without having former in-situ biomass data, both of which are generally lacking in tropical countries. The gain in accuracy was marginal when we included such data in the estimation. Finally, we demonstrate that logging and other biomass changes can be accurately mapped. A biomass change map based on interferometry corresponded well to a very accurate map derived from repeated scanning with airborne laser.

Conclusions: Satellite based, stereo imaging with advanced RADAR instruments appears to be a promising method for REDD. Interferometric processing of the RADAR data provides maps of forest height changes from which we can estimate temporal changes in biomass and carbon.

Keywords: Forest monitoring; Biomass; Carbon; InSAR

Background

Management of forest carbon (C) stocks is increasingly addressed due to its impact on the global greenhouse gas cycle and climate. Deforestation contributes to a significant fraction of the total anthropogenic C emissions [1,2]. The C loss from land use change, i.e. mainly deforestation, is currently about 900 Mt/yr [2]. C loss from forest damage and mortality is negligible in comparison, estimated to be only 1.9 Mt/yr for disturbances in the Amazon [3], and only 13.5 Mt/yr in the extremely extensive mountain pine beetle epidemic in British Columbia [4]. The suite of methods for mapping, monitoring and estimating parts of the forest C cycle is expanding rapidly, including field inventory, modelling and remote sensing. Field inventory is widely used for national forest inventories, and its feasibility for monitoring C stocks in forests such as the Niassa National Reserve miombo woodland in Mozambique has recently been demonstrated [5]. Remote sensing has a wide range of applications. For example, time-series of vegetation indices from MODIS has been used to estimate carbon fluxes in Alaskan ecosystems during 2000-2010 [6] and sustainable amounts of wood harvesting in Southeast Asia [7]. Models such as the Forest Vegetation Simulator

* Correspondence: sos@skogoglandskap.no
[1]Norwegian Forest and Landscape Institute, P.O. Box 115, 1431 Ås, Norway
Full list of author information is available at the end of the article

(FVS) can be used to compare effects of forest management alternatives, e.g. tree species selection, on future C sequestration [8].

Deforestation in the tropics is of particular significance due to its rapid speed [9], and the REDD (Reducing Emissions from Deforestation and Forest Degradation in Developing Countries) initiative aims at reducing the C losses through performance-based credits by comparison of performance against a business-as-usual reference emission level. In addition to deforestation, forest degradation and enhancement of carbon stock through forest growth are other components of the forest C changes that should influence the REDD credits. In order to realize this payment-for-ecosystem-service, the tropical countries need to document their annual changes in forest C stocks, and satellite remote sensing is likely to be a major data provider for this [10,11]. Such data would also enable detection of logging areas for possible counteractions. However, there is a need for new satellite remote sensing methods for REDD. Firstly, there is a need for methods that can overcome the limitations of today's methods, i.e. clouds, small areal coverage and failure to detect C stock changes other than deforestation. Secondly, there is a need for historical data on forest changes for the business-as-usual emission level.

Optical satellite data is the dominating remote sensing method today, having the limitation that the correlation with biomass is weak and tend to saturate at low levels. The PRODES project in Brazil represents state-of-the-art [12]. Annual, full-coverage of Brazil is obtained with about 233 Landsat images, from which clear-cuts are detected from a semi-automatic pixel-unmixing classification based on soil and shadow fractions. Although this is carried out in the most cloud-free season of August-September, clouds are preventing data in some areas. Persistent cloud cover is common in tropical forest areas [13]. A second limitation with the method is that it merely tracks land cover changes such as changes from forest to non-forest and vice versa that can be detected, while forest degradation is hardly detectable, and they make up a considerable share of cuttings in the tropics. Finally, the conversion of the annual clear-cut area into changes in forest C stocks is crudely obtained by using fixed emission factors. The recently upgraded Global Forest Watch [14] is a new and valuable forest monitoring tool; however, it is also largely based on Landsat [15] and is apparently having the same abilities and limitations as the PRODES system.

Remote sensing methods that provide 3D data have a considerable advantage in comparison with 2D, optical data. They provide measurements of forest height, and its changes, which is crucial for forest biomass changes. Airborne 3D remote sensing, i.e. airborne laser scanning (ALS) or stereo photogrammetry, is a more accurate tool

than optical satellite data and it can detect also forest degradation and growth [16]. For most countries, applications with complete ALS coverage are cost-prohibitive. The feasible application of ALS would be strip sampling, which may provide accurate estimates of C stocks changes compared to other methods [17]. In [18], carbon stocks were estimated with 1-ha resolution at the national scale of Panama, based on a sample of field plots and ALS in combination with full areal coverage of Landsat and MODIS data. The accuracy was estimated to 20.5 t/ha of C at the 1-ha pixel level.

The current study is focusing on interferometric SAR (InSAR), which can provide 3D data. Methods based on satellite SAR (Synthetic Aperture RADAR) are getting more attention, and they may resolve the limitations with optical data. SAR is an imaging RADAR system. The cloud problem is non-existent with SAR, because the longer wavelengths, commonly 3-70 cm, penetrate clouds. In Sweden, clear-cuts have been accurately detected as a decrease in the backscatter of ALOS PALSAR [19]. The idea we pursue here is based on changes in surface height obtained from InSAR, where changes in biomass and carbon stocks can be retrieved directly from changes in InSAR height. Logging leads to a reduction in height, while forest growth leads to an increase in height. Heights can be derived from SAR data in two or more ways, mainly by phase differences (interferometric SAR, InSAR) or by parallaxes (radargrammetry) in a SAR image pair. These techniques go back to the 1960s and 1970s, when they were demonstrated with stereo acquisitions in airborne SAR systems [20,21].

With the satellite based InSAR method the heights are derived from phase differences between two SAR images taken from different positions in space. This can provide accurate height measurements; however, the accuracy depends on various acquisition properties. In particular for short wavelength SAR (e.g. X-band) over forested areas the SAR imaging needs to be carried out by two satellites going together in a close formation. This is called a bistatic, or single pass, acquisition, where one satellite is submitting microwave RADAR pulses and both satellites receive the same echoes from Earth's surface. A bi-static acquisition removes temporal de-correlation, i.e. phase noise caused by differences in the position of branches and in moisture. Phase noise is also influenced by the distance between the satellites, i.e. the baseline, which should neither be too large nor too small. With increasing baseline there is an increasing volume de-correlation caused by an increasing difference in the look angle (local incidence angle) into the canopy volume. Contrary to this, when the baseline is very small the noise increases because of quantization errors, i.e. a given height correspond to a tiny fraction of a 2π cycle of phase difference. In order to compress data onboard the satellite the data is typically compressed to 5 or 8 bit [22], and a tiny fraction of such a number will correspond to a crude height measurement. Finally, random errors and phase noise will be relatively large when the backscatter signal is weak, i.e. a low signal-to-noise ratio, depending on incidence angle, polarization and topography [23].

It has recently been demonstrated that forest biomass, or the equivalent stem volume, is strongly related to InSAR height, i.e. the height above ground of the center of the SAR echo [24–28] This relationship may vary with stand structure, in particular tree number density [29,30]. However, existing studies show fairly accurate models without taking forest structure into account, likely because the variation in forest structure is moderate, e.g. [24]. In addition, it has been demonstrated that loggings, i.e. clear-cutting and forest degradation, can be detected as temporal changes in surface height [31,32]. The latter studies detected the logging-induced decreases from the 90 m SRTM C-band DSM to a recent and higher resolution X-band DSM.

One significant step further from detection of logged areas would be to estimate the corresponding changes in above ground biomass (AGB), or C stocks, as surface height changes over time. This should cover not only deforestation, but also forest degradation and forest gain and growth. In addition, the 11 year InSAR height changes from SRTM in 2000 to Tandem-X in 2011 could be used for the business-as-usual reference levels. The C-band SRTM had a near-global coverage and Tandem-X DSM has a global coverage. Hence, these two surface models can be combined and used to detect surface height changes in forests, and possibly estimate changes in forest C stocks. Besides possibly providing estimates in C stocks, this would also provide maps showing the changes, i.e. both decreases from logging and increased from forest growth.

A main question is how accurate C stock, or AGB changes can be estimated by means of InSAR, and in addition, there are several possible limitations, including:

1. The lack of a digital terrain model (DTM) making it impossible to derive AGB and C stocks,
2. The difference in SAR wavelength between the C-band SRTM and the X-band Tandem-X,
3. The coarse resolution of the C-band SRTM (90 m),
4. The lack of AGB data from field from the time of the SRTM acquisition, and
5. Unstable relationships between AGB and InSAR height.

Crucial here is that it is the changes in C stocks we are after, and not the C stocks themselves. Hence, the lack of a DTM could be overcome if AGB and InSAR height are proportional, i.e. a straight linear relationship

without saturation going through origin. This has been found to be the case for a tropical forest in Brazil [33] and for a spruce forest in Norway [24]. Also, without a DTM the relationship between AGB and InSAR height needs to be available from an external model. In the present study an external model was fit by processing the Tandem-X data against a DTM and estimating the relationship between AGB and InSAR height for a number of field plots.

The main objective of this study was to estimate how accurately AGB changes (ΔAGB) over time can be estimated from DSM changes between the SRTM C-band DSM acquired in 2000 and a Tandem-X DSM from 2011. The specific aims were to estimate

1) bias, accuracy and precision for ΔAGB on 200 m^2 plots, and
2) the accuracy in the spatial variation in ΔAGB over the study area.

In addition, we aimed at quantifying how much the performance of this method is reduced

3) because different SAR wavelengths are used at the two points of time (C-band in 2000 and X-band in 2011), and
4) if plot biomass values were not available to fit models at the first point of time (year 2000) and if a DTM was not available (which would be the general case for most REDD countries).

We had at hand a unique data set for this study, comprising repeated field inventory, repeated airborne laser scanning, and full coverage X- and C-band SRTM and Tandem-X data.

Results and discussion
AGB – InSAR height relationship

The relationship between AGB and Tandem-X InSAR height could be represented as a proportionality, where AGB inceased by 14.9 t/ha per m increase in InSAR height. We first fitted an ordinary linear regression model, which provided an estimated intercept of 16 t/ha and a slope of 13.4 t/ha/m, and had a RMSE of 54.3 t/ha. The intercept was negligible and not statistically significant from zero. We re-fitted the model without an intercept, as shown in Figure 1. The exclusion of the intercept had nearly no effect on the model performance, i.e. the RMSE showed almost no increase.

A positive correlation between biomass and InSAR height is as expected, because they both increase with tree height and stand density. In a theoretical approach [29] demonstrated that the relationship between stem

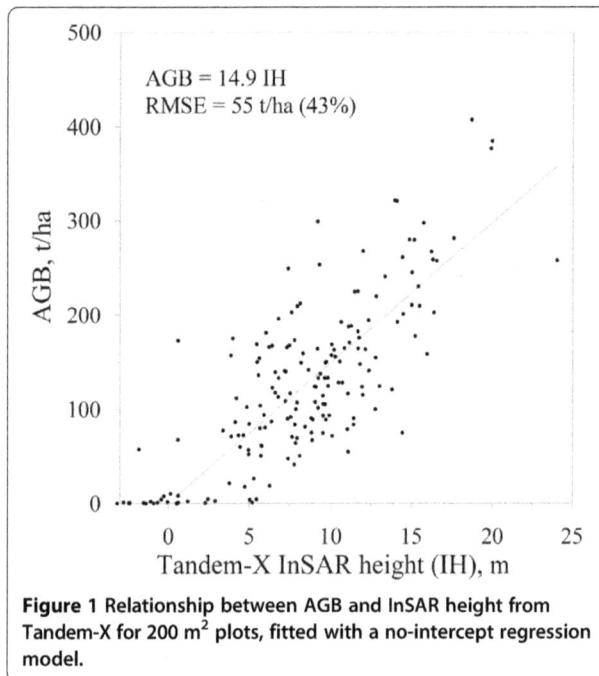

Figure 1 Relationship between AGB and InSAR height from Tandem-X for 200 m^2 plots, fitted with a no-intercept regression model.

volume, or correspondingly biomass (B), and height (H) follows a power law function:

[7] $B \propto H^a$.

If the height variable, H, represents mean tree height or top height the α takes a value in the range 1.5-2.0 [29,30,34], which implies a curvilinear relationship with a saturation effect for high biomass values. In the present study the height, H, is not representing tree height, but rather canopy height. InSAR height, as any type of canopy height variables, depends not only on tree height, but also on the shape of each tree crown and on stand density (the amount of gaps). Proportionality between biomass and InSAR height, or at least α values close to unity, has been demonstrated both in boreal and tropical forests [24,25,33].

A few outliers having zero InSAR height and AGB > 50 t/ha might be observations where logging has occurred between the field inventory in 2010 and the Tandem-X acquisition in 2011. Stand structure might affect the relationship between biomass and InSAR height [29], and further studies may reveal whether this is the case. It is unlikely that more accurate biomass estimates could be obtained for these plots being as small as 200 m^2, taking into account what is achieved with other remote sensing methods including airborne laser scanning [35]. Hence, the potential to improve the accuracy taking into account stand structure seems limited.

Some plots had negative InSAR heights, down to -3 m. This is attributable to various types of errors in the

Tandem-X DSM. These error types include remaining bias and ramp errors over the entire study area, phase discontinuities in steep topography such as forest stand edges, and remaining residual errors not removed by the multilooking and Goldstein filtering (see below).

We add here that in a real case in a tropical country without a full coverage DTM the model would have to be obtained in another way than here, e.g. from plots in some smaller study area with an accurate DTM or from scattered field plots in a sample survey having accurate terrain heights from GPS measurements on the plots.

Accuracy in estimated AGB changes at plot level

The obtained model (Figure 1) was used to predict AGB changes on the field plots from 2000 to 2011 directly from DSM changes. These predictions were clearly correlated to field measured AGB changes with r = 0.69 (Figure 2). The RMSE of these predictions was 67 t/ha, which was only slightly higher than the obtained accuracy, 59 t/ha, in the model in Figure 1. The negative changes seen in Figure 2 are mostly large and represent logging, while the positive changes are smaller and results from forest growth.

The influence of SRTM band and separate models at two points of time

The relationship between AGB and InSAR height was fairly similar with Tandem-X in 2011 and the X- and C-band SRTM in 2000 (Table 1). After correcting vertical offset and ramp errors in the SRTM DSMs based on

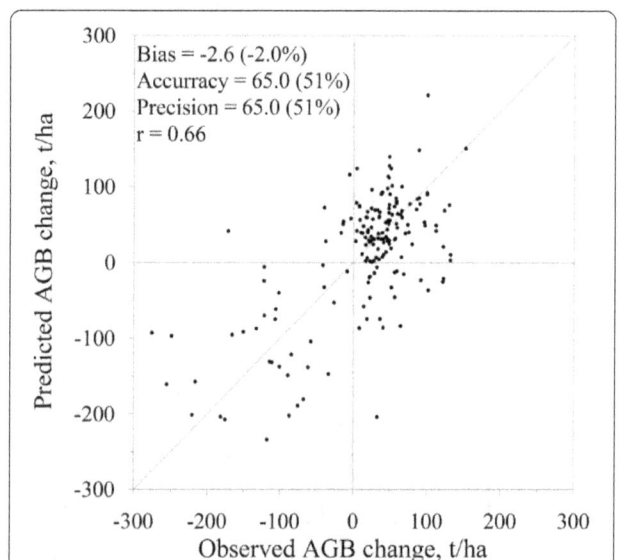

Figure 2 Predicted AGB changes for 200 m^2 plots based on InSAR height changes from the SRTM C-band DSM and a Tandem-X DSM, and the model AGB = 14.9IH, where IH is InSAR height above ground. Bias, accuracy and precision are given in t/ha as well as in % of mean biomass in 2010 (127 t/ha).

Table 1 Separate models for each year and each band: Estimating the relationship between AGB and the three alternative InSAR heights, using no-intercept regression models

InSAR data	Year	Slope, t/ha/m	RMSE, t/ha
SRTM-C	2000	13.6	59.9
SRTM-X	2000	15.1	53.9
Tandem-X	2011	14.9	54.7

ground control points, and subtracting the DTM, we obtained no-intercept models having slopes of 13.6 and 15.1, respectively for the C- and X-band SRTM. A lower slope value for C-band than for X-band was contrary to what we expected. The penetration of the C-band microwaves, having slightly longer wavelength, should be some 2–3 m deeper into the canopy than the X-band [36]. This should produce lower height values for C-band as compared to X-band in forested areas. A given biomass value would correspond to a lower C-band than X-band height, and hence a higher biomass-to-height ratio for C-band, which is contrary to what we found here. This unexpected finding must be attributed to random errors in the modelling, e.g. random errors in the GCP points used. The lower spatial resolution and lack of details in the C-band DSM increase the probability of errors in the parameter estimate. The accuracy was similar for the three InSAR data sets, varying from 53.9 to 59.9 t/ha (Table 1).

By applying these models we obtained AGB changes in alternative ways. The real AGB changes as obtained from repeated field inventory had a mean value of 14.7 t/ha, while the predicted values varied from 12.1 to 21.9 t/ha (Table 2). The major finding here is that the apparent ideal case based on a 30 m resolution X-band DSM in the year 2000, having a DTM, and separately calibrated AGB - InSAR models at both points of time (alternative 4) was only marginally better than the crude case with the 90 m C-band DSM and using only one AGB – InSAR model from Tandem-X data (alternative 1). The bias was of the same magnitude, a few t/ha, although with opposite signs, and the accuracy and correlation was only slightly better with alternative 4. The two alternatives in between, i.e. alternatives 2 and 3, had bias and accuracy values of fairly the same magnitude. Hence,

if forest C changes since 2000 are to be estimated with InSAR data, this could be based on the crude alternative 1, which would be the only feasible alternative in many tropical countries. One explanation for bias is the lack of complete synchronicity of the field and the InSAR data, i. e. 8 month difference at the beginning of the period and one year at the end. Real AGB changes due to logging and forest growth may explain some of the bias. For example, if one of the field plots had been logged between the last ALS acquisition and the Tandem-X acquisition, and the InSAR height decreased from 20 to 0 m above ground on this plot, this would generate a bias of -0.11 m when averaged over the 176 field plots. In addition, the results are sensitive to the GCPs used. In this study we placed GCPs subjectively taking into account the differential interferograms. Poorly placed GCPs could generate a bias.

A particular feature of using surface models for C change monitoring is that it is sensitive to height biases. A tiny bias on height changes over large areas would translate into a considerable bias on the estimated REDD credit. There was a negative bias of 2.6 t/ha on the predicted AGB changes with alternative 1 (Table 2). The predicted mean change on the field plots was 12.1 t/ha while the field measured value was 14.7 t/ha. Although this makes up only 2% of the mean standing AGB in 2010, it represents a 18% underestimation of the change in AGB and the above-ground C stock. In a performance based payment system this would correspond to an underestimated REDD-credit of 2,400 US$ per km^2, if we assume that C makes up 50% of AGB and that the payment is 5 US$ per ton CO_2. Based on the 176 field plots the real REDD-credit should have been about 14,000 US$ per km^2, while the estimated REDD-credit would be about 11,000 US$ per km^2. This bias corresponds to a -0.17 m bias on the change in InSAR surface height, and this illustrates that a minor bias on the derived DSMs and their changes over time can generate a considerable error if the bias occurs at large scale. However, it is likely that an error like this would vary over the area of interest, i.e. that there would be both negative and positive biases varying smoothly over the landscape due to errors varying from SAR image to SAR image.

Table 2 Comparison of performance of model alternatives for ΔAGB (t/ha)

Alt	Description	ΔAGB	Bias	Accuracy	Precision	r
1	90 m C-band SRTM as reference DSM, AGB model calibrated only in 2010	12.1	−2.6	65	65	0.66
2	30 m X-band SRTM as reference DSM, AGB model calibrated only in 2010	20.4	5.7	59	59	0.73
3	90 m C-band SRTM as reference DSM, AGB calibrated with separate models in 2000 and 2010	21.9	7.2	63	63	0.67
4	30 m X-band SRTM as reference DSM, AGB calibrated with separate models in 2000 and 2010	19.0	4.3	59	59	0.73

Field measured ΔAGB was used as the reference with a mean value of 14.7 t/ha.

Accuracy in spatial variation of AGB changes

There was clearly a correspondence in the spatial pattern of AGB changes (Figure 3). Clear-cuts with reduction in AGB and C stocks are visible as red areas, while forest growth representing increases in AGB and C is visible in green. As expected, the ALS data apparently provided more accurate data by having sharp edges on clear-cuts, while the InSAR-based changes were smoother. The changes from X-band SRTM had more distinct change features than those from the more coarse resolution C-band DSM. There were moderately strong correlations between the AGB changes obtained from ALS and those obtained from InSAR. It is notable that the C-band SRTM performed almost equally well as the X-band SRTM. The correlation coefficient was $r = 0.62$ with the X-band and $r = 0.65$ with the C-band SRTM (Table 2). ALS is here used as reference data, because it was not feasible to get field measurements covering the entire study area. The real accuracy of the InSAR based changes is unknown. If we assume that the errors on ΔAGB from ALS and errors on ΔAGB from InSAR are uncorrelated, then the spatial variation of InSAR-based changes are more accurate than these results indicate.

It is likely that some loggings occurred between the first ALS acquisition on 8th and 9th June 1999 and the SRTM acquisition in February 2000 (8 months), and between the last ALS acquisition on 2nd July 2010 and the Tandem-X acquisition on 14th July 2011 (1 year). This would include both thinning and clear-felling, and would generate errors in the relationships. The real relationships between changes in biomass and changes in InSAR height are likely to be more accurate than what was obtained here. We had no logging records available and could not quantify the effect of the slight mismatch of the timing of the data acquisitions.

For the entire study area the mean height change for all pixels from C-band SRTM to Tandem-X was a decrease ΔDSM of 0.046 m which corresponded to a predicted mean decrease of ΔAGB = 0.71 t/ha. Estimating the accuracy of this estimate was not within the scope of the present study, and would require further and careful analyses.

We see the presented method as a possible approach to provide MRV data for REDD. First, it can provide complete coverage. This is not a requirement in REDD; however it would enable statistics for various spatial scales. Both the national and sub-national level is defined in REDD, in addition to project areas [37]. This means that an entire country can participate in REDD and receive a REDD credit. However, also a part of a country can be an entity in REDD. This can be an administrative unit such as a state within a federation, it can be a forest type region such as a mangrove forest area, or it can be an area covered by a project addressing forest C stocks. In addition, a given country may want to distribute its REDD credit internally, i.e. setting up a performance based payment system within their country.

Secondly, the method were here tested in a Norwegian forest; however, we believe that the method would be appropriate also in a tropical forest. The two requirements for using the method are that the relationship between AGB and InSAR height (i) is straight linear, and (ii) is stable over time. A straight linear relationship has also been found in a virgin tropical forest in Brazil by [33]. They used the airborne OrbiSAR system, which provided X- and P-band InSAR data. The InSAR height for the forest was derived as the difference between the X- and P-band DSMs. The relationship had a slope (proportionality) of about 13.5 t/ha, which is close to what we have obtained for spruce forests in Norway [24,25,38]. However, the straightness of the relationship might vary with forest type. In a regular Eucalyptus plantation in Brazil the relationship was found to be curvilinear [39]. Secondly, the stability of the relationship across weather conditions is also promising. It has been shown in a study in Siberia that frost lowered the L-band InSAR height with 4 m [40]. In a Finnish study by [41], the X-band InSAR height in a pine forest decreased during the autumn, however, that was attributed to needle

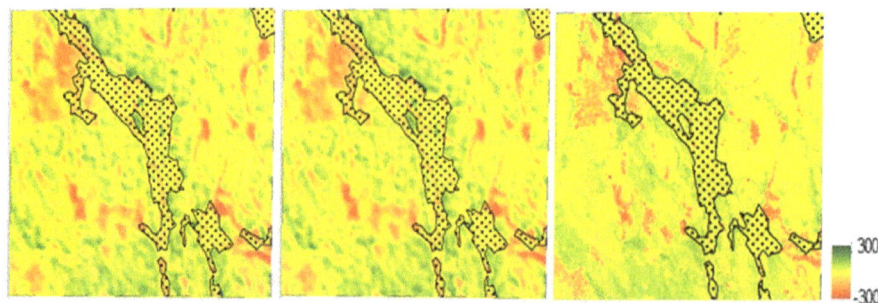

Figure 3 Predicted AGB change (t/ha) over the study area based on height changes from C-band SRTM to Tandem-X and the model AGB = 14.9IH (left), based on SRTM X-band (middle), and based on repeated ALS acquisitions (right). Agricultural fields are outlined and marked with black dots.

fall. In a study in Norway we found a considerable decrease in InSAR height for frozen conditions, as compared to non-frozen. However, we found little or no difference between acquisitions during unfrozen conditions spring-summer-autumn [42].

Thirdly, the data requirements should be feasible. Tandem-X has covered the Earth already two times or more, i.e. in 2011 and 2012, and apparently the two satellites will continue to go in Tandem-X formation for some more years with the remaining fuel, which will enable systematic cover of tropical countries in some more years. At the end of the Tandem-X lifetime, a continuation might be covered by best-available-alternatives. BIOMASS is a tailored mission for forest biomass probably operational from 2021 or 2022 [43]. In the meantime we would have to use the best available technology. This could be X-band radargrammetry which can provide almost equally accurate biomass estimates as Tandem-X [44] or optical stereo imaging. Possible new SAR missions in the next years include SAOCOM-CS and Tandem-L. In addition, the SRTM DSM would be invaluable for deriving data on business-as-usual, from 2000 up to Tandem-X in 2011. Field inventory would be necessary to calibrate the relationship between biomass and InSAR height. Such relationships need to be calibrated for various forest types. As indicated by the present study, it is apparently sufficient to do this calibration at one point of time.

In this study we are addressing above-ground biomass only, while an MRV system would require monitoring also of below ground biomass, dead wood, litter and soil. Hence, our proposed InSAR method doesn't solve all the needs for an MRV system; however, it is a possibly important step in the right direction. Below-ground biomass C can be estimated as it normally makes up an amount proportional to above ground biomass C. Dead wood, litter and soil C are more difficult to estimate, and sophisticated models such as the Yasso model [45] seems to be the only feasible way to derive litter and soil C.

Conclusions

In conclusion, AGB changes could be estimated fairly accurately from the DSM changes between the SRTM C-band DSM in 2000 and a Tandem-X DSM in 2011. Estimated AGB changes for 200 m^2 plots (and for 10 m × 10 m pixels) were close to unbiased, with accuracy and precision around 50%. The derived ΔAGB values from InSAR varied consistently with those obtained from ALS over the study area. The performance of the method was slightly better with X-band SRTM data than with C-band. The accuracy was negligibly improved in the case where separate AGB – InSAR height models were fitted

for both points of time, which required AGB data for both points of time as well as an accurate DTM.

Methods
Study area
The study area was covered by field inventory, ALS and InSAR data at two points of time, i.e. approximately in 2000 and 2011. It was located in a forest area in the municipality of Våler (59°30'N 10°55'E, 70–120 m above sea level), southeast Norway, of about 852.6 ha. The main tree species was Norway spruce (*Picea abies* (L.) Karst.). Scots pine (*Pinus sylvestris* L.) was also quite frequent and there were some scattered broadleaves, in particular birch (*Betula pubescens* Ehrh.). The forest has been actively managed for timber production, including commercial thinning, clear-felling followed by planting in the spruce dominated stands, and selective logging followed by natural regeneration in the pine dominated stands. The study area including the field and ALS data have earlier been used and described in detail by [46] and [47].

Field plot data
We established 176 circular 200 m^2 plots for field measurements of above ground biomass. The study area was stratified into four predefined forest types; (1) recently regenerated forest (age ≥ 20 years), (2) young forest, (3) spruce dominated mature forest and (4) pine dominated mature forest. This was done based on an existing stand map. We laid out plots in systematic grids in each stratum. Each plot center was initially located in the field using 1:5000 topographic maps, and their positions were later determined accurately using Differential Global Positioning System (GPS) and Global Navigation Satellite System (GLONASS) having an accuracy of < 0.5 m [46]. The field inventory was carried out in the summers of 1998 and 1999, and then redone in the fall of 2010 and spring of 2011. The field inventory comprised recording of tree species, callipering of all trees with diameter at breast height (dbh) ≥4 cm, and height measurements with a Vertex hypsometer on sample trees selected with a probability proportional to stem basal area. The number of trees with height measurements ranged from 3 to 43 per plot with an average of 18. Non-measured tree heights were later obtained from species specific diameter-height relationships.

AGB was estimated as the sum of the individual components stump, stem, bark, dead and living branches and foliage of individual trees predicted using previously fitted species-specific allometric models having tree species, *dbh* and height as input variables [48]. AGB was estimated for both points of time, from which we also obtained AGB changes at the plot level (Table 3).

Table 3 Above-ground biomass on the plots (t/ha)

	Mean	min - max
AGB 1999	112	2 - 349
AGB 2010	127	0 – 407
ΔAGB	15	−275 - 153

ALS data

The study area was covered by complete cover ALS data acquired under leaf-on conditions on 8th - 9th June 1999 and 2nd July 2010 with a Piper PA-31-310 Navajo aircraft at a speed of 70 – 80 m/s at an elevation of 700 – 900 m above the ground (Table 4).

A DTM was derived from the ALS acquisitions, and the heights of the echoes were recalculated to heights above the ground.

Model-predicted AGB changes from the repeated ALS acquisitions were used as a reference representing the true spatial variation of ΔAGB. These ALS-based predictions were produced with separate models for six strata (Table 5). The strata mainly represent the age and the dominating tree species, while one stratum contains stands having thinning during the study period. For each plot we extracted height distribution metrics of the two main ALS echo types (first and last), including height percentiles, cumulative canopy density above certain thresholds, coefficient of variation and mean. We did this for both the 1999 and the 2010 laser scans, and extracted the temporal changes in these metrics. Simple linear models for predicting ΔAGB were developed by regressing field measured ΔAGB against these ALS height metric changes. We applied a stepwise regression model with forward selection of explanatory variables.

The number of selected metrics varied from one to four. As an example, the model for change in above ground biomass in young stands, stratum 4, was -56.3 + 32.2 $\delta meanf$, where $\delta meanf$ was the change in the mean height of the first echoes. We used these regression models to predict ΔAGB for each 10 m × 10 m pixel over the entire study area. For land cover types other than forest land we applied the model for stratum 1, clear-cut. In order to compare the predicted changes

Table 5 ΔAGB models based on airborne laser scanning

Stratum	Model[1]	RMSE, t/ha
1. Clear-cut	$\Delta AGB = 1.27 + 385\ \delta dl5$	37.1
2. Thinned	$\Delta AGB = -78.1 + 14.9\ \delta pf0 + 26.1\ \delta pf20 + 3.65\ \delta cvf -16.3\ \delta pl0$	13.3
3. Recently regenerated forest	$\Delta AGB = 2.41 + 17.9\ \delta pf20$	26.3
4. Young forest	$\Delta AGB = -56.3 + 32.2\ \delta meanf$	36.5
5. Spruce dominated mature forest	$\Delta AGB = 0.879 - 233\ \delta df3 + 6.53\ \delta pl10 + 13.7\ \delta pl90$	26.4
6. Pine dominated mature forest	$\Delta AGB = 8.28 + 15.1\ \delta pf50 -89.1*\delta df8-7.53\ \delta pl60 + 8.67\ \delta pl80$	22.2

[1]p = height percentile of vegetation echoes (0, 10,…, 90); d = cumulative canopy density above vegetation threshold (0, 1, …, 9); cv = coefficient of variation of height of vegetation echoes; mean = arithmetic mean of height of vegetation echoes; f = first echo; l = last echo.

from InSAR and from ALS we selected 784 pixels as a systematic grid over the area.

InSAR data

For calibrating the model of AGB against InSAR height we processed the Tandem-X data against the DTM from ALS, while for obtaining the temporal changes in AGB from temporal changes in surface height we processed them against the SRTM DSMs (Figure 4).

Temporal changes in AGB were to be detected as changes from the SRTM DSMs to a Tandem-X DSM, corresponding to changes in canopy height.

We used both the X- and C-band SRTM data, which were acquired during 12th – 20th February 2000. Both the X- and C-band SRTM data have full areal coverage in the study area. We obtained these data as DSMs, i.e. the X-band DSM from the German Aerospace Centre (DLR) in 2002, and the C-band DSM was derived from a procedure built into the ENVI/Sarscape 5.0 software, which downloads data as SRTM-3 version 4 (http://srtm.csi.cgiar.org/index.asp). The data were received in geographic (lat/lon) projection with a spatial resolution of 1 arc sec (15 m x 31 m) for the X-band and 3 arc sec (46 m x 93 m) for the C-band, which we resampled with bilinear interpolation to UTM32 and 10 m x 10 m pixels.

Table 4 Key parameters for the airborne laser scanning campaigns

Parameter	1999	2010
Instrument	Optech ALTM 1210	Optech ALTM Gemini
Pulse repetition frequency	10 kHz	100 kHz
Scan frequency	21 Hz	55 Hz
Scan half-angle (after processing)	14.0°	13.8°
Pulse density on ground	1.2 m^{-2}	7.3 m^{-2}

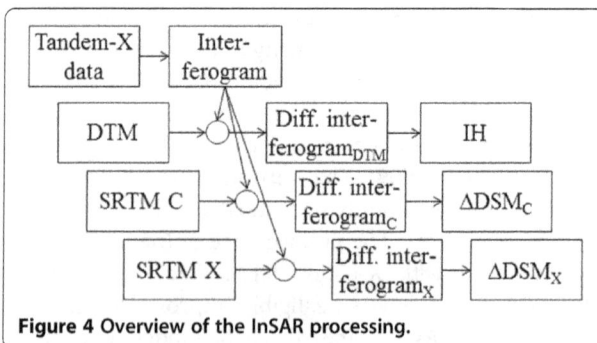

Figure 4 Overview of the InSAR processing.

Table 6 Technical properties of the Tandem-X InSAR data acquisition, incidence angle θ$_i$, normal baseline B⊥, and height of ambiguity HoA

Date	Time	Orbit	Polarization	θ$_i$, degrees	B⊥, m	HoA, m
14th July 2011	05:32	Descending	HH-HH	46	55	147

The Tandem-X data were from an ascending, right looking TanDEM-X stripmap image pair acquired in the morning on 14th July 2011 (Table 6). The data were received from DLR as co-registered Single Look Complex (SLC) data in CoSSC format. The basic idea with InSAR processing is to extract the phase component from the complex data in each of the two SAR images, i.e. to derive the phase difference for each pixel in the form of an interferogram, from which topographic information can be extracted. The InSAR processing was carried out with the ENVI/Sarscape software. The processing was done three times, i.e. separately with the X-band SRTM DSM as a reference, with the C-band SRTM DSM as a reference, and with the ALS DTM as a reference. In each the interferogram was processed into a differential interferogram, being the part of the phase differences in the Tandem-X image pair which represented either the 11 year changes in canopy height (from SRTM) or the canopy height (from DTM). We carried out phase unwrapping using the Minimum Cost Flow method, and converted the unwrapped phases into elevation data and transformed from satellite slant-range geometry to a geocoded DSM. During the processing we used a multilooking of 5 azimuth x 5 range, which corresponded to about 10 m x 10 m, which was also the spatial resolution of the geocoded DSMs obtained by bilinear interpolation (Figure 5).

It is important for the accuracy of this method to minimize height errors. In a DSM there is typically one height error which varies from pixel to pixel as random noise. In the processing we applied the Goldstein filter [49] to reduce this noise. In the Tandem-X World-DEM^{TM} specification this error is quantified as relative vertical accuracy of < 2 m or < 4 m depending on the slope. The second type of error is quantified as absolute

vertical accuracy and should be < 10 m. This is a type of error that varies gradually over larger distances, and in the present study area of limited size such an error might appear as a bias or a ramp. In the processing we have largely removed bias and ramp errors by using ground control points (GCPs). We selected points manually in the data where two DSMs were expected to have identical values, i.e. where they could be tied together. They were carefully selected in order to be useful for all DSM corrections, and should represent locations without forests (zero canopy height) and without any temporal changes. This was accomplished by taking into account three differential interferograms, i.e. from the InSAR processing of Tandem-X against the DTM; against the C-band SRTM DSM; and against the X-band SRTM DSM. They were placed in sites having a low fringe density, and where the phase values clearly indicated no canopy height and no temporal change. We fitted Equation (1) to these GCPs:

$$\Delta\phi = k_0 + k_1 \, \text{RG} + k_2 \, \text{AZ} \qquad (1)$$

where $\Delta\phi$ was the phase difference at each GCP, k_0, k_1 and k_2 were correction factors, and RG and AZ were the range and azimuth co-ordinates (Table 7). After correcting the differential interferogram with these factors the RMSE (root mean square error) of the GCP heights was in the range of 1.3 – 1.5 m. From these corrections we also derived ramp corrections for the SRTM DSMs, which enabled us to extract regression models of AGB against InSAR height for the two SRTM DSMs.

Weather conditions influence the penetration of SAR microwaves into the vegetation. In dry or frozen conditions the penetration is deeper than in moist conditions because the amount of liquid water in the canopy is low,

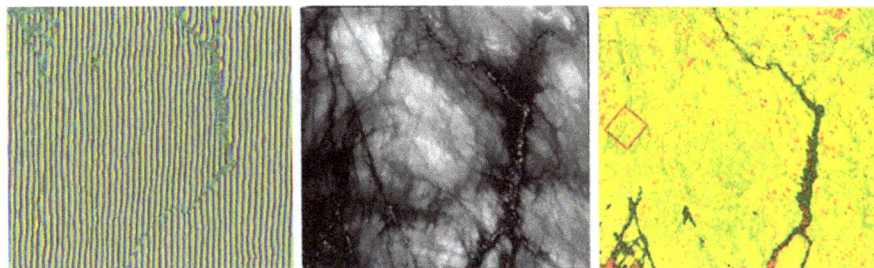

Figure 5 Processing from interferogram (left) into DSM (middle) and height change from SRTM C-band to Tandem-X (right). Clear-cuts during the 11 years are clearly visible as red areas. The SAR image covers about 35 km x 35 km. The study area is indicated as a red rectangle.

Table 7 Correction factors for phase offset and phase ramp errors (radians), see Equation (1), and the final accuracy (RMSE) for the 39 Ground Control Points (GCP)

Reference DSM	k_0	k_1	k_2	RMSE
ALS DTM	0.1603781470	−0.0000644062	0.0000598075	1.54 m
SRTM-C	0.1027309745	−0.0000652833	0.0000286684	1.29 m
SRTM-X	0.2173415607	−0.0000852327	0.0000247047	1.27 m

and the dielectric constant is low [50,51]. According to this we might expect an InSAR DSM to vary between the acquisitions. Apparently, this has played a minor role in this case, because the obtained relationship between AGB and InSAR height was very stable from the February SRTM acquisitions to the July Tandem-X acquisition. It is possible that this effect is minor in general, or that the dielectric properties were similar although the SRTM was in the winter while the Tandem-X was in the summer. The weather in the study area during 12th – 20th February 2000 was unstable and varied with temperatures around and slightly above zero, and with some precipitation coming as moist snow and sleet. For the Tandem-X acquisition the mean temperature was 14°C and no precipitation from 08.00 13th July – 08.00 14th July.

Analyses

We derived the relationship between AGB and InSAR height from field inventory on 200 m^2 plots in fall 2010 and spring 2011, each linked to one 10 m x 10 m InSAR height pixel from the combination of a Tandem-X DSM from July 2011 and a DTM from ALS. This model was based on all field plots without taking tree species into account, because the field plots were dominated by spruce and in most cases contained a mixture of species. Spruce was present in 90% of the plots, and made up 59% of AGB on the plots. We could have estimated species specific models, and they would likely be different [25]. However, the number of plots for each model would then have been small and increasing the random error of the parameter estimates. Similarly, for the temporal changes we fitted a model ΔAGB from field inventory against height changes of the surface models, i.e. ΔDSM. We applied the following statistical measures where N is the number of observations, P_i is the predicted value for observation i, and O_i is the observed value for observation i:

$$Bias = N^{-1} \sum_i (P_i - O_i) \tag{2}$$

$$Accurracy = RMSE = \left[N^{-1} \sum_i (P_i - O_i)^2 \right]^{0.5} \tag{3}$$

$$Precision = RMSE_s$$
$$= \left[N^{-1} \sum_i (P_i - O_i - bias)^2 \right]^{0.5} \tag{4}$$

The evaluation of the spatial accuracy on temporal C changes was based on a visual examination of mapped changes and correlation analyses. Estimated AGB changes from the repeated ALS acquisitions were used as a reference representing the "true" spatial variation of ΔAGB. We are aware that also ALS based changes have errors; however, this was the best available data set for spatial variation in changes. We laid out a 10 m x 10 m grid over the study area and predicted AGB changes for each grid cell with InSAR and ALS. The ΔAGB was predicted directly from the DSM change from SRTM to Tandem-X multiplied by the slope of the relationship between AGB and InSAR height. We selected 784 pixels distributed systematically over the area and extracted the predicted AGB changes from ALS and from InSAR, and used Pearson correlation analyses to represent the spatial correspondence.

Abbreviations

AGB: Above-ground biomass; ALS: Airborne laser scanning; C: Carbon; dbh: Diameter at breast height; DSM: Digital surface model; DTM: Digital terrain model; GCP: Ground control point; IH: InSAR height; InSAR: Interferometric, synthetic aperture RADAR; MODIS: Moderate resolution imaging spectroradiometer; REDD: Reduced Emissions from Deforestation and forest Degradation in Developing Countries; RMSE: Root mean square error; SAR: Synthetic aperture RADAR; SRTM: Shuttle RADAR topography mission.

Competing interests

The authors declare that they have no competing interests.

Authors' contributions

All authors have made substantial contributions and have given final approval of the version to be published. Næsset, Gobakken and Bollandsås have contributed with field and ALS data together with variables derived from them and their documentation, while Solberg has carried out the InSAR processing and done the majority of the writing.

Authors' information

All authors have a master degree in forestry, and are working mainly on remote sensing in forestry at a university campus at Ås, Norway. S.S. is a senior researcher working on forest disturbance and remote sensing since 1990 at the Norwegian Forest and Landscape Institute, formerly the Norwegian Forest Research Institute. The three other authors are working closely together mainly on airborne laser scanning in forestry in the forest mensuration group at the Norwegian University of Life Sciences. The group is led by E.N., who is professor and a world leading researcher in the field. T. G. is professor and O.M.B. is a researcher in this group.

Acknowledgements

We acknowledge the German Aerospace Centre (DLR) for providing Tandem-X data through the AO XTI_VEGE0315, and the European Space Agency for providing PRODEX funding.

Author details

[1]Norwegian Forest and Landscape Institute, P.O. Box 115, 1431 Ås, Norway. [2]Norwegian University of Life Sciences, P.O. Box 5003, 1432 Ås, Norway.

References

1. Anon: *Global Forest Watch*. Washington: World Resources Institute; 2014.
2. Askne JIH, Dammert PBG, Ulander LMH, Smith G: **C-band repeat-pass interferometric SAR observations of the forest.** *Ieee Transact Geosci Remote Sensing* 1997, **35**:25–35.
3. Asner GP, Mascaro J, Anderson C, Knapp DE, Martin RE, Kennedy-Bowdoin T, Van Breugel M, Davies S, Hall JS, Muller-Landau HC, Potvin C, Sousa W, Wright J, Bermingham E: **High-fidelity national carbon mapping for resource management and REDD+.** *Carbon Bal Manage* 2013, **8**:7–7.
4. Deutscher J, Perko R, Gutjahr K, Hirschmugl M, Schardt M: **Mapping tropical rainforest canopy disturbances in 3D by COSMO-skymed spotlight InSAR-stereo data to detect areas of forest degradation.** *Remote Sens* 2013, **5**:648–663.
5. Ene LT, Næsset E, Gobakken T, Gregoire TG, Stahl G, Holm S: **A simulation approach for accuracy assessment of two-phase post-stratified estimation in large-area LiDAR biomass surveys.** *Remote Sens Environ* 2013, **133**:210–224.
6. Espírito-Santo FDB, Gloor M, Keller M, Malhi Y, Saatchi S, Nelson B, Junior RCO, Pereira C, Lloyd J, Frolking S, Palace M, Shimabukuro YE, Duarte V, Mendoza AM, López-González G, Baker TR, Feldpausch TR, Brienen RJW, Asner GP, Boyd DS, Phillips OL: **Size and frequency of natural forest disturbances and the Amazon forest carbon balance.** *Nat Commun* 2014, **5**.
7. Galvez FB, Hudak AT, Byrne JC, Crookston NL, Keefe RF: **Using climate-FVS to project landscape-level forest carbon stores for 100 years from field and LiDAR measures of initial conditions.** *Carbon Bal Manage* 2014, **9**:1.
8. Gama FF, Dos Santos JR, Mura JC: **Eucalyptus biomass and volume estimation using interferometric and polarimetric SAR data.** *Remote Sens* 2010, **2**:939–956.
9. Goldstein RM, Werner CL: **Radar interferogram filtering for geophysical applications.** *Geophys Res Lett* 1998, **25**:4035–4038.
10. Graham LC: **Synthetic interferometer radar for topographic mapping.** *Proc IEEE* 1974, **62**:763–768.
11. Hansen MC, Potapov PV, Moore R, Hancher M, Turubanova SA, Tyukavina A, Thau D, Stehman SV, Goetz SJ, Loveland TR, Kommareddy A, Egorov A, Chini L, Justice CO, Townshend JRG: **High-resolution global maps of 21st-century forest cover change.** *Science* 2013, **342**:850–853.
12. Hansen MC, Shimabukuro YE, Potapov P, Pittman K: **Comparing annual MODIS and PRODES forest cover change data for advancing monitoring of Brazilian forest cover.** *Remote Sens Environ* 2008, **112**:3784–3793.
13. IPCC: *Summary for Policymakers in T. F. Stocker, D. Qin, G.-K. Plattner, M. Tignor, S. K. Allen, J. Boschung, A. Nauels, Y. Xia, and V. Bex, editors. Climate Change 2013: The Physical Science Basis*, Contribution of Working Group I to the Fifth Assessment Report of the Intergovernmental Panel on Climate Change. 2013.
14. Krieger G, Fiedler H, Hajnsek I, Eineder M, Werner M, Moreira A, TanDEM-X: **mission concept and performance analysis: Pages 4890–4893 in** *IEEE International Geoscience and Remote Sensing Symposium, IGARSS '05*. Proceedings of IEEE International Geoscience and Remote Sensing Symposium: Seoul; 2005.
15. Krieger G, Moreira A: **Spaceborne Interferometric and Multistatic SAR Systems.** In *Bistatic Radar: Emerging Technology*. Edited by Cherniakov M. West Sussex: Wiley & Sons; 2008:95–158.
16. Kurz WA, Dymond CC, Stinson G, Rampley GJ, Neilson ET, Carroll AL, Ebata T, Safranyik L: **Mountain pine beetle and forest carbon feedback to climate change.** *Nature* 2008, **452**:987–990.
17. La Prade G: **An analytical and experimental study of stereo for radar.** *Photogramm Eng* 1963, **29**:294–300.
18. Le Toan T, Quegan S, Davidson MWJ, Balzter H, Paillou P, Papathanassiou K, Plummer S, Rocca F, Saatchi S, Shugart H, Ulander L: **The BIOMASS mission: mapping global forest biomass to better understand the terrestrial carbon cycle.** *Remote Sens Environ* 2011, **115**:2850–2860.
19. Liski J, Palosuo T, Peltoniemi M, Sievanen R: **Carbon and decomposition model Yasso for forest soils.** *Ecol Model* 2005, **189**:168–182.
20. Lynch J, Maslin M, Balzter H, Sweeting M: **Choose satellites to monitor deforestation.** *Nature* 2013, **496**:293–294.
21. Marklund LG: *Biomassafunktioner för tall, gran och björk i Sverige*. Sveriges: Landbruksuniversitet; 1988:1–73.
22. Mette T, Papathanassiou KP, Hajnsek I: **Biomass estimation from polarimetric SAR interferometry over heterogeneous forest terrain.** In *IEEE Intl. Geosci. Remote Sensing Symp, Anchorage.*; 2004:511–514.
23. Neeff T, Dutra LV, Dos Santos JR, Freitas CD, Araujo LS: **Tropical forest measurement by interferometric height modeling and P-band radar backscatter.** *For Sci* 2005, **51**:585–594.
24. NVE: *senorge.no. Norges vassdrags- og energidirektorat*.
25. Næsset E: **Predicting forest stand characteristics with airborne scanning laser using a practical two-stage procedure and field data.** *Remote Sens Environ* 2002, **80**:88–99.
26. Næsset E: **Accuracy of forest inventory using airborne laser scanning: evaluating the first Nordic full-scale operational project.** *Scand J For Res* 2004, **19**:554–557.
27. Næsset E, Bollandsas OM, Gobakken T, Gregoire TG, Stahl G: **Model-assisted estimation of change in forest biomass over an 11 year period in a sample survey supported by airborne LiDAR: a case study with post-stratification to provide "activity data".** *Remote Sens Environ* 2013, **128**:299–314.
28. Olander LP, Gibbs HK, Steininger M, Swenson JJ, Murray BC: **Reference scenarios for deforestation and forest degradation in support of REDD: a review of data and methods.** *Environ Res Lett* 2008, **3**:025011.
29. Patenaude G, Milne R, Dawson TP: **Synthesis of remote sensing approaches for forest carbon estimation: reporting to the Kyoto Protocol.** *Environ Sci Pol* 2005, **8**:161–178.
30. Persson H, Fransson J: **Forest variable estimation using radargrammetric processing of TerraSAR-X images in boreal forests.** *Remote Sens* 2014, **6**:2084–2107.
31. Potter C, Klooster S, Genovese V: **Alaska ecosystem carbon fluxes estimated from MODIS satellite data inputs from 2000 to 2010.** *Carbon Bal Manage* 2013, **8**:12.
32. Potter C, Klooster S, Genovese V, Hiatt C: **Forest production predicted from satellite image analysis for the Southeast Asia region.** *Carbon Bal Manage* 2013, **8**:9.
33. Praks J, Demirpolat C, Antropov O, Hallikainen M: *On forest height retrival from spaceborne X-band inferometic SAR images under variable seasonal conditions*. Otaniemi: XXXII Finnish URSI convention on radio science and SMARAD seminar, 24–25 April 2013; 2013:115–118.
34. Ribeiro NS, Matos CN, Moura IR, Washington-Allen RA, Ribeiro AI: **Monitoring vegetation dynamics and carbon stock density in miombo woodlands.** *Carbon Bal Manage* 2013, **8**:11–11.
35. Santoro M, Pantze A, Fransson JES, Dahlgren J, Persson A: **Nation-wide clear-cut mapping in Sweden using ALOS PALSAR strip images.** *Remote Sens* 2012, **4**:1693–1715.
36. Sarabandi K: **Δk-radar equivalent of interferometric SAR's: a theoretical study for determination of vegetation height.** *Geosci Remote Sensing, IEEE Transact on* 1997, **35**:1267–1276.
37. Soja MJ: *Modelling and Retrieval of Forest Parameters from Synthetic Aperture Radar Data*. Gothenburg: Chalmers university of technology; 2012.
38. Solberg S, Astrup R, Bollandsas OM, Naesset E, Weydahl DJ: **Deriving forest monitoring variables from X-band InSAR SRTM height.** *Can J Remote Sens* 2010, **36**:68–79.
39. Solberg S, Astrup R, Breidenbach J, Nilsen B, Weydahl D: **Monitoring spruce volume and biomass with InSAR data from TanDEM-X.** *Remote Sens Environ* 2013, **139**:60–67.
40. Solberg S, Astrup R, Gobakken T, Næsset E, Weydahl DJ: **Estimating spruce and pine biomass with interferometric X-band SAR.** *Remote Sens Environ* 2010, **114**:2353–2360.
41. Solberg S, Astrup R, Weydahl DJ: **Detection of forest clear-cuts with Shuttle Radar Topography Mission (SRTM) and Tandem-X InSAR Data.** *Remote Sens* 2013, **5**:5449–5462.
42. Solberg S, Riegler G, Noni P: **Estimating Forest Biomass From TerraSAR-X Stripmap Radargrammetry,** *Ieee Transactions on Geoscience and Remote Sensing*. 2015, **53**(1):154–161.
43. Solberg S, Weydahl DJ, Astrup R: *Temporal stability of X-band single-pass InSAR heights in a spruce forest: Effects of acquisition properties and season*, IEEE transaction on Geoscience and Remote sensing Accepted. ; 2014.
44. Thiel C, Schmullius C: **Investigating the impact of freezing on the ALOS PALSAR InSAR phase over Siberian forests.** *Remote Sensing Lett* 2013, **4**:900–909.
45. UNFAO: *Global forest resources assessments 2005*. 2005.
46. UNFCCC: *Decision 4/CP.15*, Methodological guidance for activities relating to reducing emissions from deforestation and forest degradation and the role of conservation, sustainable management of forests and enhancement of forest carbon stocks in developing countries. Bonn, Germany: United

Nations Framework Convention on Climate Change; 2009. http://unfccc.int/resource/docs/2009/cop15/eng/11a01.pdf.

47.   Vastaranta M, Holopainen M, Karjalainen M, Kankare V, Hyyppa J, Kaasalainen S: **TerraSAR-X stereo radargrammetry and airborne scanning LiDAR height metrics in imputation of forest aboveground biomass and stem volume.** *Ieee Transact Geosci Remote Sensing* 2014, **52:**1197–1204.

48.   Way J, Paris J, Kasischke E, Slaughter C, Viereck L, Christensen N, Dobson MC, Ulaby F, Richards J, Milne A, Sieber A, Ahern FJ, Simonett D, Hoffer R, Imhoff M, Weber J: **The effect of changing environmental-conditions on microwave signatures of forest ecosystems - preliminary-results of the March 1988 Alaskan aircraft SAR experiment.** *Int J Remote Sens* 1990, **11:**1119–1144.

49.   Weydahl DJ, Sagstuen J, Dick OB, Ronning H: **SRTM DEM accuracy assessment over vegetated areas in Norway.** *Int J Remote Sens* 2007, **28:**3513–3527.

50.   Woodhouse IH: **Predicting backscatter-biomass and height-biomass trends using a macroecology model.** *Ieee Transact Geosci Remote Sensing* 2006, **44:**871–877.

51.   Yu XW, Hyyppa J, Kaartinen H, Maltamo M: **Automatic detection of harvested trees and determination of forest growth using airborne laser scanning.** *Remote Sens Environ* 2004, **90:**451–462.

## 12

Evaluation of modelled net primary production using MODIS and landsat satellite data fusion

Steven Jay[1*], Christopher Potter[2], Robert Crabtree[1], Vanessa Genovese[3], Daniel J. Weiss[1,4] and Maggi Kraft[1]

Abstract

Background: To improve estimates of net primary production for terrestrial ecosystems of the continental United States, we evaluated a new image fusion technique to incorporate high resolution Landsat land cover data into a modified version of the CASA ecosystem model. The proportion of each Landsat land cover type within each 0.004 degree resolution CASA pixel was used to influence the ecosystem model result by a pure-pixel interpolation method.

Results: Seventeen Ameriflux tower flux records spread across the country were combined to evaluate monthly NPP estimates from the modified CASA model. Monthly measured NPP data values plotted against the revised CASA model outputs resulted in an overall R^2 of 0.72, mainly due to cropland locations where irrigation and crop rotation were not accounted for by the CASA model. When managed and disturbed locations are removed from the validation, the R^2 increases to 0.82.

Conclusions: The revised CASA model with pure-pixel interpolated vegetation index performed well at tower sites where vegetation was not manipulated or managed and had not been recently disturbed. Tower locations that showed relatively low correlations with CASA-estimated NPP were regularly disturbed by either human or natural forces.

Keywords: Net primary production, MODIS, Landsat, EVI, Ameriflux

Background

Net photosynthetic accumulation of carbon by plants, also known as net primary production (NPP), captures solar energy and drives most biotic processes on Earth. Climate controls on NPP fluxes on land are an issue of central relevance to humanity, in part due to possible limitations on the extent to which NPP in managed ecosystems can provide adequate food and fiber for growing populations [26].

Measurement of NPP presents many challenges in any ecosystem, and particularly in heterogeneous environments such as wetland, cultivated, ex-urban, and mountainous landscapes. Traditionally, NPP has been calculated by harvesting and measuring dry biomass or from eddy flux towers estimates [35]. Measuring dry biomass is labor- and time-intensive and logistically impossible to perform at scales other than the small plot (generally < 1 ha) [14]. Eddy flux towers measure the amount of CO_2 being exchanged with the atmosphere across a landscape. This technique can cover a larger area than using small plot biomass measurements. However, eddy flux tower measurements are affected by wind direction and atmospheric conditions, and logistical limitations have led to under-representation of tower sites in remote, disturbed, or degraded ecosystems [8].

Advances in modeling techniques and the integration of satellite multi-spectral data can greatly increase our capacity to estimate global and continental NPP. Common gridded approaches using satellite imagery to estimate NPP have assumed a constant land cover type within each pixel [10, 20, 32, 43]. By making this generalization, the influence of land cover types covering small

*Correspondence: jay@yellowstoneresearch.org
[1] Yellowstone Ecological Research Center, 2048 Analysis Dr. Ste. B, Bozeman, MT 59718, USA
Full list of author information is available at the end of the article

fractions of a pixel is potentially lost when pixels are classified. This can have detrimental effects on the pixel's estimate of NPP if some highly productive systems such as wetlands, or conversely, low productive areas such as bare ground, are ignored. In an attempt to improve NPP estimates across North America, we have developed an image fusion technique to incorporate high resolution land cover data into a modified version of the Carnegie Ames Stanford Approach (CASA) ecosystem model [20, 26] to improve NPP estimates at a 0.004 degree resolution.

One widely used estimate of global productivity is the MODIS MOD17 algorithm [43]. The MODIS MOD17 product currently estimates gross primary production (GPP) and NPP as a fraction of GPP at a 1-km spatial scale and an 8-day temporal scale. Turner et al. [36] found generally strong agreement of ground based measurements with MODIS productivity products using the Big-Foot [34, 35] validation procedure. However, the MODIS products tended to overestimate NPP at low productivity sites and underestimated NPP at highly productive sites outside the tropical zone [43]. This overestimation has been mostly attributed to high light interception values during the annual maxima as well as anomalous values in non-growing seasons. The low estimation tended to be a function of using incorrect light use efficiency terms for NPP [36].

The CASA ecosystem model [20, 26] uses freely available geographic data layers for climate and soils, and satellite imagery to estimate monthly NPP. CASA has been applied and tested around the world in hundreds of published studies (e.g., [2, 7, 9, 11, 18]). The CASA model estimates NPP at optimal metabolic rates, adjusting these rates based on scalars related to the effects of climate, soil moisture and texture, and land use [20]. For this study, CASA NPP is calculated based on a constant maximum light use efficiency concept [15], the MODIS vegetation index, solar radiation, temperature, and a soil moisture scalar.

Previous NPP estimates using the CASA model have shown strong correlation in timing with eddy flux towers; Potter et al. [26] reported an $R^2 = 0.77$ using a small set of monthly AmeriFlux tower NPP estimates. The CASA model was also capable of predicting annually summed NPP with an $R^2 = 0.90$ on a global scale using the NOAA Advanced Very High Resolution Radiometer (AVHRR) VI and regressing against over 1900 observed NPP data points [22]. The model has been used successfully in measuring the global effects of deforestation on NPP [21, 26], particularly in tropical Amazonian and Asian carbon fluxes [23, 27, 41].

We have developed a new approach for this study to improve the spatial resolution and potentially the accuracy of the CASA NPP model by using an image fusion technique, whereby high resolution, Landsat-derived land cover is fused with MODIS MOD13A1 Enhanced Vegetation Index (EVI) data. The proportion of each land cover type within each pixel area was used to influence the ecosystem model result. This technique produced CASA model predictions of monthly NPP at 0.004 degree (approx. 500-m) resolution for North America from 2000–2010 that were compared to tower flux estimates of NPP for evaluation.

Methods

The CASA model requires several input data sets in order to successfully estimate NPP values. Land cover specific interpolated vegetation indices, gridded temperature and precipitation data, soil texture, solar radiation, and elevation are all used to estimate NPP. The vegetation indices, climate data, and solar radiation datasets were compiled for each month from 1999 to 2010 and the soil texture and elevation data remains static. These data parameterize the CASA model which is programmed in Python and integrates with ArcGIS software using the ArcPy processing module. Data inputs and CASA processing are described in more detail below.

Interpolated vegetation index

An interpolated vegetation index was created using a combination of MODIS MOD13A1 16-day EVI data and both 2006 NLCD [3] land cover data and Canadian land cover data [6, 12, 17]. A five step process was developed to create land cover specific CASA NPP estimates, which could then be aggregated to estimate total landscape NPP (Fig. 1).

Step one of the processing chain created "fractional cover" estimates. Fractional cover estimates were created by combining monthly 0.004 degree resolution, cloud-filled [40] MODIS EVI data with 0.0003 degree (approx. 30-m) resolution 2006 NLCD and 2001 Canadian land cover data to estimate the proportion of each land cover contained in a MODIS pixel. High resolution land cover data was aggregated to 11 general land cover classes. The 11 classes used in modeling were; water, evergreen forest, wetlands, lichen, mixed forest, woodlands, grasslands, croplands, deciduous forest, and brushland. The high resolution land cover data was then resampled to 0.004 degree resolution, while maintaining the proportion of each land cover type contained within the pixel to create a fractional cover layer.

The next step required the identification of "pure-pixels". Pure pixels are pixels that contain more than 90 % of a single class. These pure pixels are used to extract MODIS EVI values using a point-intercept method. Pure pixels are identified by applying a mask to each fractional

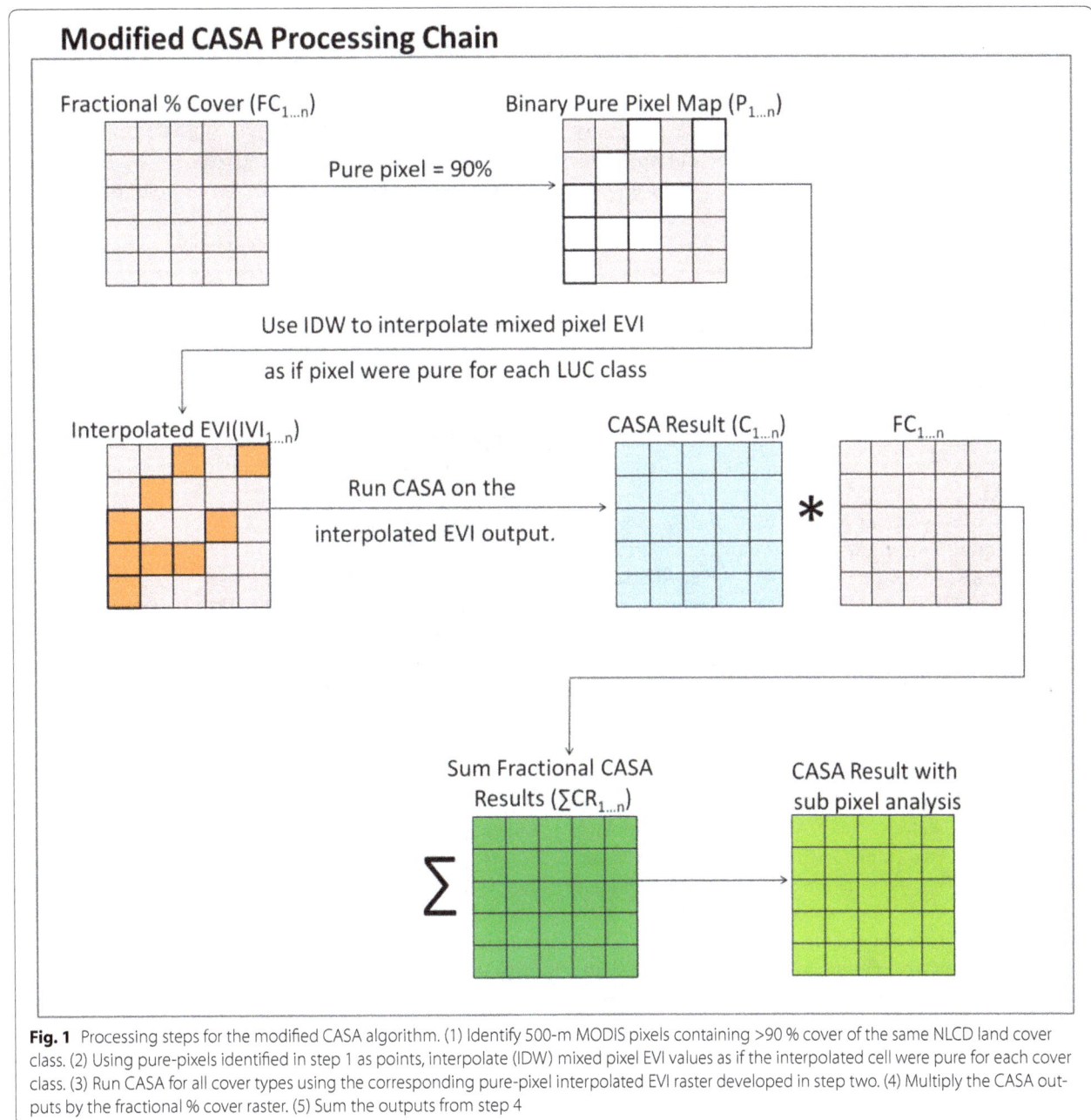

Fig. 1 Processing steps for the modified CASA algorithm. (1) Identify 500-m MODIS pixels containing >90 % cover of the same NLCD land cover class. (2) Using pure-pixels identified in step 1 as points, interpolate (IDW) mixed pixel EVI values as if the interpolated cell were pure for each cover class. (3) Run CASA for all cover types using the corresponding pure-pixel interpolated EVI raster developed in step two. (4) Multiply the CASA outputs by the fractional % cover raster. (5) Sum the outputs from step 4

land cover data set and remove all pixels with a value below 0.9. These pixels are then converted to a point representing the center of its corresponding pixel. These points a then overlaid with the monthly MODIS EVI data and the intersecting pixel values are extracted for each month from 2000 to 2010.

The third step requires using the pure pixel values to produce a land cover specific vegetation index by performing an inverse distance weighted (IDW) algorithm. The land cover specific pure pixels tended to be highly clustered making the IDW method appropriate because of Tobler's Law [33]. The IDW process was run using ESRI ArcGIS 10.2 software. The optimal power function for each land cover was calculated using the Geostatistical Analyst extension in ArcGIS 10.2. The optimal power function is identified using cross-validation to find a minimum root mean square prediction (RMSPE) value [19, 39] and a variable search radius was used.

Next, the interpolated vegetation indices (IVI) are used as the vegetation input for the CASA model. The CASA model uses the IVI along with solar radiation and temperature and wetness scalars to estimate NPP [20]. Each land cover is run independently through the CASA model to produce NPP estimates specific for each cover type. These results are then multiplied by the corresponding fractional cover value calculated in step 1. This step produces NPP estimates proportional to the amount of each land cover type per pixel.

In the final step, the proportional NPP estimates are summed to create a total NPP estimate per pixel. This total NPP value is now influenced by the proportion of the pixel covered by specific land cover type. This sub pixel analysis provides the opportunity to estimate NPP values without using the assumption that a MODIS pixel is a homogenous land cover as previous estimates have done [22, 26].

Climate data

Two consecutive years of precipitation and temperature data are required for CASA model initialization in order to model the soil moisture reservoir (explained in detail below). For model years 2000 and 2001, National Centers for Environmental Prediction (NCEP) North American Regional Reanalysis (NARR) climate datasets were used as input data for the CASA model. These gridded datasets had a spatial resolution of 0.3 degree (approximately equal to 25 km resolution for the continental USA) and were developed at NOAA's National Center for Atmospheric Research (NCAR). The NARR data is an extension of the NCEP global reanalysis. The NARR model uses the high resolution NCEP data along with an advanced data assimilation model. This assimilation technique improves the accuracy of gridded temperature and precipitation estimates. NCEP data was used for the years 2000 and 2001 because the higher resolution MODIS land surface temperature data was not available prior to 2001 and CASA initialization requires data from 1999 and 2000 to properly initialize. Model year 2001 used NCEP data so as to avoid mixing two different data sources during modeling.

For model years 2002–2010, MODIS MOD11 Terra land surface temperature data was used. The MODIS land surface temperature product is a 0.05 degree (approximately 3 km) resolution, daily global product (MOD11C1). These values were derived from the daily MODIS day/night LST/E product from pairs of 7 daytime and nighttime MODIS TIR bands (20, 22, 23, 29, and 31–33). These data inputs were used for years after 2001, because they did not exist prior for years prior to 2000. MODIS land surface data requires some pre-processing in order to obtain monthly average values. Raw data was converted from raw digital numbers to degrees Celsius by multiplying the raw value by a pre-determined scaling

factor of 0.02 and subtracting 273.15 [37]. All temperature data was then resampled to 0.004 degrees for input into the CASA model.

The Global Land Data Assimilation System (GLDAS) Noah Land Surface Model L4 monthly 1.0 degree precipitation models were used as the CASA input. This 1.0 degree (approximately 75 km) resolution data set assimilates measurements from several ground-based weather stations to interpolate climatic variables. Average monthly precipitation data sets were downloaded from the NASA Goddard Earth Sciences Data and Information Services Center (http://disc.sci.gsfc.nasa.gov/services/grads-gds/gldas). GLDAS precipitation data was obtained as daily precipitation averages. To convert the data to total monthly averages, the daily average was multiplied by the number of days in its associated month. The data was then resampled to 0.004 degrees for input into the CASA model.

Solar radiation

Monthly solar radiation values were modeled for North America. Surface solar irradiance was estimated using the ESRI ArcGIS solar radiation model. Total radiation for each pixel on a topographic surface is calculated by estimating the sum of the direct and diffuse radiation across all of North America [4, 5, 31]. Direct and diffuse radiation simulate shadow patterns at discrete intervals through time across the landscape [5].

Soil texture and elevation

Soil texture data was obtained from the SSURGO database (NRCS) for the Continental United States and Alaska. Canadian soil texture data was obtained from the Canadian Soil Database using the Soil Landscapes of Canada (SLC) data. These datasets were than simplified into basic soil texture classes based on the content of clay, which can be interpreted by the CASA model. The seven classes are: organic soils, 0–5 % clay, 5–15 % clay, 15–30 % clay, $\geq$30 % clay, and lithosols. This data is resampled to 0.004 degree resolution for input into the CASA model. Elevation data was obtained from the USGS National Elevation Dataset (NED) at 0.008 degree resolution and resampled to 0.004 degree resolution for CASA model input.

CASA ecosystem model

Monthly flux of the net fixation of CO_2 by vegetation is computed by the CASA model using light use efficiency (LUE) [25]. Monthly NPP is estimated using surface solar irradiance, Sr, an interpolated vegetation index, a constant light use efficiency term (e_{max}), and scalar values for temperature (T) and moisture

(W). These terms are used to calculate NPP using this equation

$$NPP = Sr \times IVI \times e_{max} \times T \times W$$

The e_{max} term is set to a constant 0.55 gC/MJ PAR, this value was determined by using predicted annual NPP values compared to previous field estimates [20, 22, 24]. Previous studies have found this value to produce good results when initializing CASA with MODIS EVI data [16, 22, 26]. The T scalar is calculated using a derivation of optimal temperatures (T_{opt}) for vegetative growth. This setting varies by latitude and longitude, ranging from 0 °C in Polar Regions to 30–35 °C in low latitude deserts. Monthly water deficits define the W scalar by comparing precipitation and soil water to potential evapotranspiration (PET) using the Priestly and Taylor method [30]. CASA model initialization requires at least two consecutive years of data of vegetation data, temperature and precipitation data. The 2 years are required to model the previous years' soil moisture reservoir and estimate the current year's water balance.

The CASA model couples evapotranspiration to water content in the soil profile by using a series of algorithms. The model's algorithms use three soil layers (surface organic matter, topsoil, and subsoil to rooting depth), which can have different textures, moisture holding capacity, and carbon–nitrogen dynamics [25]. These soil layers are used to calculate a water balance using precipitation and soil parameters versus evapotranspiration and drainage. Inputs from precipitation recharge the soil layers until field capacity is reached and excess is then defined as drainage and leave the system as runoff.

The CASA model is run independently on each land cover's IVI with all other inputs remaining the same. These land cover specific modeling results are then scaled proportionally to amount of each respective land cover present in the pixel. If a particular land cover is not present in a pixel, the CASA result is not included in the final composite estimate. This ensures that only the influence of cover types present is used to calculate the final NPP value.

NPP evaluation procedure

A total of 51 Ameriflux tower sites with over 2000 combined data points were reviewed for CASA NPP validation. Sites meeting the criteria outlined by Potter et al. [26] were selected for CASA model validation. In order to meet the selection criteria, sites needed to have a minimum of 3 years of flux data and the data must span the entire year in order to compare dormant season results. Of the 51 sites reviewed, 17 sites met these criteria (with locations shown in Fig. 2) and were included in this analysis. Of the Ameriflux locations selected for validation,

the majority of the sites (11 of the 17) were located in some type of forest land cover, 7 were in evergreen needle leaf forests, 3 in deciduous broadleaf forests, and 1 in mixed forest. The remaining six sites were cropland (3 sites), grassland (2 sites), and the final site was a savanna woodland site. For sites that met these criteria, Ameriflux data sets were downloaded from the Carbon Dioxide Information Analysis Center (CDIAC; http://public. ornl.gov/ameriflux/dataproducts.shtml). We selected the Level 4 gap-filled and ustar-filtered records, which contain estimated gross primary productivity (GPP) and total ecosystem respiration.

Monthly records of NPP were estimated using an uncertainty range of 40–50 % of the GPP carbon flux defined for temperate ecosystems. Waring et al. [38] evaluated a constant ratio of NPP/GPP for forested sites and found the ratio to be 0.47 ± 0.04 SD. Zhang et al. [42] found the global NPP/GPP ratio to be 0.52 with minimal variation. However, they did find densely vegetated regions to have a lower NPP/GPP ratio than sparsely vegetated regions. Comparison between a NPP/GPP ratio of 40 and 50 % and found no significant difference (p < 0.01). The NPP/GPP ratio of 0.40 had slightly better correlation than 0.50. Very little inter-annual analysis of the NPP/GPP ratio has been performed and only limited to certain specific species, however Campioli et al. [1] found inter-annual fluctuation of the NPP/GPP ratio in Beech stands to not be significant throughout the year.

Using a point intersect technique, CASA model results were extracted for the years matching the Ameriflux site data. We also extracted the neighboring 3 × 3 pixel area around the Ameriflux site to test a 1.5 × 1.5 km footprint around the validation site. The 3 × 3 pixel area was averaged to produce a single CASA model NPP value that was then compared to the Ameriflux site data. An expanded footprint was tested because air around the tower is mixing and moving due to wind and the tower measurement is indicative of the local area rather than a single point in space. Furthermore, because flux towers are typically positioned in areas surrounded by similar land cover, it is reasonable to summarize the CASA results over a local neighborhood that is less slightly variable over time than the response at a single pixel.

Next, a series of linear regressions was performed comparing Ameriflux site data with the annual CASA results. Regression analysis was performed comparing the results of the point intersection and the 3 × 3 pixel average against both the 40 % estimate of GPP and 50 % estimate of annual GPP. For revised model evaluation, seasonal CASA output was compared to the seasonal flux of the Ameriflux sites. Comparing the seasonal flux of both the ground measurements and the model estimates provides

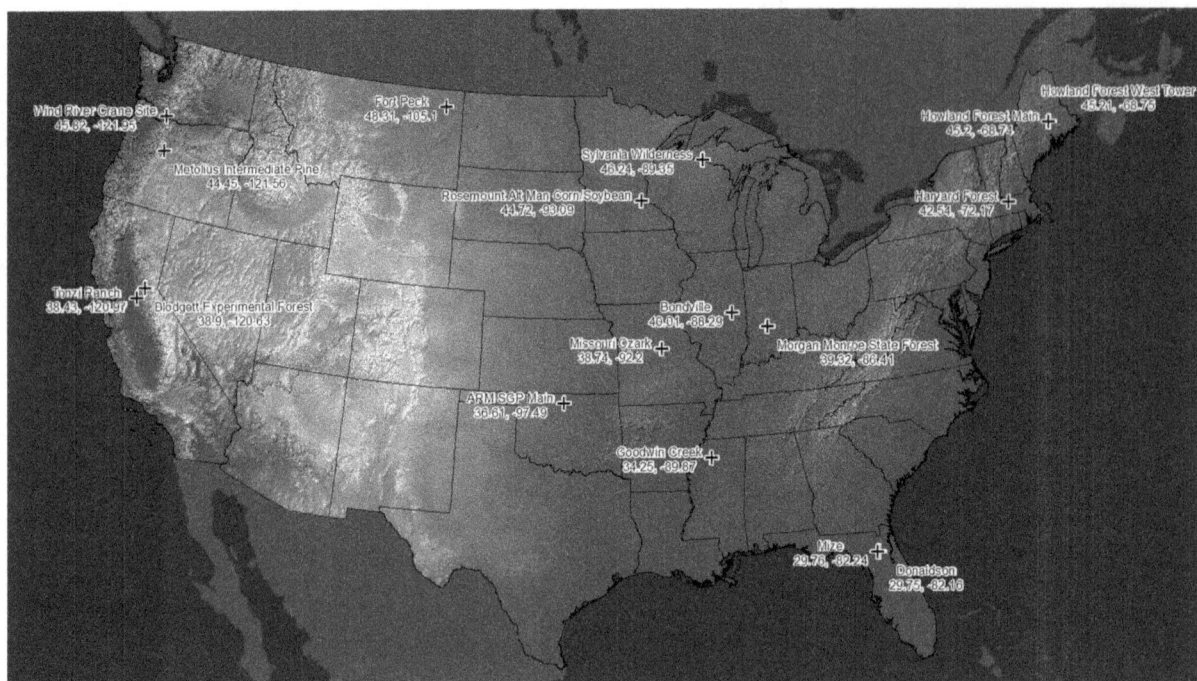

Fig. 2 Tower flux site locations used for comparisons to CASA model NPP estimates

insight into how well the CASA model estimates timing of yearly productivity maxima and minima.

Results

Comparison of MODIS EVI values to IVI values

Four mountain wetland areas were selected to demonstrate to differences between IVI values and MOD13 composite EVI values. These locations were selected based on the authors' previous experience working in these landscapes, and the relatively high proportion (15–53 %) of the pixel containing wetlands. Results showed a significant difference between pure-pixel EVI and MODI13 composite EVI values for three of the four sites compared (Table 1). For example, at the Yellowstone Lake location (Fig. 3), MOD13 EVI values approach zero in March and April, in contrast to the pure-pixel interpolated EVI values which had a much more muted or non-existent dip during these warming months. This difference may be a result of the pure-pixel interpolated

EVI capturing annual lake level changes and vegetation greening during spring run-off periods. The remaining sites (Fig. 3) showed that the major differences between pure-pixel and MOD13 EVI occurred in the winter months of December to February, and that the pure-pixel EVI values were slightly lower in the peak summer months of July and August.

Combined flux tower NPP comparisons

Seventeen Ameriflux towers spread across the Continental United States (CONUS) were combined to evaluate monthly NPP estimates from the modified CASA model. Using a NPP:GPP ratio of 40 % in the tower flux measurements, 1030 monthly data values plotted against the CASA model outputs resulted in an overall R^2 of 0.72 (Fig. 4). Averaging NPP for a three cell by three cell buffer around the flux tower location resulted in a slight decrease in correlation between modeled NPP and flux tower measurements ($R^2 = 0.71$). Removing flux tower

Table 1 Comparison of pure-pixel interpolated EVI values and MOD13 EVI values using a paired t-test

Location	Latitude	Longitude	Percent wetland (%)	t-value	df	p-value
Yellowstone Lake	44.406	−110.252	53	−13.54	132	<0.05
Headwaters State Park	45.919	−111.49	15	3.58	132	<0.05
Red Rocks NWR	44.623	−111.806	52	−1.70	132	0.09
Beartooth Pass	44.931	−109.523	30	−2.81	132	<0.05

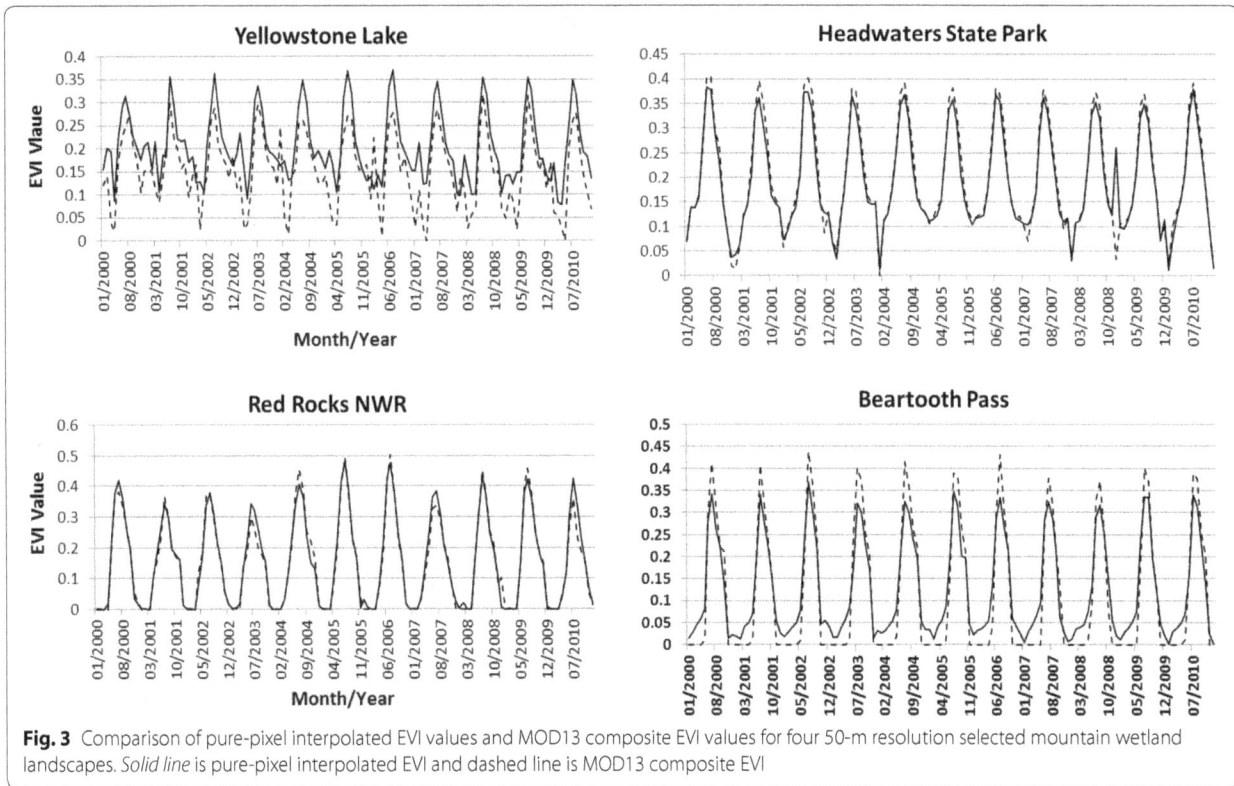

Fig. 3 Comparison of pure-pixel interpolated EVI values and MOD13 composite EVI values for four 50-m resolution selected mountain wetland landscapes. *Solid line* is pure-pixel interpolated EVI and dashed line is MOD13 composite EVI

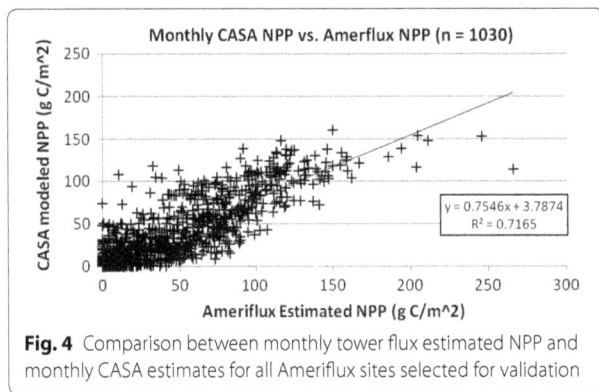

Fig. 4 Comparison between monthly tower flux estimated NPP and monthly CASA estimates for all Ameriflux sites selected for validation

highest levels of correlation (Table 2), with the Sylvania Wilderness flux tower in northern Michigan returning a $R^2 = 0.93$ and the Morgan Monroe State Forest flux tower in Indiana a correlation of $R^2 = 0.93$. Southern and western CONUS tower sites matched most poorly with CASA model estimates, with the Donaldson flux tower in Florida showing the lowest correlation of $R^2 = 0.01$ and the ARM-SGP main tower in Oklahoma returning an $R^2 = 0.08$. When grouped by land cover types, the combination of deciduous broadleaf forests monthly NPP measurements matched most closely with CASA model estimates, resulting a correlation of $R^2 = 0.88$, followed by croplands ($R^2 = 0.73$), grasslands ($R^2 = 0.65$), and evergreen needleleaf forests ($R^2 = 0.57$) (Table 3).

Individual flux tower NPP comparisons

The ARM-SGP site located in north-central Oklahoma showed a low correlation coefficient between flux tower NPP measurements and CASA model estimates (Fig. 6). This site (and several others shown in Fig. 5) was located in a region dominated by a mix of cropland and grassland cover types, which at ARM-SGP consisted of wheat (*Triticum aestuvum* L.), corn (*Zea Mays* L.) and soybean (*Glycine Willd.*). During flux measurements from January 2003 through October 2006, multiple rotations of different crops occurred. During 2003–2004, common wheat

sites that have been recently disturbed or managed the overall R^2 increases to 0.82. The revised CASA monthly NPP estimates were found to be 25 % lower overall than the tower-based NPP measurements, for reasons explained below.

On a seasonal basis, the modified CASA model most closely matched the tower flux NPP during summer ($R^2 = 0.58$) and autumn ($R^2 = 0.72$), and most poorly in the winter ($R^2 = 0.22$). Correlations were also relatively low in the spring, showing a $R^2 = 0.48$. Tower sites in the Northeast and northern Midwest showed the

Table 2 CASA annual NPP validation results by Ameriflux tower site

Site	Land cover	R^2
Donaldson	Evergreen needleleaf forest	0.01
ARM-SGP	Croplands	0.08
Tonzi	Woodlands	0.23
Mize	Evergreen needleleaf forest	0.25
Blodgett	Evergreen needleleaf forest	0.34
Ft Peck	Grasslands	0.54
Goodwin Creek	Grasslands	0.54
Rosemount Alt	Croplands	0.65
Wind River	Evergreen needleleaf forest	0.70
Metolius Intermediate	Evergreen needleleaf forest	0.72
Missouri Ozark	Deciduous broadleaf forest	0.80
Bondville	Croplands	0.82
Howland West	Evergreen needleleaf forest	0.86
Howland Main	Evergreen needleleaf forest	0.86
Harvard Forest	Deciduous broadleaf forest	0.90
Morgan Monroe	Deciduous broadleaf forest	0.91
Sylvania Wilderness	Mixed forest	0.93
Overall		0.72

Table 3 CASA NPP validation results by land cover type within tower site

Land cover	R^2	n
Evergreen needleleaf forest	0.57	401
Deciduous broadleaf forest	0.88	212
Croplands	0.73	154
Grasslands	0.65	117
Woody savannas	0.23	81
Mixed forest	0.93	65

was planted, in 2005 corn was planted instead of wheat, and soybeans were planted in 2006. Comparing monthly average NPP for each separate year, 2003 showed an $R^2 = 0.1$, 2004 showed an $R^2 = 0.36$, 2005 showed an $R^2 = 0.17$, and 2006 showed an $R^2 = 0.30$.

A summary of the monthly flux comparisons (Fig. 7) revealed that the CASA model does not yet account for either irrigation schedules that boost NPP of crops as intended by the farmers, or for crop harvests that are typically occurring in June–August of each year. This lack of sensitivity to crop-specific irrigation likely explains the low monthly NPP correlation observed at this site as well as the 25 % underestimation of NPP shown for all tower sites combined in Fig. 4. There was also a consistent increase in tower-measured NPP flux in September and October for most crops at the ARM-SGP site, which

was not estimated by the CASA model during these cooler months outside the main corn and soybean growing seasons.

Four example Ameriflux sites (two forests, one cropland, and one grassland) where the CASA model NPP matched closely with tower-measured monthly NPP (Fig. 8) resulted in R^2 values between 0.82 and 0.93. The Goodwin Creek grassland sight (R^2 of 0.54), however, was an exception to this pattern. In most years, the revised CASA model closely tracked the seasonality of NPP and the peak summer NPP values in these tower flux measurements. It is worth noting that the Bondville tower site had the lowest level of mixed land cover types (with surrounding grasslands and pastures) of any of the cropland locations shown in Fig. 5.

The seasonal pattern in tower-measured NPP at Goodwin Creek grasslands was inconsistent from year-to-year (Fig. 8), which differed from the CASA model NPP, which suggested some management or local disturbance impacts on the tower fluxes that are not unaccounted for by the CASA model. The Ameriflux data source (available at http://ameriflux.ornl.gov/) states that grass surrounding the base of the tower was mowed periodically to maintain a height consistent with the regional grasslands, which confirms the sudden and unpredictable loss of tower-measured NPP at this location.

Another site that showed a low correlation with CASA estimated NPP was the Donaldson flux tower in Florida. This site is a managed slash pine (*Pinus elliottii*) plantation ecosystem in north-central Florida [29]. The stand is even-aged with the overstory comprising of 100 % slash pine with assorted native species in the understory. Review of the measured NPP pattern shows consistent CASA mode overestimation in January. This is likely a result of the site being located in a warm climate, where evapotranspiration remains high all year but precipitation is inconsistent, leading to spikes that result from rapid soil wetting and drying. Graphing the soil water scalar term against NPP (Fig. 9) shows spikes in soil water stress echoed within CASA NPP. Management and disturbance also played a role in the poor result at the Donaldson site. A 100-year drought was reported from 2000 through the summer of 2002. The CASA algorithm was able to detect this drought as indicated by the similar amplitude observed in the NPP time-series flux; however, CASA was not able to match peaks and troughs measured at the site. Following the drought, fertilizer was applied in 2002 to the plantation. Interestingly, CASA slightly over-predicted the tower NPP measurements despite the fertilizer application, and did not detect a large decline in productivity in June 2002. No major disturbances were reported during 2003 and this year resulted in the highest correlation between the CASA-modeled and the measured NPP

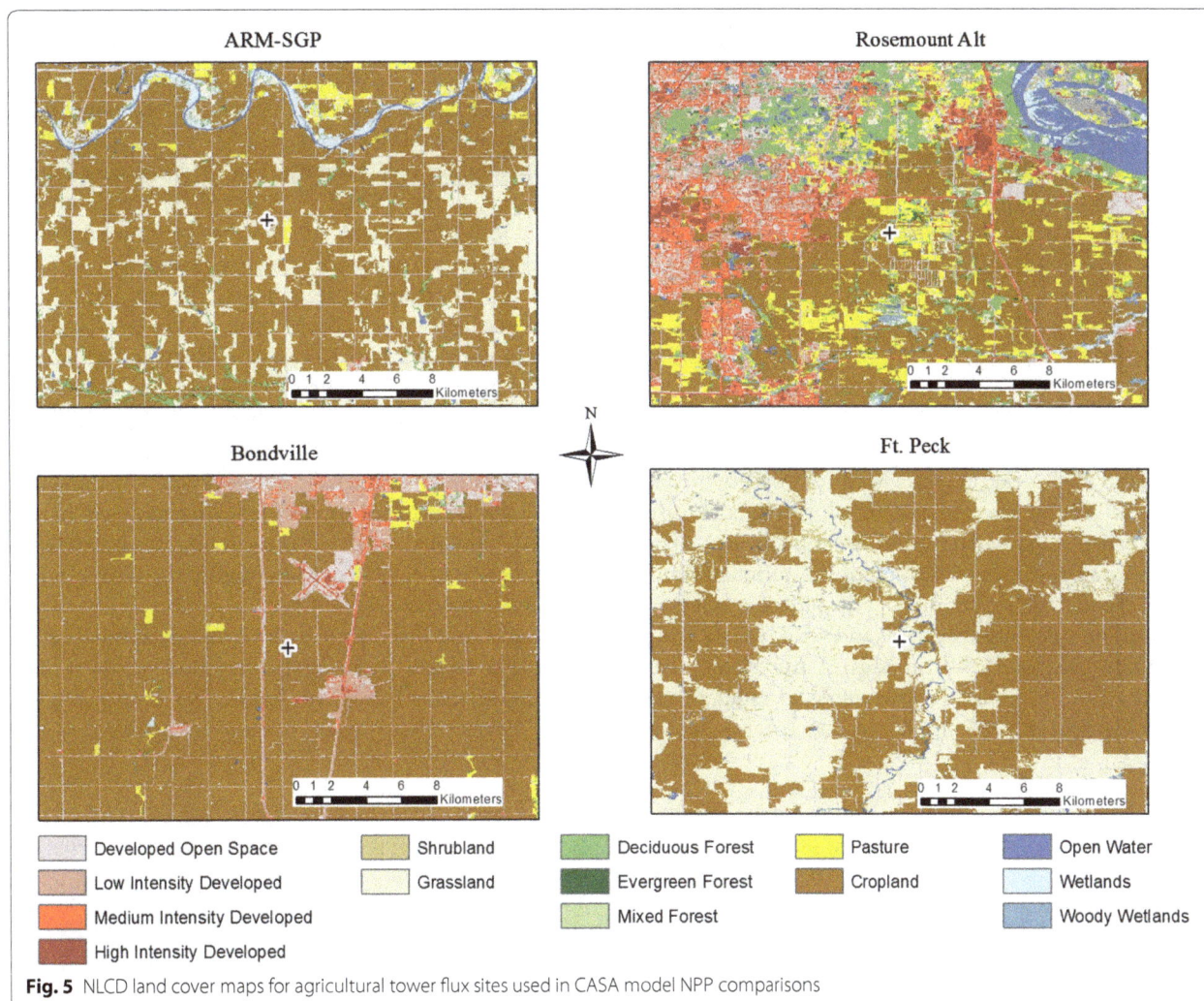

Fig. 5 NLCD land cover maps for agricultural tower flux sites used in CASA model NPP comparisons

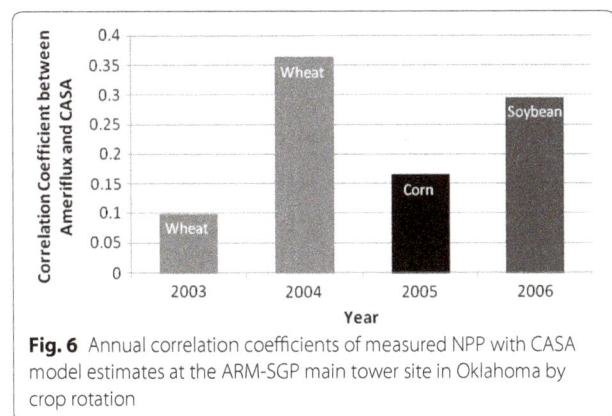

Fig. 6 Annual correlation coefficients of measured NPP with CASA model estimates at the ARM-SGP main tower site in Oklahoma by crop rotation

fluxes. A series of tropical storms struck the site during the summer of 2004, but the CASA algorithm did not detect the loss of productivity of the forest overstory due

to the site being inundated in September following this series of storms.

Discussion

Generally, we found that the revised CASA model with pure-pixel interpolated EVI performed well at tower sites where vegetation was not manipulated or managed and had not been recently disturbed. Sites such as the Sylvania Wilderness and Morgan Monroe, which are large, relatively undisturbed tracts of temperate forest, both had very high correlations with CASA-estimated NPP. Sites such as Goodwin Creek grasslands and the Donaldson managed pine plantation locations both had relatively low correlations with CASA-estimated NPP, which we attribute to the disturbances experienced by these sites.

Cropland-dominated tower site comparisons exhibited several instances of mistiming and underestimation of peak monthly NPP by the revised CASA model. The southern Plains states fluxes typically measured

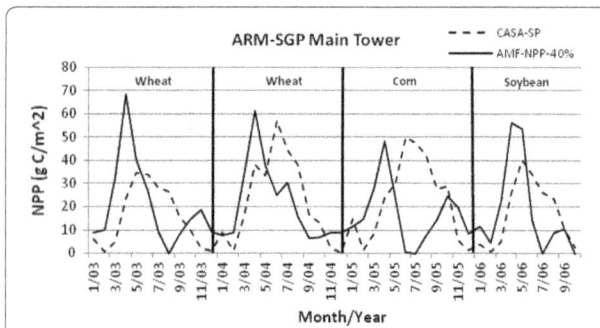

Fig. 7 Comparisons of monthly NPP flux in cropland types at the ARM-SGP main tower in Oklahoma

a cropland NPP peak in April-May, followed by a steep decline in NPP in June-August, which presumably result from, respectively, early springtime rainfall and irrigation and mid-summer harvesting of the various rotating crops. The CASA model did not include the effects of irrigation on boosting crop NPP and yield, which is very commonly practiced in farmlands of the Plains states [28]. The CASA model did not detect a regular June-August decline in NPP measured due to crop harvest

practices, indicating the pure-pixel interpolated EVI may have remained high in summer due to inclusion of surrounding grassland and pasture areas that were not harvested at that time and were shown to surround most central tower locations in Fig. 5.

Crop rotation also played a role at some agricultural tower sites. At the ARM-SGP tower site in 2005, corn was planted instead of wheat, and the change from a C3 plant to a C4 plant may explain the drop in correlation with the CASA model estimates, since the CASA algorithm uses a constant light-use efficiency term. However, C4 plants (such as corn) tend to have higher light-use efficiency terms than C3 plants (such as wheat). Soybeans were planted in 2006 and again we observed a slight increase in correlation in the rotation from a C4 crop to a C3 crop. This shortcoming in the CASA model could be mitigated by using different light-use efficiency terms based on the crop type found at a particular location. This is one technique that Yu et al. [41] used to modify the CASA model to estimate productivity in China.

Changes and discrepancies in land cover used in the study also explain some of the variation in the results. This model relies heavily on land cover inputs and the broad cover classes used may not capture regional

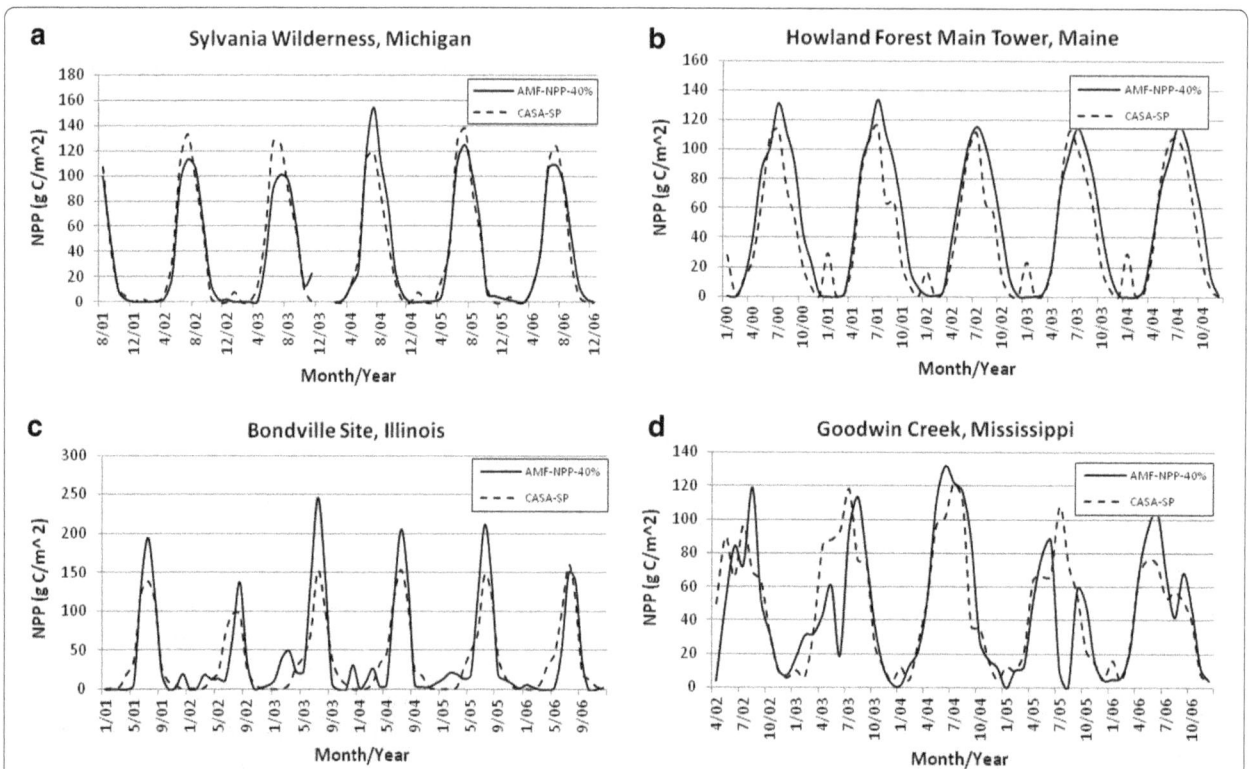

Fig. 8 Comparison of monthly NPP patterns between Ameriflux measured NPP and CASA modeled NPP. **a** Sylvania Wilderness, Michigan, mixed forest land cover, **b** Howland Forest Main Tower, Maine, evergreen needleleaf forest, **c** Bondville, Illinois, cropland, **d** Goodwin Creek, Mississippi, grassland

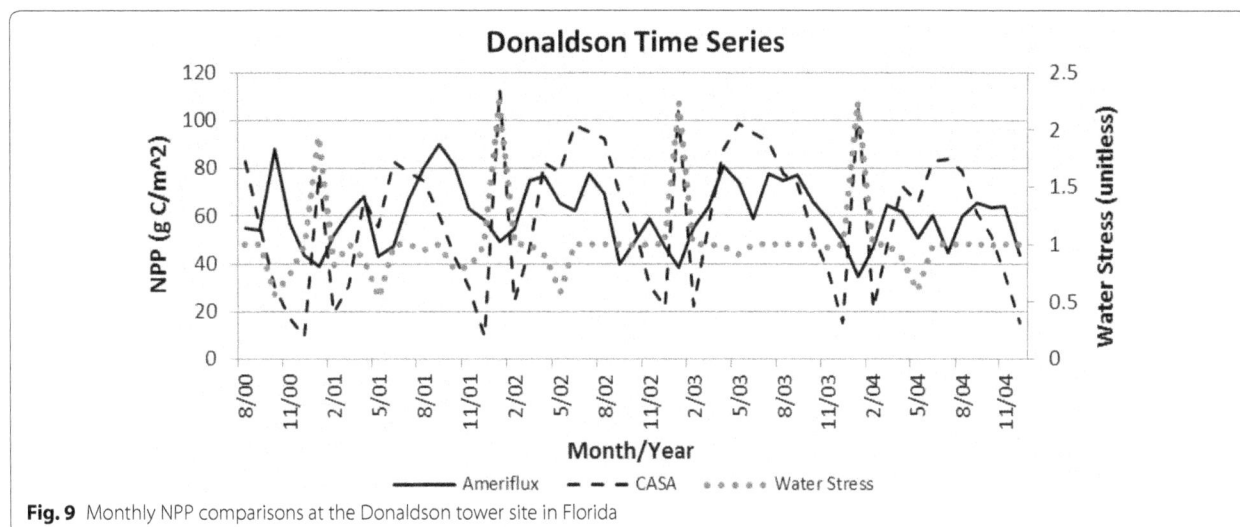

Fig. 9 Monthly NPP comparisons at the Donaldson tower site in Florida

differences in vegetation. For example, evergreen forests of the Pacific Northwest are not comprised of the same species as the Slash Pine forests in Florida yet the model treats these land covers as the same. Additionally, the model also assumes that land cover is consistent year to year, which is not what is happening in reality. The static nature of land cover limits the ability of the model to capture land cover changes and adjust NPP estimates accordingly. This can be observed by the low R^2 achieved at sites with crop rotations.

Previous comparison studied by Li et al. [13] concluded that variations in NPP in woodland locations of drought-prone climate zones, such as the central California Tonzi Ranch tower site, cannot be matched closely by CASA unless soil water availability was modified in the model structure. This site was a savanna consisting of scattered blue oak trees (*Quercus douglasii*), with occasional gray pine trees (*Pinus sabiniana* L.), surrounded by grazed grassland. Oak trees in this region were able to continue to transpire into the summer months, albeit at low rates, under very dry soil conditions and maintain basal levels of carbon metabolism, because tree roots were able to access sources of water in the soil unavailable to grass roots. Consequently, for CASA model applications in oak woodlands of California, an adjustment should be in the available water storage content for the deeper rooting layer of shrubs and trees that may be present at such sites. This adjustment made available 80% more soil water for transpiration by shrubs and trees than is commonly set for other moist forested climate zones of the western United States, and resulted in an R^2 match of 0.89 between monthly tower measurements and CASA estimated NPP.

Conclusions

Using image fusion of Landsat and MODIS satellite data products to enhance the CASA ecosystem model shows promise for accurate monthly NPP estimation, especially in heterogeneous but relatively undisturbed landscapes. However, more validation and work will be required to fully understand how well the CASA model can perform at managed cropland sites. This is due to the fact that the model inputs and algorithms are not yet sensitive to some management practices, such as irrigation and crop rotations. Further validation in wetlands and mountainous landscapes with new tower flux measurements will be required to fully document the advantages of image fusion to improve model NPP estimates.

Abbreviations

AVHRR: Advanced Very High Resolution Radiometer; CASA: Carnegie Ames Stanford Approach; CDIAC: Carbon Dioxide Information Analysis Center; CONUS: Continental United States; eMax: light utilization efficiency term; EVI: Enhanced Vegetation Index; GLDAS: Global Land Data Assimilation System; GPP: gross primary productivity; IVI: interpolated vegetation index; NARR: North American Regional Reanalysis; NCEP: National Centers for Environmental Prediction; NLCD: National Land Cover Database; NPP: net primary productivity.

Authors' contributions

SJ carried out the CASA model runs, performed the data validation, and drafted the manuscript; CP developed the original CASA model, provided insight and expertise on model runs and results, and edited the manuscript; RC developed the conceptual data fusion model, provided ecological insight and expertise, and edited the manuscript; VG programmed the CASA model, DW obtained and processed MODIS EVI data, MK obtained and standardized climate inputs. All authors read and approved the final manuscript.

Author details

[1] Yellowstone Ecological Research Center, 2048 Analysis Dr. Ste. B, Bozeman, MT 59718, USA. [2] CASA Systems 2100, LLC, PO Box 1631, Los Gatos, CA 95030, USA. [3] Science and Environmental Policy, California State University, Monterey Bay, 100 Campus Center, Seaside, CA 93955, USA. [4] Department of Zoology, University of Oxford, The Tinbergen Building, S Parks Rd, Oxford, OX1 3PS, UK.

Availability of data and materials

The datasets supporting the conclusions of this article are available in the COASTERdata repository, http://www.coasterdata.net.

Competing interests

The authors declare that they have no competing interests.

Funding

This research was funded by NASA.

References

1. Campioli M, Gielen B, Göckede M, Papale D, Bouriaud O, Granier A. Temporal variability of the NPP-GPP ratio at seasonal and interannual time scales in a temperate beech forest. Biogeosciences. 2011;8(9):2481–92.
2. Crabtree R, Potter C, Mullen R, Sheldon J, Huang X, Harmsen J, Rodman A, Jean C. A modeling and spatio-temporal analysis framework for monitoring environmental change using NPP as an ecosystem indicator. Remote Sens Environ. 2009;113:1486–96.
3. Fry J, Xian G, Jin S, Dewitz J, Homer C, Yang L, Barnes C, Herold N, Wickham J. Completion of the 2006 National Land Cover Database for the Conterminous United States. PE&RS. 2011;77(9):858–64.
4. Fu P. A geometric solar radiation model with applications in landscape ecology. Ph.D. thesis, Department of Geography, University of Kansas, Lawrence, Kansas, USA; 2000.
5. Fu P, Rich P. A geometric solar radiation model with applications in agriculture and forestry. Comput Electron Agric. 2002;37:25–35.
6. Gibson J, Nedelcu S, Pavlic G, Budkewitsch P. A complete orthorectified Landsat-7 mosaic of the Canadian Arctic Archipelago. Natural Resources Canada. 2012: ISBN 978-1-100-19023-5.
7. Golubyatnikov L, Denisenko E. Modeling the values of net primary production for the zonal vegetation of European Russia. Biol Bull. 2001;28:293–300.
8. Goulden L, Munger J, Fan S, Daube B, Wofsy S. Measurements of carbon sequestration by long-term eddy covariance: methods and a critical evaluation of accuracy. Glob Change Biol. 1996;2:169–82.
9. Huanga C, Asner G, Barger N. Modeling regional variation in net primary production of pinyon–juniper ecosystems. Ecol Model. 2012;227:82–92.
10. Heinsch F, Zhao M, Running S, Kimball J, Nemani R, Davis K, Bolstad P, Cook B, Desai A, Ricciuto D, Law B, Oechel W, Kwon H, Luo H, Wofsy S, Dunn A, Munger J, Baldocchi D, Xu L, Hollinger D, Richardson A, Stoy P, Siqueira M, Monson R, Burns S, Flanagan L. Evaluation of remote sensing based terrestrial productivity from MODIS using regional tower eddy flux network observations. IEEE Trans Geosci Remote Sens. 2006;44:1908–25.
11. Jinguo Y, Zheng N, Chenli W. Vegetation NPP distribution based on MODIS data and CASA model—a case study of northern Hebei Province. Chin Geograph Sci. 2006;16:334–41.
12. Latifovic R, Pouliot D, Olthof I. North American Land Change System: Canadian Perspective. In: 30th Canadian symposium on remote sensing, Lethbridge, Alberta. 2009.
13. Li S, Potter C, Hiatt C. Monitoring of net primary production in california rangelands using landsat and MODIS satellite remote sensing. Nat Resour. 2012. doi:10.4236/nr.2012.
14. Long S, Garcia Moya E, Imbamba S, Kamnalrut A, Piedade M, Scurlock J, Shen Y, Hall D. Primary productivity of natural grass ecosystems of the tropics: a reappraisal. Plant Soil. 1989;115:155–66.
15. Monteith J, Moss C. Climate and the efficiency of crop production in Britain. Philos Trans R Soc Lond Biol Sci. 1977;281(980):277–94.
16. Olson R, Scurlock J, Cramer W, Parton W, Prince S. From sparse field observations to a consistent global dataset on net primary production. IGBP-DIS Working Paper No. 16. 1997. IGBP-DIS. Toulouse, France.
17. Olthof I, Latifovic R, Pouliot D. Northern Land Cover of Canada. Natural Resources Canada: Canada Center for Remote Sensing. 2011; ISBN: 978-1-100-15034-5.
18. Ott L, Pawson S, Collatz G, Gregg W, Menemenlis D, Brix H, Rousseaux C, Bowman K, Liu J, Eldering A. Assessing the magnitude of CO_2 flux uncertainty in atmospheric CO_2 records using products from NASA's Carbon Monitoring Flux Pilot Project. J Geophys Res Atmos. 2015. doi:10.1002/2014JD022411.
19. Philip G, Watson D. A precise method for determining contoured surfaces. Aust Pet Explor Assoc J. 1982;22:205–12.
20. Potter C, Randerson J, Field C, Matson P, Vitousek P, Mooney H, Klooster S. Terrestrial ecosystem production: a process model based on global satellite and surface data. Global Biogeochem Cycles. 1993;7(4):811–41.
21. Potter C. Terrestrial biomass and the effects of deforestation on the global carbon cycle: results from a model of primary production using satellite observations. Bioscience. 1999;49(10):769–78.
22. Potter C, Klooster S, Myneni R, Genovese V, Tan P, Kumar V. Continental-scale comparisons of terrestrial carbon sinks estimated from satellite data and ecosystem modeling 1982–1998. Global Planet Change. 2003;39:201–13.
23. Potter C, Klooster S, Steinbach M, Tan P, Kumar V, Shekhar S, Carvalhos C. Understanding global teleconnection of climate to regional model estimates of Amazon ecosystem carbon fluxes. Glob Change Biol. 2004;10:693–703.
24. Potter C, Klooster S, Huete A, Genovese V. Terrestrial carbon sinks for the United States predicted from MODIS satellite data and ecosystem modeling. Earth Interact. 2007;(13):1–21.
25. Potter C, Klooster S, Hiatt C, Genovese V, Castilla-Rubio JC. Changes in the carbon cycle of Amazon ecosystems during the 2010 drought. Environ Res Lett. 2011;6(3):034024.
26. Potter C, Klooster S, Genovese V. Net primary production of terrestrial ecosystems from 2000–2009. Clim Change. 2012. doi:10.1007/s10584-012-0460-2.
27. Potter C, Klooster S, Genovese V, Hiatt C. Forest production predicted from satellite image analysis for the Southeast Asia region. Carbon Balance Manage. 2013;8:9.
28. Potter C. Combined use of climate and satellite image data to predict the timing and origin of dust bowl storms in the Plains states, USA. Int J Environ Sci. 2015;5:6.
29. Powell T, Gholz H, Clark K, Starr G, Croer W, Martin T. Carbon exchange of a mature, naturally regenerated pine forest in north Florida. Glob Change Biol. 2008;14(11):2523–38.
30. Priestly C, Taylor R. On the assessment of surface heat flux and evaporation using large-scale parameters. Mon Weather Rev. 1972;100:81–92.
31. Rich P, Dubayah R, Hetrick W, Saving S. Using viewshed models to calculate intercepted solar radiation: applications in ecology. American Society for Photogrammetry and Remote Sensing Technical Papers. American Society of Photogrammetry and Remote Sensing. 1994. p. 524–529.
32. Running S, Ramakrishna R, Heinsch F, Zhao M, Reeves M, Hashimoto H. A continuous satellite-derived measure of global terrestrial primary production. Bioscience. 2004;54(6):547–60.
33. Tobler WR. A computer movie simulating urban growth in the Detroit region. Econ Geogr. 1970;46:234–40.
34. Turner D, Ritts W, Cohen W, Gower S, Zhao M, Running S, Wofsy S, Urbanski S, Dunn A, Munger J. Scaling gross primary production (GPP) over boreal and deciduous forest landscapes in support of MODIS GPP product validation. Remote Sens Environ. 2003;88(3):256–70.
35. Turner D, Ritts W, Cohen W, Maeirsperger T, Gower S, Kirschbaum A, Running S, Zhao M, Wofsy S, Dunn A, Law B. Site-level evaluation of satellite-based global terrestrial gross primary production and net primary production monitoring. Glob Change Biol. 2005;11(4):666–84.
36. Turner D, Ritts D, Cohen W, Gower S, Running S, Zhao M, Costa M, Kirschbaum A, Ham J, Saleska S, Ahl D. Evaluation of MODIS NPP and GPP products across multiple biomes. Remote Sens Environ. 2006;102:282–92.
37. Wan Z. MODIS land surface temperature products user's guide. Institute for Computational Earth System Science, University of California, Santa Barbara, CA. 2006. http://www.icess.ucsb.edu/modis/LstUsrGuide/usr-guide.html.
38. Waring R, Landsberg J, Williams M. Net primary production of forests: a constant fraction of gross primary production? Tree Physiol. 1998;18(2):129–34.

39. Watson D, Philip G. A refinement of inverse distance weighted interpolation. Geoprocessing. 1985;2:315–27.

40. Weiss D, Atkinson P, Bhatt S, Mappin B, Hay S, Gething P. An effective approach for gap-filling continental scale remotely sensed time-series. ISPRS J Photogramm Remote Sens. 2014;98:106–18.

41. Yu D, Peijun S, Shao H, Zhu W, Pan Y. Modelling net primary productivity of terrestrial ecosystems in East Asia based on an improved CASA ecosystem model. Int J Remote Sens. 2009;30(18):4851–66.

42. Zhang Y, Xu M, Chen H, Adams J. Global pattern of NPP to GPP ratio derived from MODIS data: effects of ecosystem type, geographical location and climate. Glob Ecol Biogeogr. 2009;18(3):280–90.

43. Zhao M, Heinsch F, Nemani R, Running S. Improvements of the MODIS terrestrial gross and net primary production global data set. Remote Sensing Environ. 2005;95:164–76.

## 13

Community assessment of tropical tree biomass: challenges and opportunities for REDD+

Ida Theilade[1], Ervan Rutishauser[2]* and Michael K Poulsen[3]

Abstract

Background: REDD+ programs rely on accurate forest carbon monitoring. Several REDD+ projects have recently shown that local communities can monitor above ground biomass as well as external professionals, but at lower costs. However, the precision and accuracy of carbon monitoring conducted by local communities have rarely been assessed in the tropics. The aim of this study was to investigate different sources of error in tree biomass measurements conducted by community monitors and determine the effect on biomass estimates. Furthermore, we explored the potential of local ecological knowledge to assess wood density and botanical identification of trees.

Results: Community monitors were able to measure tree DBH accurately, but some large errors were found in girth measurements of large and odd-shaped trees. Monitors with experience from the logging industry performed better than monitors without previous experience. Indeed, only experienced monitors were able to discriminate trees with low wood densities. Local ecological knowledge did not allow consistent tree identification across monitors.

Conclusion: Future REDD+ programmes may benefit from the systematic training of local monitors in tree DBH measurement, with special attention given to large and odd-shaped trees. A better understanding of traditional classification systems and concepts is required for local tree identifications and wood density estimates to become useful in monitoring of biomass and tree diversity.

Keywords: Community monitoring, Tree biomass, Indonesia, Wood density, Species identification, MRV, REDD+

Background

Programs aiming at curbing deforestation and forest degradation in tropical regions (REDD+) rely upon cost-efficient techniques to monitor, report and verify forest carbon stocks. A complete enumeration of all living plants in a given landscape is impossible, and most studies rely upon a "sample plot" approach in which all trees are measured. However, the representativeness of a plot network for an entire landscape remains challenging to ascertain [1], but recommendations on the shape, size or number of sample plots have recently been proposed (e.g. [2–4]).

While professional foresters or scientists are generally in charge of establishing such sample plots, several

*Correspondence: er.rutishauser@gmail.com
[2] CarboForExpert, 1248 Hermance, Switzerland
Full list of author information is available at the end of the article

REDD+ projects have recently shown how local communities might represent a cheap and efficient alternative to external professionals [5–7]. In South East Asia, community monitoring was able to measure forest carbon stocks with similar accuracy as that of professional foresters [5]. Error in plot-level biomass estimates carried out by non-professional ranged between ±10% [5, 8]. At plot-level, error in biomass estimates can be divided into: (1) model error, such as the choice of a particular allometric model, prediction errors or error on the model parameters [9, 10], and (2) measurement error on the tree growth variables (e.g. tree diameter or height) or omission of trees. To mitigate these errors, standardized protocols and practices have been developed [11, 12] and generic allometric models to estimate tree biomass are now widely applied.

However, a significant difference in community vs forester's estimates of biomass (381 vs 449 Mg ha^{-1} respectively) was found in Indonesia by Danielsen and colleagues [5]. This discrepancy is exclusively due to measurement errors, as tree biomass was computed using the same model for both observers. In dense tropical forests, errors of measurement may be due to the presence of buttresses, irregular-shaped trunks, misplacement of the tape measure on the trunk, misreading of the actual measure or error of transcription on the tally sheet. Most REDD+ pilot programs use temporary sample plots to assess carbon stocks. The lack of repeated measurements prevents the assessment of measurements' accuracy and precision. Indeed, tree diameter could be measured accurately (mean of replicates close to the true value), but imprecisely (high variance among replicates), or precisely (low variance of replicates) but inaccurately (e.g. measured with an instrument calibrated with an incorrect standard) [13]. As a consequence, both imprecision or inaccuracy may inflate the uncertainty surrounding tree biomass estimates.

Large tropical trees are known to be more challenging to measure due to large buttresses or odd-shape stems [14], while they account for a large fraction of aboveground biomass [15]. Hence, forests with numerous large trees are more prone to be affected by errors of measurement and to large uncertainties in their biomass estimates. Due to lack of time, data precision and accuracy are barely assessed and reported in forest carbon monitoring. However, assessing main sources of error will help identifying areas where more investment in explanations and training are needed.

Another source of uncertainty relates to tree wood density (WD) that may vary at tree, species and landscape scales [16, 17]. In low accuracy estimation of carbon stocks (Tier 1), WD are approximated by an average regional default value [18, 19]. More sophisticated tree biomass estimates (Tiers 2 and 3) rely upon allometric models based on WD, tree height and diameter at breast height (DBH) [20]. Hence, botanical identification of trees is an important investment for REDD+ activities to accurately estimate tree biomass and monitor biodiversity. Due to the low number of tropical tree taxonomy experts, it has been proposed that para-taxonomists (people who lack formal education, but who are trained to undertake taxonomic tasks) can provide information at a greater rate and at a lower cost compared to expert botanists and conventional approaches [21]. Even though some communities seems to name trees consistently [22, 23], a previous study from Central Kalimantan, Indonesia resulted in poor matching between vernacular names and actual taxa, possibly due to the variety of dialects encountered [24]. On the other hand, wood densities have been found

to be relatively homogeneous within Indonesian tree genera [25], and a congruent identification of the common genera by local monitors could replace the use of average WD with genus-specific values and reduce uncertainties in corresponding forest carbon stock estimates.

The present study addresses the following questions:

1. How accurate and precise are tree diameter measurements carried out by community monitors?
2. Does prior experience from logging inventories reduce measurement errors?
3. Is local ecological knowledge useful for tree identifications?
4. How do different sources of error propagate into tree biomass estimates?

Results
Source of errors in tree diameter measurements
Tree girth of 103 trees were measured by eleven local monitors, with 95% of all measurements comprised between −5.73 and 5.83 cm around the actual DBH value. Only 86 measurements out of 1,749 felt out of this confidence interval, designated hereafter to as "large errors". Large errors were more frequent and of greater magnitude (i.e. larger SD) among trees with large DBH (Figure 1). Errors were biased positively, and stand-level biomass was generally overestimated (range −4 to +20%; mean +7%). Half (52.3%) of this errors ($|DBH_{mes} - DBH_{mean}| > 6$ cm) were found among trees designated as having "odd shape" by local monitors, while these trees made up only 16% of the sample. A fifth of the measurements done on trees with odd shape was affected by large errors, significantly more than those carried on more regular stems (16 vs 3% respectively, $\chi^2 = 81.3$, df $= 1$, P $< 10^{-5}$).

Prior experience in measuring trees did not significantly decrease the likelihood of doing a large error ($\chi^2 = 2.5$, df $= 1$, P $= 0.11$). But when the repeatability of measurement was investigated, experienced monitors performed better. Difference in paired DBH measurements significantly differed (Pairwise Student test: t $= -2.34$, df $= 146.4$, P $= 0.02$) among experienced and inexperienced monitors, averaging 0.9 and 2.4 cm respectively (Figure 2).

Estimating wood hardness
For each tree, local monitors were also asked to estimate the wood density on a 3-classes scale (i.e. very light, light and heavy). While this simple classification returned generaly poor results (Figure 3), experienced monitors were able to discriminate trees with low wood densities (Figure 3, ANOVA: $F_{2,613} = 11.76$, P $< 10^{-4}$) while inexperienced monitors could not (ANOVA: $F_{2,511} = 0.424$, P $= 0.655$).

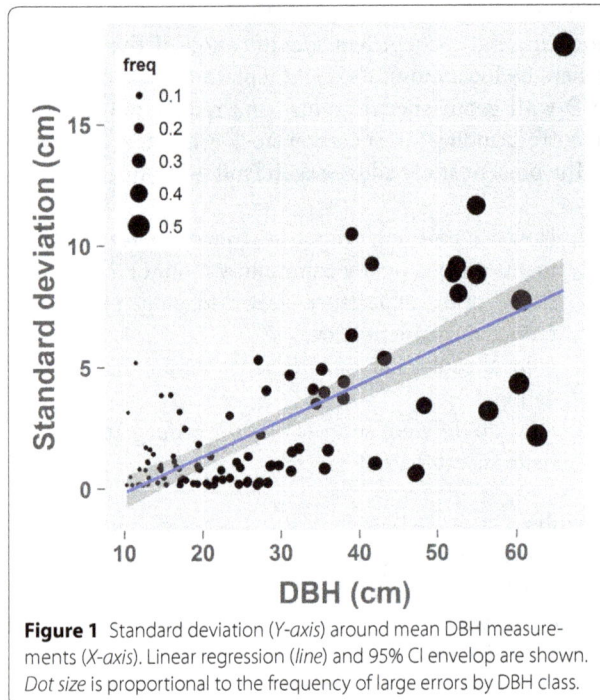

Figure 1 Standard deviation (*Y-axis*) around mean DBH measurements (*X-axis*). Linear regression (*line*) and 95% CI envelop are shown. *Dot size* is proportional to the frequency of large errors by DBH class.

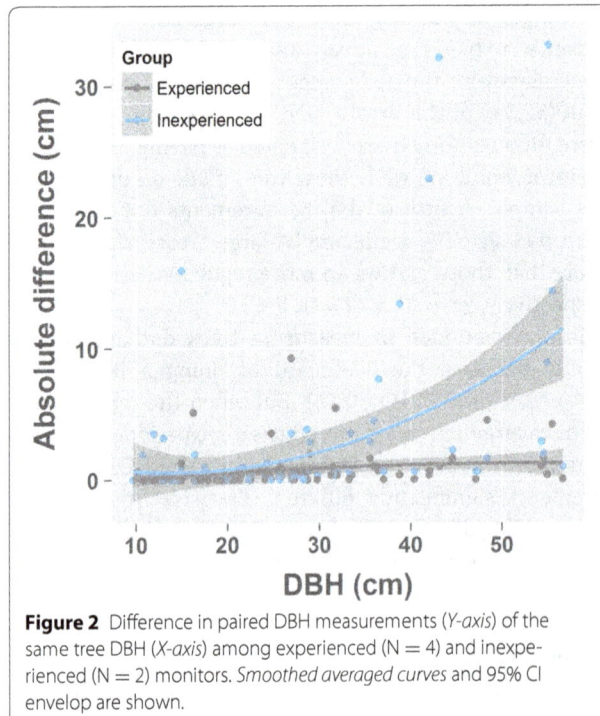

Figure 2 Difference in paired DBH measurements (*Y-axis*) of the same tree DBH (*X-axis*) among experienced (N = 4) and inexperienced (N = 2) monitors. *Smoothed averaged curves* and 95% CI envelop are shown.

Vernacular identification

The third information collected in the field was the vernacular name of each tree. Overall, there was very little agreement among observers in naming trees (Figure 4).

For instance, the number of vernacular names averaged nine per taxa. More consistency was found among Dipterocarp trees, which were better identified by experienced monitors than inexperienced ones (ANOVA: $F_{1,42} = 10.55$, P = 0.002).

Propagating error of DBH measurement and wood hardness into tree biomass estimates

For both experienced and inexperienced monitors, the bias increased with tree biomass (Figure 5). When accounting for DBH measurements and average wood density per wood hardness class, experienced monitors performed better and generated lower bias compared to their inexperienced counterparts (Figure 5, Estimates 1). When all trees were assigned the same wood density, biases lowered but remained high for large trees (Figure 5, Estimates 2).

Discussion

Tree diameter measurements

Overall, local monitors had good ability to measure trees, with 95% of the measurements found within 6 cm around the actual DBH. Large errors were not randomly distributed, but increased in frequency (i.e. number of occurence) and magnitude (i.e. breath of SD) with DBH (Figure 1). Half of these errors were found among odd-shaped trees, while these trees made up only 16% of the sample. A fifth of the repeated measures done on odd-shaped trees was affected by at least one large error, significantly more than among regular stems (16 vs 3%). When averaged out at stand level, we found a significant bias towards larger DBH measurements that resulted in an stand-level biomass overestimation of 7%. This error remain low and of similar magnitude as that reported in other studies [25, 26]. We have decided to use the most recent allometric models to calculate tree biomass, as generic models were shown to perform better at our site [27]. However, we acknowledge that the choice of a particular allometric model may result in greater inaccuracies than the physical measurements described above [9].

Beyond tree measurements

As botanical identification is mandatory to determine specific WD and calculate tree biomass, two methods were tested to see whether local knowledge could help towards this task. The introduction of a simple 3-scales wood hardness classification returned unconvincing results (Figure 3), as inexperienced monitors were not able to distinguish between hardwood classes while experienced monitors were able to distinguish very light wood only. Likewise, more consistency was found among experienced monitors to name Dipterocarp trees (Figure 4), i.e. the main commercial timber family

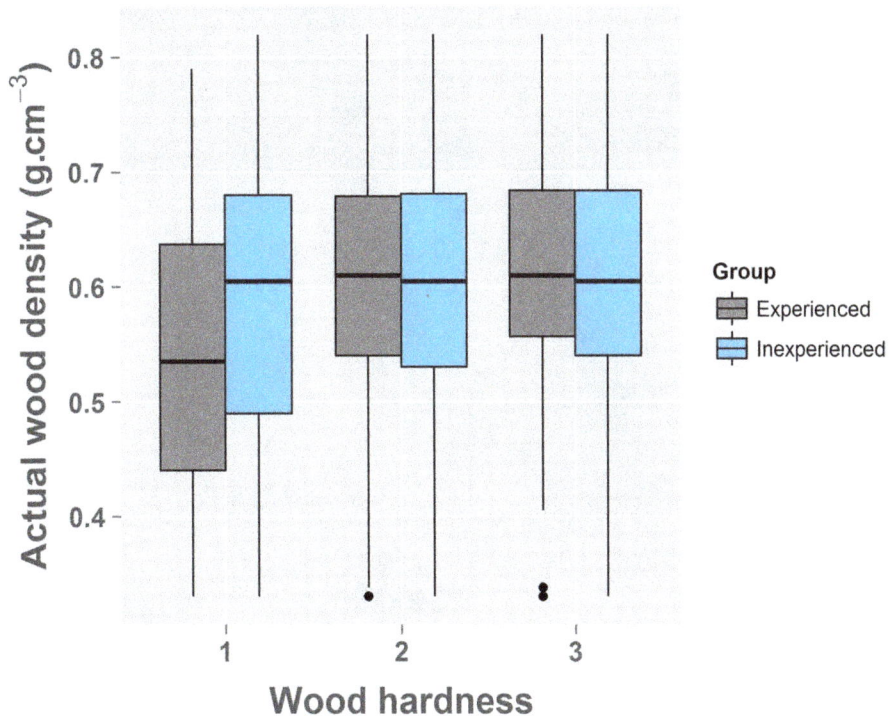

Figure 3 Boxplot of wood densities by wood hardness class estimated by experienced (*grey*) and inexperienced (*blue*) observers.

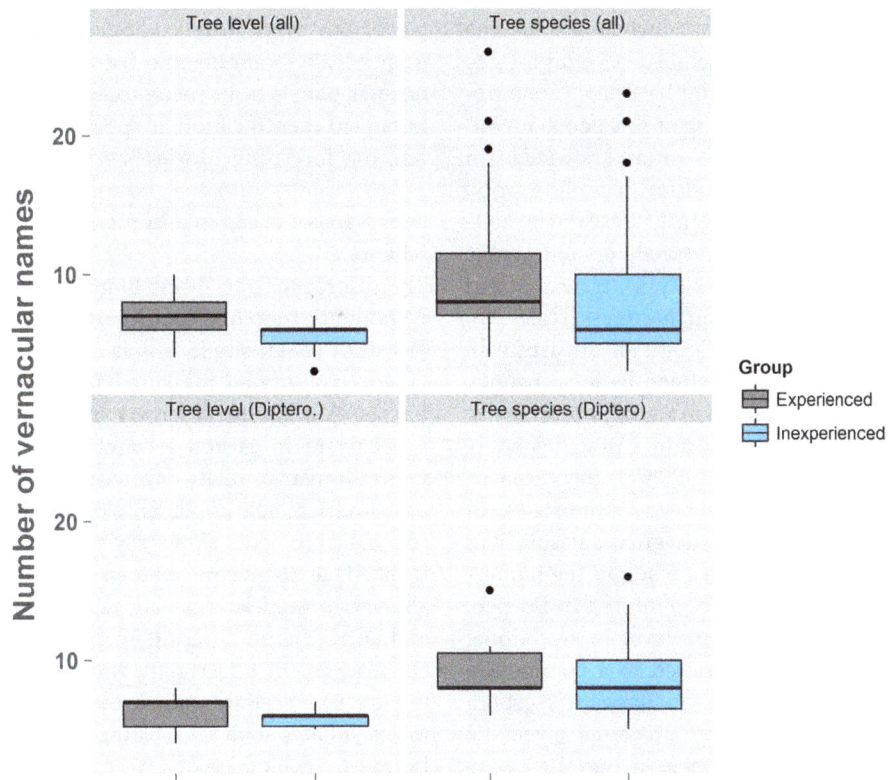

Figure 4 Number of vernacular names (*boxplots*) at tree and species by experienced and inexperienced monitors for all trees (*top*) and Dipterocarps only (*bottom*).

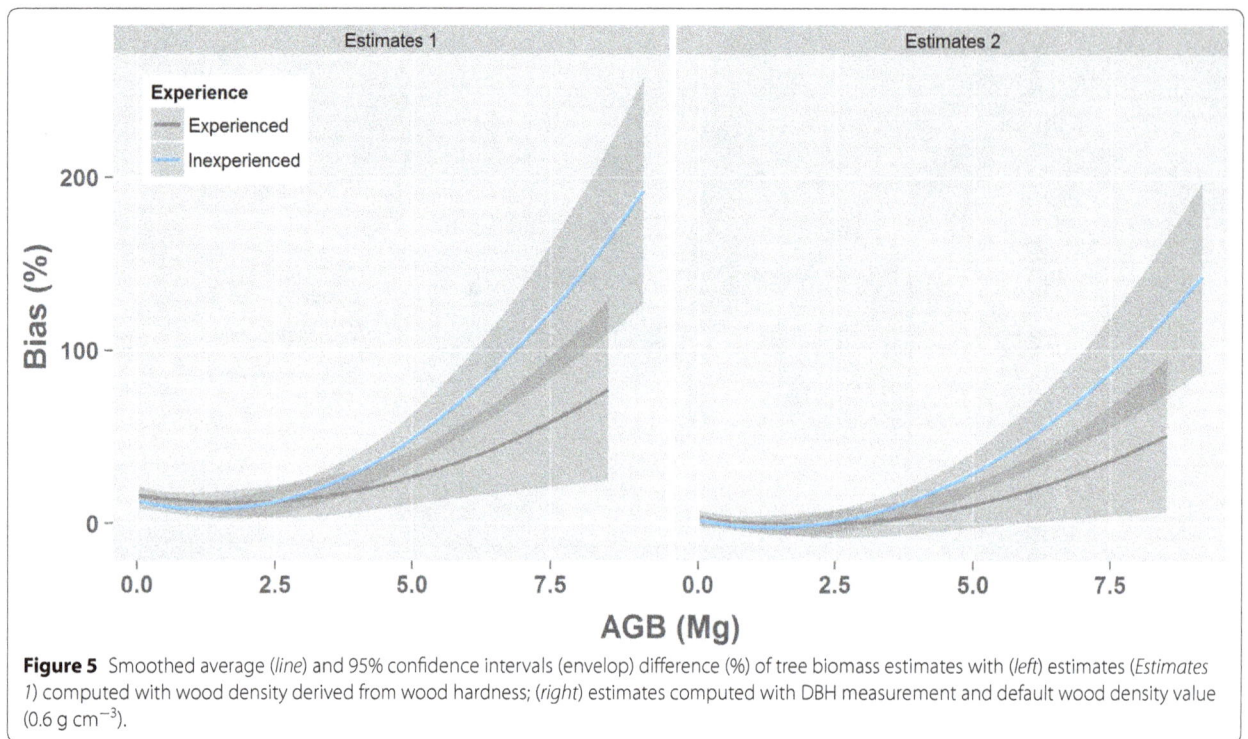

Figure 5 Smoothed average (*line*) and 95% confidence intervals (envelop) difference (%) of tree biomass estimates with (*left*) estimates (*Estimates 1*) computed with wood density derived from wood hardness; (*right*) estimates computed with DBH measurement and default wood density value (0.6 g cm^{-3}).

in the region. This is not surprising as their experience consists mainly in identifying commercial hard wood species, including Dipterocarps, during pre-logging inventories [28]. The overall inability of monitors to classify trees based on coarse wood hardness categories may arise from a misunderstanding of this peculiar concept. Local people usually possess sound knowledge on what different species can be used for, including wood properties such as workability, termite resistance, suitability for tools, firewood or boat-making. A possible explanation may lie in their inability to 'translate' this knowledge into this simple wood hardness scale. We suggest that future studies take point of departure in emic categories, i.e. categories defined by local people. Overall, there was little agreement among observers in naming trees. This result corroborate a previous study carried in Borneo, where only 10–20% of the vernacular names employed by Dayak para-taxonomists could be related to a given taxa [24]. The great variability in vernacular names in the region is a result of the numerous ethnic groups and dialects encountered in Borneo. Locally, trees are named based on local or traditional usage and names might be restricted to a community or even a group of villagers. Different species or genera having similar properties or usage are often given the same vernacular name. For instance, at our site, some trees were given names that can be translated as "big tree". Hence collection and interpretation of vernacular

names remains challenging. However, vernacular names remain employed in the logging industry and timber trade, but with little consistency with scientific taxonomy [23]. Refining the list of commonly used vernacular names of Bornean trees, and the corresponding botanical identification at species or genus level would improve forest inventories based on vernacular names.

Improvement of community monitoring in a REDD+ scheme

The discrepancy in forest biomass stocks measured by community monitors and foresters reported in a previous study at our site [5, 28], is likely to be due to the difficulty to accurately measure large trees in dense tropical forests. Measurement errors among odd-shaped trees is recurrent in carbon accounting studies. As tree biomass allometries relate dry mass with a theoretical taper or cylindrical bole diameter, biomass estimation requires tree measurements above any major irregularities of the trunk. Due to the polynomial form of current generic allometric models, a linear relationship between error and DBH (Figure 1) results mechanically in an exponential inflation of uncertainty when expressed in biomass (Figure 5). We have shown that error in biomass estimates inflates with tree biomass and inexperience. For instance, the biomass of a typical tree of 7.5 ton might be over/underestimated by 47 or 80% by an experienced or inexperienced observer respectively (Figure 5, Estimates

2). This difference goes up to 55 and 120% respectively, when estimated WD are included into biomass computation (Figure 5, Estimates 1).

This issue becomes more acute when monitoring forest biomass over time, as rapid radial increments of buttresses will compound the overestimation of biomass increase [29]. While local monitors accurately measured DBH of most trees, much attention and training should be paid on large trees (>60 cm DBH). Prior experience in measuring trees did not lower the likelihood of doing large errors, but increased accuracy of repeated measurements. Thereby, trained monitors are less prone to systematic bias, a key feature in terrestrial carbon monitoring where true biomass value is sought. Accuracy will also be requested to estimate changes in forest carbon stocks over repeated censuses. Indeed, error of measurements and data correction might prevent the detection of any directional change in biomass stock [30].

In a multi-country comparison of the efficiency (i.e. costs and accuracy) of local communities to monitor tree biomass stocks, Brofeldt and collaborators [28] relied at the second census upon a few community members trained initially, while the rest of team received a brief training only. Based on this study, we recommend that all community monitors involved in REDD+ programmes receive a complete training on tree measurement with special attention on dealing with large and odd-shaped trees. When multi-census has to be carried out, points of measurements should be clearly marked in the field (i.e. paint mark on the trunk). Technical improvements to increase accuracy of community-based measurements of carbon stock will likely facilitate the uptake and scaling up of local information as part of the national forest monitoring system (NFMS) and the associated monitoring, reporting, and verification (MRV) system for REDD+ [31]. This is in line with current United Nations Framework Convention on Climate Change (UNFCCC) texts and guidance documents on the technical aspects of REDD+ which outline explicit roles for indigenous people and local communities in implementing REDD+ [32–34].

Conclusion

Several REDD+ studies have recently shown how community monitors represent a cost-efficient and reliable alternative to external professionals. In this study, we have investigated different sources of error in tree diameter measurements conducted by community monitors and propagated those at both tree and stand levels biomass estimates.

Local monitors had good ability to measure tree DBH with 95% of all measurements found within a confidence interval of 6 cm around the actual DBH. Large errors were more frequent and of greater magnitude among trees with a large DBH (>60 cm DBH) and odd-shaped trunks. Monitors with experience from logging inventories performed better and generated lower bias compared to inexperienced monitors although the likelihood of large errors was identical among both groups. Overall, we found a directional bias towards overestimated DBH among monitors that led to a slight inflation of stand-level biomass (7%).

We suggest that future REDD+ programmes may benefit from the systematic training of local monitors in measuring tree DBH with special attention given to large and odd-shaped trees. A better understanding of traditional classification systems and concepts, possibly combined with a basic training of local monitors in taxonomy, is required for tree identifications to become useful in monitoring either forest biomass, or tree diversity.

Methods
Study site and community monitors
The study area is located in the district of Kutai Barat District, East Kalimantan, Indonesia. Monitoring plots were established in the customary forest surrounding the Dayak village of Batu Majang. The tropical lowland rainforest at 300 m.a.s.l. is characterised by species of the Dipterocarp family such as *Shorea* sp., *Dipterocarpus* sp., *Anisoptera* sp., and *Hopea* sp. among other high quality timber species. Despite the customary harvest of a few trees and other non-wood forest products, the forest structure is similar to that of a primary forest. The local community is committed to conserve the forest for various reasons, such as protecting the watershed and hunting/harvesting resources. Several permanent forest plots were established in 2012, in which all trees >10 cm DBH were tagged, measured and identified to species level [27].

Representatives of the local Dayak community helped select eleven participants (referred hereafter to as community monitors) based on their interest and experience with forest resources, to measure the girth, estimate wood density, and identify trees in the permanent plots. All community monitors were male, had attended primary school, and received 3 h of specific training on tree measurement in the field. Six monitors had a prior employment in timber companies, doing surveys (i.e. mapping harvestable stems) for logging operations. This group is referred to as "experienced", while others (n = 5) with no previous experience are referred to as "inexperienced".

Data collected
In 2014, 103 trees were randomly chosen among two permanent monitoring plots and measured by local monitors. While creating a tree-walk and numbering the trees,

the community monitors were trained at measuring tree girth and estimate wood hardness. Girth measurement was done at 130 cm height using classical tapes with centimeter units. Monitors were instructed carefully to avoid common mistakes such as a twisted or lax tape, a thumb placed under the tape, and measuring below breast height.

When measurement was hampered by the presence of buttresses, lianas, or trunk deformities, i.e. extra efforts had to be made to measure tree, monitors were asked to record the tree as "odd shaped". Wood properties of common tree species is often known by local communities. To test whether such information could be used to refine tree biomass estimates, each monitor was asked to assess wood hardness using a simple classification: "1" for very light wood, "2" for floater (light wood) and "3" for sinker (heavy wood). These categories are used in the logging industry and are well-known to local people. Finally, monitors were asked to name each tree using Dayak common names. Community members worked in teams of two people, monitor A measuring the girth, assessing wood density and naming trees along the full tree-walk, and monitor B writing down information on a pre-prepared form.

Statistical analysis

Overall precision

We investigated the distribution of error measurements on a per-tree basis. As each tree was measured at least once by the different community members, we computed the differences between each measurement and the average DBH for each tree. We further used the 5th and 95th percentiles of these differences to identify large errors. For each tree, we defined the actual DBH (DBH_{mean}), as the average of all measurements comprised within the 5th and 95th percentiles. The minimum number of measurements used to compute the actual DBH is 12 (max = 17).

The precision of measurements of a given tree diameter refers to the variance of the different measurements. We used the standard deviation to estimate how the different measures spread out from the mean value. The bigger the error, the larger the standard deviation.

$$SD(\sigma) = \sqrt{\frac{1}{n} \sum \left(DBH_{mes,i} - DBH_{mean,i} \right)^2}$$

Repeatability of measurement

102 trees were measured twice by six observers. We estimated the repeatability of girth measurements among those observers, by calculating the absolute difference among both measurements.

Comparison of wood hardness and botanical estimation

In 2012, all trees were identified at species level by a professional botanist [27]. Trees were identified directly in the field to the lowest taxonomical level. Among the 102 trees accounted for in the present study, 70% were identified at species level and 30% at genus level (Additional file 1). From these identifications, wood densities were extracted from the Global Wood Density Database [35] and considered as actual wood densities (WD). The capacity of local observers to group trees in three classes of wood hardness was further assessed with a one-way ANOVA by wood hardness classes and observers experience.

Error propagation in tree biomass estimates

We integrated information gathered in the field by local monitors (i.e. wood hardness and DBH measurements) into biomass estimates. Wood hardness was associated to the 25, 50 and 75th percentile of actual wood densities respectively (1 = 0.55, 2 = 0.63, 3 = 0.73 g cm^{-3}). Tree biomass (Estimates 1) was computed using a generic allometric model [20], as follow:

$$AGB_{est} = \exp[-1.803 - 0.976 \times E + 0.976 \times \ln(WD)$$
$$+2.673 \times \ln(DBH) - 0.0299 \times \ln(DBH)^2]$$

where E is a synthetic index of temperature seasonality, maximum climatological water deficit, and precipitation seasonality (E = −0.09162301 at our site), WD is the wood density (g cm^{-3}), and DBH, the diameter at breath height (cm).

Alternatively, tree biomass (Estimates 2) was computed using a default WD value for Bornean forests (WD = 0.6, 37) to estimate a "Tier 1" level of uncertainty. Both estimates were further compared to the best tree biomass estimate (AGB_0), computed with actual WD and DBH (DBH_{mean}) as recommended by Tier 3 standard [19]. Differences in tree biomass are expressed as bias (e.g. [estimate$_1$ − AGB_0]/AGB_0). To check if errors could cancel each other at stand level (i.e. no directional bias), tree biomass were summed for each monitor and the relative bias (%) per monitor was computed as follow:

$$bias_j(\%) = \frac{\sum AGB_{ij} - \sum AGB_0}{\sum AGB_0},$$

where i = the ith tree, j = the jth monitor and AGB_0 = best tree biomass estimate.

Additional file

Additional file 1: Table S1. List of local names commonly used by Dayaks in Batu Majang, Kutai Barat, East Kalimantan, Indonesia.

Authors' contributions

IT, ER, MKP equally contributed in designing the protocol and writing the manuscript; MKP conducted data collection and ER did the statistical analysis. All authors read and approved the final manuscript.

Author details

[1] Faculty of Science, Institute of Food and Resource Economics, University of Copenhagen, Rolighedsvej 25, 1958 Frederiksberg C, Denmark. [2] Carbo-ForExpert, 1248 Hermance, Switzerland. [3] Nordic Agency for Development and Ecology (NORDECO), Skindergade 23, 1159 Copenhagen K, Denmark.

Acknowledgements

We are most grateful to Pak Yosep, head of village of Batu Majang, Agus, Anse Latus, Lusang, Prin, Samuel Ajang, Simon, Sius, Syahdan, Vincen Idum, Vincensius Yen and Sarjuni for conducting field work. We thanks Yuyun Karniawan and Itong Sarjuni for facilitating liaison and transportation, and Pak Yosep for housing Itong Sarjuni during the field visit. We also thank Kristell Hergoualch (CIFOR) and Andreas de Neergaard (University of Copenhagen) for providing tree botanical identifications.

Compliance with ethical guidelines

Competing interests

The authors declare that they have no competing interests.

References

1. Chave J, Condit R, Aguilar S, Hernandez A, Lao S, Perez R (2004) Error propagation and scaling for tropical forest biomass estimates. Philos Trans R Soc B Biol Sci 359:409–420
2. Baraloto C, Molto Q, Rabaud S, Hérault B, Valencia R, Blanc L et al (2013) Rapid simultaneous estimation of aboveground biomass and tree diversity across neotropical forests: a comparison of field inventory methods. Biotropica 45:288–298
3. Wagner F, Rutishauser E, Blanc L, Herault B (2010) Assessing effects of plot size and census interval on estimates of tropical forest structure and dynamics. Biotropica 42:664–671
4. Walker SM, Pearson T, Casarim FM, Harris H, Petrova S, Grais A et al (2012) Standard operating procedures for terrestrial carbon measurement. Winrock International, USA
5. Danielsen F, Adrian T, Brofeldt S, van Noordwijk M, Poulsen MK, Rahayu S et al (2013) Community monitoring for REDD+: international promises and field realities. Ecol Soc 18:41
6. Larrazábal A, McCall MK, Mwampamba TH, Skutsch M (2012) The role of community carbon monitoring for REDD+: a review of experiences. Curr Opin Environ Sustain 4:707–716
7. Butt N, Slade E, Thompson J, Malhi Y, Riutta T (2013) Quantifying the sampling error in tree census measurements by volunteers and its effect on carbon stock estimates. Ecol Appl 23:936–943
8. Molto Q, Rossi V, Blanc L (2013) Error propagation in biomass estimation in tropical forests. Methods Ecol Evol 4:175–183
9. Picard N, Boyemba Bosela F, Rossi V (2014) Reducing the error in biomass estimates strongly depends on model selection. Ann For Sci. doi:10.1007/s13595-014-0434-9
10. GOFC-GOLD (2012) A sourcebook of methods and procedures for monitoring and reporting anthropogenic greenhouse gas emissions and removals caused by deforestation, gains and losses of carbon stocks in forests remaining forests, and forestation. Wageningen, Global Observation of Forest Cover and Land Dynamic (GOFC-GOLD)
11. IPCC (2014) 2013 Revised supplementary methods and good practice guidance arising from the kyoto protocol
12. Clark DB, Kellner JR (2012) Tropical forest biomass estimation and the fallacy of misplaced concreteness. J Veg Sci 23:1191–1196
13. Clark DA (2002) Are tropical forests an important carbon sink? Reanalysis of the long-term plot data. Ecol Appl 12:3–7
14. Slik J, Paoli G, McGuire K, Amaral I, Barroso J, Bastian M et al (2013) Large trees drive forest aboveground biomass variation in moist lowland forests across the tropics. Glob Ecol Biogeogr 22:1261–1271
15. Chave J, Muller-Landau HC, Baker TR, Easdale TA, Ter Steege H, Webb CO (2006) Regional and phylogenetic variation of wood density across 2,456 neotropical tree species. Ecol Appl 16:2356–2367
16. Henry M, Besnard A, Asante W, Eshun J, Adu-Bredu S, Valentini R et al (2010) Wood density, phytomass variations within and among trees, and allometric equations in a tropical rainforest of Africa. For Ecol Manag 260:1375–1388
17. Brown S (1997) Estimating biomass and biomass change of tropical forests: A primer. FAO Forestry Paper, vol 134. UN FAO, Rome (FAO [series editor]: Forestry Paper)
18. IPCC (2006) Guidelines for national greenhouse gas inventories, vol 4. Institute for Global Environmental Strategies (IGES), Hayama (Eggelstons S, Buendia L, Miwa K, Todd N, Tanabe K [series editors])
19. Chave J, Réjou-Méchain M, Búrquez A, Chidumayo E, Colgan MS, Delitti WB et al (2014) Improved allometric models to estimate the aboveground biomass of tropical trees. Glob Chang Biol 20:3177–3190
20. Sheil D, Lawrence A (2004) Tropical biologists, local people and conservation: new opportunities for collaboration. Trends Ecol Evol 19:634–638
21. Jinxiu W, Hongmao L, Huabin H, Lei G (2004) Participatory approach for rapid assessment of plant diversity through a folk classification system in a tropical rainforest: case study in Xishuangbanna, China. Conserv Biol 18:1139–1142
22. de Lacerda AEB, Nimmo ER (2010) Can we really manage tropical forests without knowing the species within? Getting back to the basics of forest management through taxonomy. For Ecol Manag 259:995–1002
23. Wilkie P, Saridan A (1999) The limitations of vernacular names in an inventory study, Central Kalimantan, Indonesia. Biodivers Conserv 8:1457–1467
24. Slik JWF (2006) Estimating species-specific wood density from the genus average in Indonesian trees. J Trop Ecol 22:481
25. Venter M, Venter O, Edwards W, Bird MI (2015) Validating community-led forest biomass assessments. PLoS One 10:e0130529
26. Butt N, Epps K, Overman H, Iwamura T, Fragoso JMV (2015) Assessing carbon stocks using indigenous peoples' field measurements in Amazonian Guyana. For Ecol Manag 338:191–199
27. Rutishauser E, Noor'an F, Laumonier Y, Halperin J, Rufi'ie, Hergoualch K, Verchot L (2013) Generic allometric models including height best estimate forest biomass and carbon stocks in Indonesia. For Ecol Manag 307:219–225
28. Brofeldt S, Theilade I, Burgess ND, Danielsen F, Poulsen MK, Adrian T et al (2014) Community monitoring of carbon stocks for REDD+: does accuracy and cost change over time? Forests 5:1834–1854
29. Sheil D (1995) A critique of permanent plot methods and analysis with examples from Budongo Forest, Uganda. For Ecol Manag 77:11–34
30. Muller-Landau HC, Detto M, Chisholm RA, Hubbell SP, Condit R (2014) Detecting and projecting changes in forest biomass from plot data. In: Coomes DA, Burslem DFRP, Simonsen WD (eds) Forests and global change. Cambridge University Press, Cambridge, pp 381–416
31. Torres A (2014) Potential for integrating community-based monitoring into REDD+. Forests 5:1815–1833
32. UNFCCC (2011) Framework convention on climate change, subsidiary body for scientific and technological advice (SBSTA), methodological guidance for activities relating to reducing emissions from deforestation and forest degradation and the role of conservation, sustainable management of forests and enhancement of forest carbon stocks in developing countries. Draft conclusions proposed by the Chair, Thirty-fifth session Durban, 28 November to 3 December 2011. UNFCCC, Bonn
33. UNFCCC (2011) Outcome of the work of the ad hoc working group on long-term cooperative action under the convention. Draft decision [-/CP.17]. UNFCCC, Bonn
34. UNFCCC (2009) Methodological guidance for activities relating to reducing emissions from deforestation and forest degradation and the role of conservation, sustainable management of forests and enhancement of forest carbon stocks in developing countries. Decision 4/CP.15, FCCC/CP/2009/11/Add.1. United Nations Framework Convention on Climate Change, Copenhagen
35. Zanne AE, Lopez-Gonzalez G, Coomes DA, Ilic J, Jansen S, Lewis SL et al (2009) Global wood density database. http://datadryad.org/repo/handle/10255/dryad.235. Accessed 28 May 2013

## 14

Water-air CO_2 fluxes in the Tagus estuary plume (Portugal) during two distinct winter episodes

Ana P Oliveira[1*†], Marcos D Mateus[2†], Graça Cabeçadas[1] and Ramiro Neves[2]

Abstract

Background: Estuarine plumes are frequently under strong influence of land-derived inputs of organic matter. These plumes have characteristic physical and chemical conditions, and their morphology and extent in the coastal area depends strongly on physical conditions such as river discharge, tides and wind action. In this work we investigate the physical dynamics of the Tagus estuary plume and the CO_2 system response during two contrasting hydrological winter periods. A hydrodynamic model was used to simulate the circulation regime of the study area, thus providing relevant information on hydrodynamic processes controlling the plume.

Results: Model simulations show that for the studied periods, the major cause of the plume variability (size and shape) was the interaction between Tagus River discharge and wind. The freshwater intrusion on Tagus shelf exerted considerable influence on biochemical dynamics, allowing identification of two regions: a high nutrient region enriched in CO_2 inside the estuarine plume and another warmer region rich in phytoplankton in the outer plume.

Conclusions: The Tagus estuarine plume behaved as a weak source of CO_2 to the atmosphere, with estimated fluxes of 3.5 ± 3.7 and 27.0 ± 3.8 mmol C m^{-2} d^{-1} for February 2004 and March 2001, respectively.

Background

Coastal regions are significantly influenced by land-derived discharges emanating from estuaries, with estuarine plumes mediating the fluxes of natural terrestrial compounds and pollutant into shelf seas [1,2]. The extent and morphology of estuarine plumes are a direct consequence of river discharge, but are also strongly dependent on other physical conditions such as tide and wind stress.

An essential characteristic of estuarine plumes may be defined by a significant salinity gradient, although the boundary of the plume is often difficult to define given the highly dynamic nature of such systems [1]. Furthermore, highly stratified plumes lead to well defined density fronts along their boundaries, where turbidity is relatively low and chlorophyll *a* relatively high, even in winter [3].

Some studies concerning nutrients, fluxes of organic constituents and phytoplankton have been undertaken in estuaries and/or their associated plumes [1,4-8]; some highlight the seasonality CO_2 source/sink behaviour of

the estuarine plumes [9-12], and only a few refer the estuarine plume dynamics and the carbonate system response [13,14]. However, the CO_2 uptake capacity of the estuarine plumes in several continental shelf zones is already extensively reported [2,3,15-20], suggesting that other estuarine plumes might counteract inner estuary CO_2 emissions. The processes controlling the CO_2 dynamics in the estuarine plume are linked to various factors such as spring/summer phytoplankton blooms, thermodynamic effects, winter floods from the inner estuary or stratification/mixing of the plume water column. On an annual basis, these processes together with the complexity of near shore ecosystems, can significantly impact water-air CO_2 exchanges in estuarine plumes [13,14].

Significant drawdown of CO_2 partial pressure (pCO_2), biological uptake of dissolved inorganic carbon (DIC) and an associated enhancement of dissolved oxygen and pH within plumes occur due to enhanced biological activity, as reported for the Mississippi River plume in the USA [21,22], the Scheldt plume in Belgium [23], and the Pearl River estuary in China [8]. For the Changjiang Estuary plume (China), the pCO_2 drawdown and DO enhancement in the warm seasons (from April to October) appeared to be controlled by primary productivity and

* Correspondence: aoliveira@ipma.pt
†Equal contributors
[1]Instituto Português do Mar e da Atmosfera (IPMA), I.P., Avenida de Brasília, 1449-006 Lisboa, Portugal
Full list of author information is available at the end of the article

water-air exchange, while mixing dominated the aqueous pCO_2 in the cold seasons extending from November to March of the following year [15]. Mixing of river water with Gulf of Maine waters as also been pointed as responsible for the carbon variability in this system [24], although biological processes were significantly intense during the spring and summer seasons. Biological activity also lowers Amazon River plume pCO_2, and contributes to a CO_2 deficit in the northern western tropical North Atlantic Ocean that outlasts the plume's physical structure [25].

This paper aims to characterize the dynamics of water-air CO_2 flux in the Tagus estuarine plume (Figure 1) during two contrasting winter periods, based on the pCO_2 dynamics derived from field data. Underlying controlling mechanisms have been investigated based on the river discharge, the role of temperature and the biological activity. This study merges field data retrieved by experimental methods with information derived from the results of a numerical model on the spatial and temporal variability of the physical structure of the plume.

Results and discussion

The sampling programs were carried out in winter, from 7 to 19 March 2001 and from 5 to 9 February 2004.

Environmental settings

The first three months of 2001, with mean air temperature of 13.0°C, were slightly warmer than the same period in 2004, with mean air temperature of 11.5°C. The winter 2001 was characterised by exceptional rain events, with precipitation values significantly higher than during the same period in 2004. The effect of the different rainy regimes is seen in the Tagus flow, with a mean value of 1893 $m^3 s^{-1}$ in March 2001 and 481 $m^3 s^{-1}$ in February 2004 (Figure 2). Atmospheric CO_2 ($pCO_{2,air}$) was slightly lower (mean value of 373 µatm) in 2001, when compared with 2004 (mean value of 380 µatm). Both periods were characterised by absence of upwelling, seen in the positive Bakun index mean values of 725 $m^3 s^{-1} km^{-1}$ and of 344 $m^3 s^{-1} km^{-1}$ for March 2001 and February 2004, respectively. Significant shifts in wind direction and intensity were observed in March 2001 (Figure 3A), with dominant direction from the SW quadrant and intensities between 7 – 10 $m s^{-1}$. In February 2004 the Tagus coastal area was under the influence of persistent south winds followed by stronger north winds (7 – 10 $m s^{-1}$ in intensity), as shown in the wind rose in Figure 3B.

The winter periods were considered statistically different (t-test, $p < 0.05$, n = 27) for all physical (T, S) and

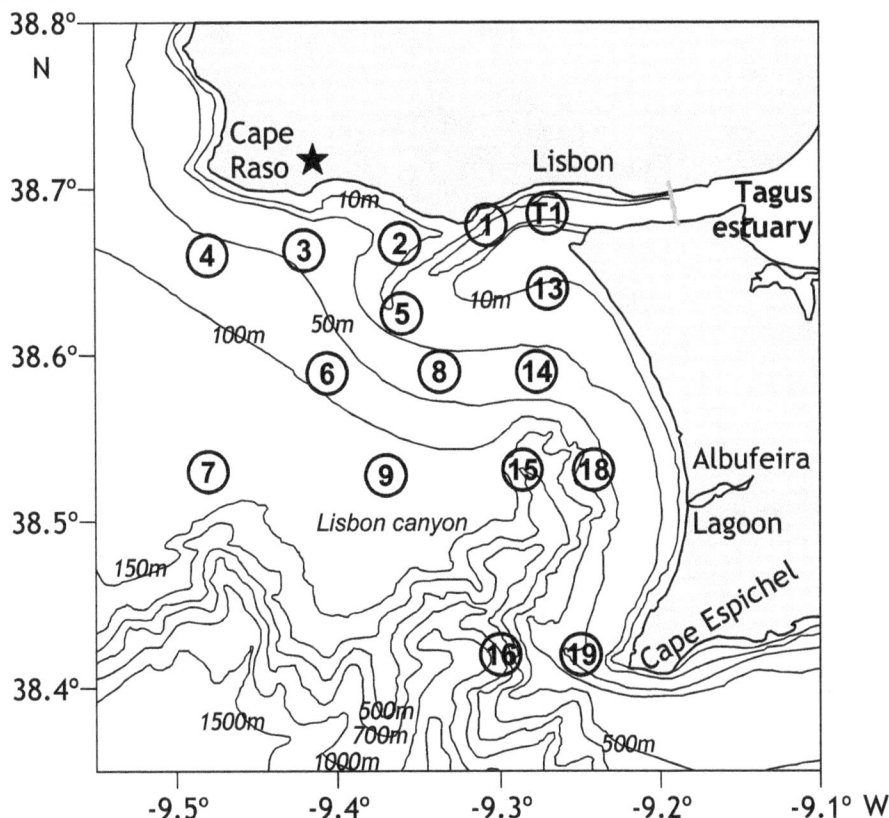

Figure 1 Location of the study site. Location of the sampling stations in the mouth of the Tagus estuary (SW Portugal) and adjacent coastal area. The position of the Guia meteorological station (38°41'27" N, 9°27'34" W) is marked with a star.

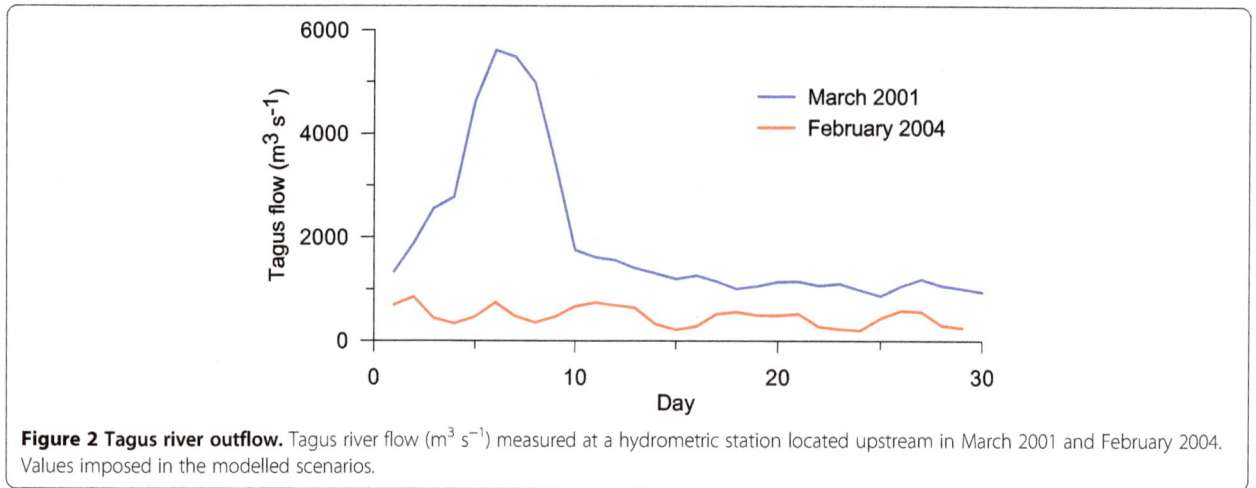

Figure 2 Tagus river outflow. Tagus river flow ($m^3\ s^{-1}$) measured at a hydrometric station located upstream in March 2001 and February 2004. Values imposed in the modelled scenarios.

biogeochemical parameters ($Si(OH)_4$, AOU, SPM, Chl a, pH, TA, pCO_2), except for DO, NO_3, NH_4 and PO_4 (Table 1). Higher values for all parameters occurred in March 2001, except for S and pH, denoting the influence of the river plume. Salinity differences were also a consequence of the river flow in the two periods.

TA values in March 2001 are considerably high, but fall within empirically established boundaries. They are within the range reported for Tagus estuary adjacent coastal waters in previous works [26,27]. Also, the Portuguese National Information System for Hydric Resources – SNIRH (data available at http://snirh.apambiente.pt/)

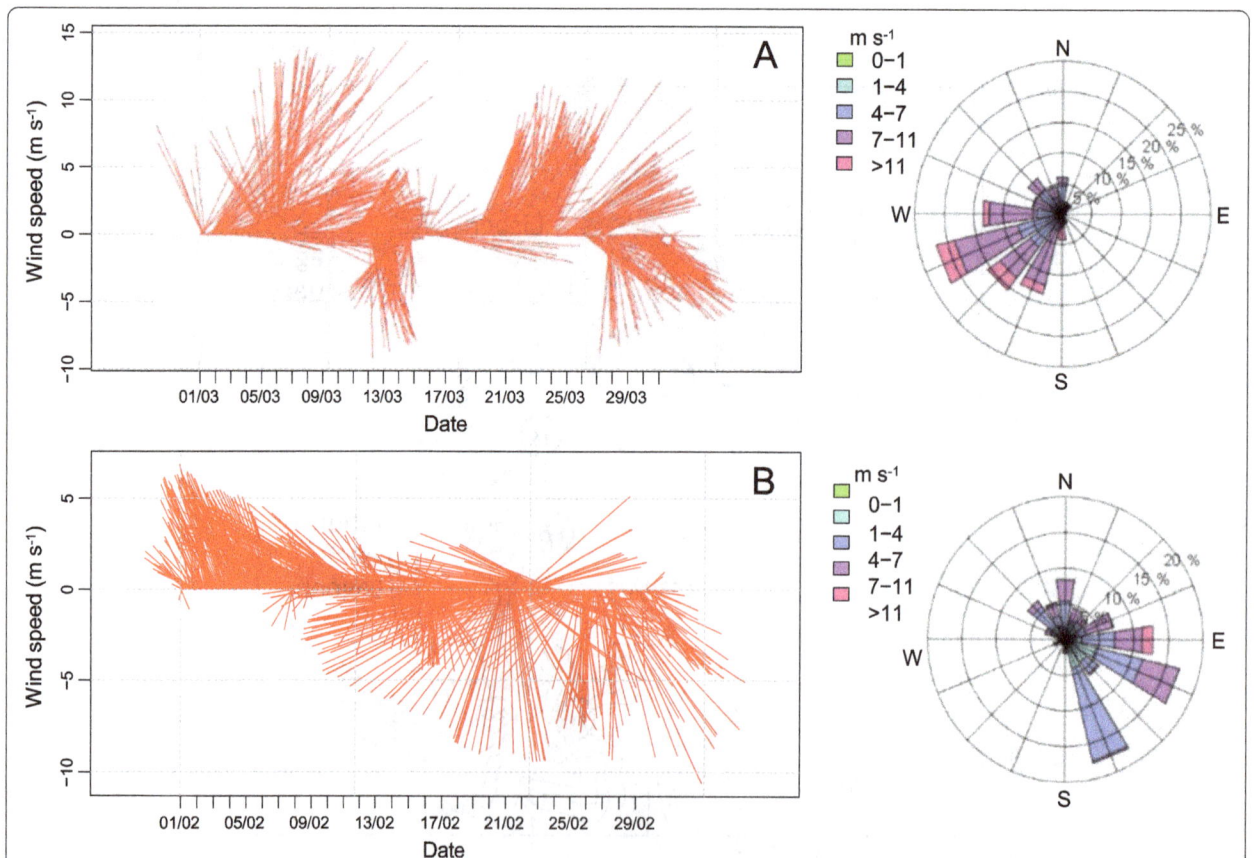

Figure 3 Wind regime. Stick diagram and wind rose of the wind regime measured at Guia meteorological station (38°41′27″ N, 9°27′34″ W) during **(A)** March 2001 and **(B)** February 2004. The wind intensity and direction showed here was used to force the model in each scenario. Wind rose shows the cumulative frequency in which wind speeds increase from the center to the outside.

Table 1 Mean values for monitored parameters

	MARCH 2001		FEBRUARY 2004	
	Range	Mean value (SD[a])	Range	Mean value (SD[b])
T (°C)	14.8 – 15.9	15.4 (0.3)	14.4 – 14.9	14.7 (0.2)
S	31.4 – 34.6	32.3 (0.8)	27.9 – 35.7	32.7 (2.1)
NO_3 (µmol l^{-1})	0.9 – 13.7	8.4 (3.5)	1.1 – 15.8	8.4 (5.9)
NH_4 (µmol l^{-1})	0.5 – 3.3	1.9 (1.0)	0.4 – 4.4	2.6 (1.5)
PO_4 (µmol l^{-1})	0.2 – 1.0	0.5 (0.2)	0.2 – 1.1	0.7 (0.3)
$Si(OH)_4$ (µmol l^{-1})	4.8 – 26.5	17.1 (7.1)	1.1 – 16.9	9.6 (6.1)
DO (mg l^{-1})	8.3 – 8.9	8.5 (0.2)	7.6 – 9.3	8.2 (0.6)
AOU (µmol kg^{-1})	−25.9 – -1.7	−9.3 (6.8)	−35.6 – 25.8	2.0 (20.4)
SPM (mg l^{-1})	3.0 – 19.2	7.2 (4.6)	2.1 – 9.6	4.1 (2.2)
Chl a (mg m^{-3})	0.7 – 1.6	1.1 (0.3)	0.2 – 1.1	0.7 (0.3)
pH	7.88 – 7.96	7.93 (0.02)	8.06 – 8.17	8.10 (0.05)
TA (µmol kg^{-1})	3026 – 3770	3519 (282)	2357 – 2767	2489 (103)
pCO_2 (µatm)	990 – 1467	1207 (140)	431 – 654	531 (70)
Wind speed (m s^{-1})	3.4 – 3.7	-	1.0 – 4.2	-
Piston velocity (cm h^{-1})	3.32 – 4.18	-	0.04 – 5.54	-

[a]standard deviation (SD) (n = 13).
[b]standard deviation (SD) (n = 14).
Seawater surface range of data and mean values for Tagus coastal area during winter 2001 and 2004. Shaded area indicates the parameters that are statistically different (t-test, $p < 0.05$, n = 27) between the 2001 and 2004 winter sampling periods.

reports TA values within the range 3000–8200 µmol kg^{-1} in the lower part of the estuary under the influence of freshwater. TA values in 2001 winter can be attributed to carbonate dissolution, which is confirmed by the significant decrease of particulate inorganic carbon from the estuary mouth (station T1, see Figure 1) to the plume. Although anaerobic degradation processes, such as denitrification and sulphate reduction, can also impact alkalinity increase, there are no evidences that such processes occurred during the March 2001 sampling period. Carbon loads to the plume were also quite different in both winter periods, being the value in the 2001 winter (2931 t C d^{-1}) ~2.2 times higher than the value in 2004 winter (1340 t C d^{-1}).

The estuarine plume boundary

The boundary of Tagus plume can be inferred by the salinity gradient resulting from the fresh water intrusion in the coastal area. As such, the size and extension of the plume is strongly related with riverine discharges and, thus, with rainfall. This is observed in the studied periods (Figure 4). Using the salinity isopleth 34.5 to set up the limit of the plume, it is possible to notice a larger plume in March 2001 as a result of higher river flow. During this period the plume is more pronounced, extending south to Albufeira Lagoon reaching the Espichel Cape limit (Figure 4A), ~30 km from the estuary mouth. In February 2004, the plume remains closer to the Tagus mouth, extending ~14 km north-west

along the coast (Figure 4B). The T-S (Figure 4C, D) and AOU-S (Figure 4E, F) diagrams reflect the impact of Tagus water input on the coastal area adjacent to the estuary in terms of salinity, temperature and oxygen. During March 2001 the Tagus Bay was under the influence of Tagus discharge, as noticed by the salinity values below 34.5, temperature higher than 15°C, and by the oxygen super-saturated water (AOU < 0) (Figure 4C, E).

Model results show a significant variation in Tagus plume dispersion pattern between 20 and 30 March, as seen in Figure 5A and B. During this period, the horizontal current structure of the plume changes the northwest direction from the river mouth to the south. From 26 to 31 March the wind is consistently from the Northern quadrant with a relatively high intensity (~5 m s^{-1}) (Figure 3), inducing marked offshore/southward advection of the estuary plume. Current velocity intensifies as a result of the river flow increase in this period. Model results for the salinity provide insights on the plume size and shape (Figure 5B), and show significant variation in its limits in response to the wind regime. It is also noticed an evolution from its original position trapped along the northern side of the river mouth, in 23 March, to a south-west transport off the Tagus mouth.

The 2004 winter period was characterised by water temperatures below 15°C (Figure 4D). Oxygen saturation showed the presence of the undersaturated plume (Stations 1 to 6, 8, 13 and 14), and an outer oversaturated area (Stations 7, 9, 15, 16 and 18) (Figure 4F). This is reinforced

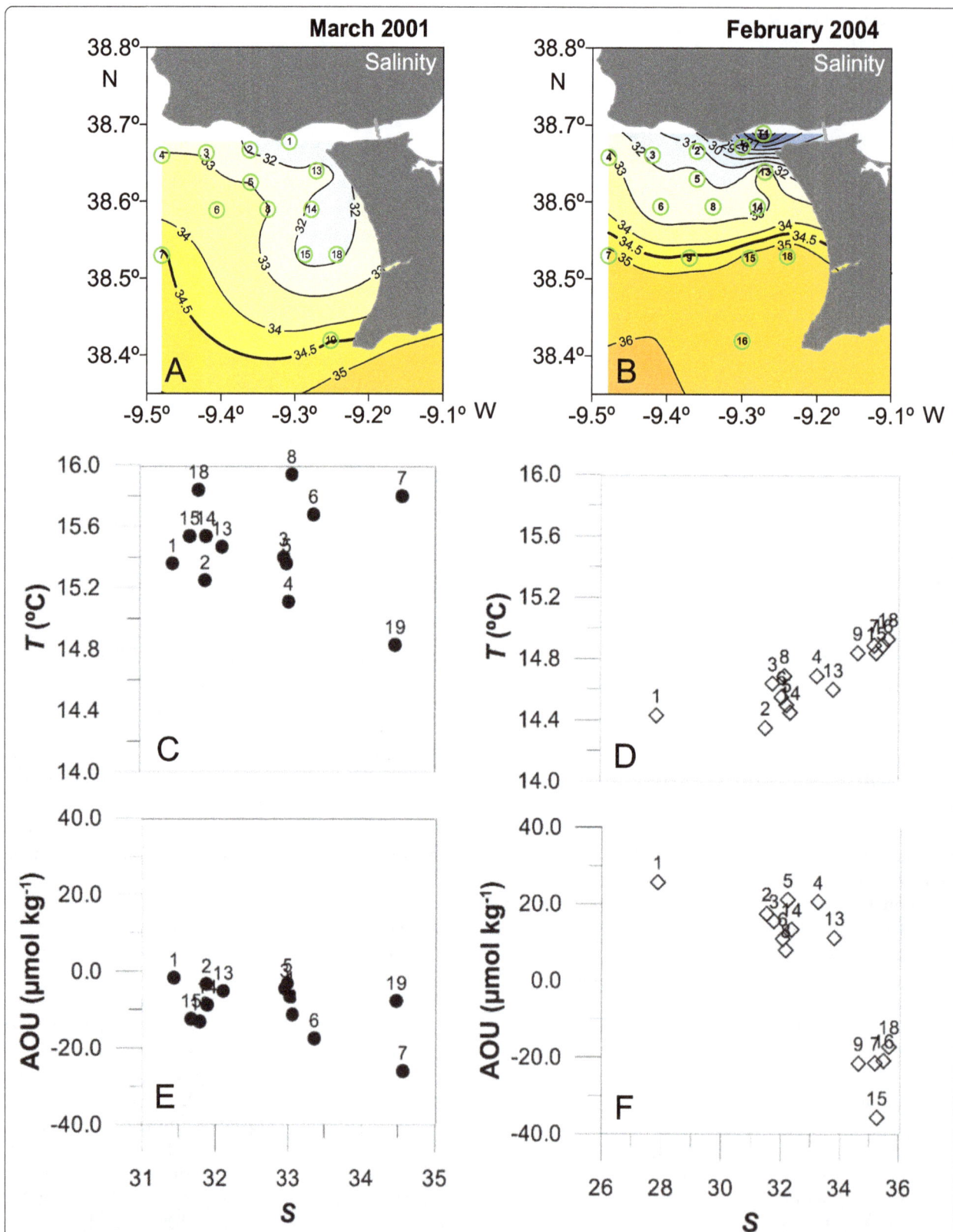

Figure 4 Tagus plume characterization. Surface salinity distribution during **(A)** March 2001 and **(B)** February 2004. T-S diagrams for **(C)** March 2001 and **(D)** February 2004 samplings, illustrating stations at the estuarine plume. AOU-S diagrams in **(E)** March 2001 and **(F)** February 2004 illustrating stations at the estuarine plume. The plume limit is represented by the 34.5 isopleth (bold line).

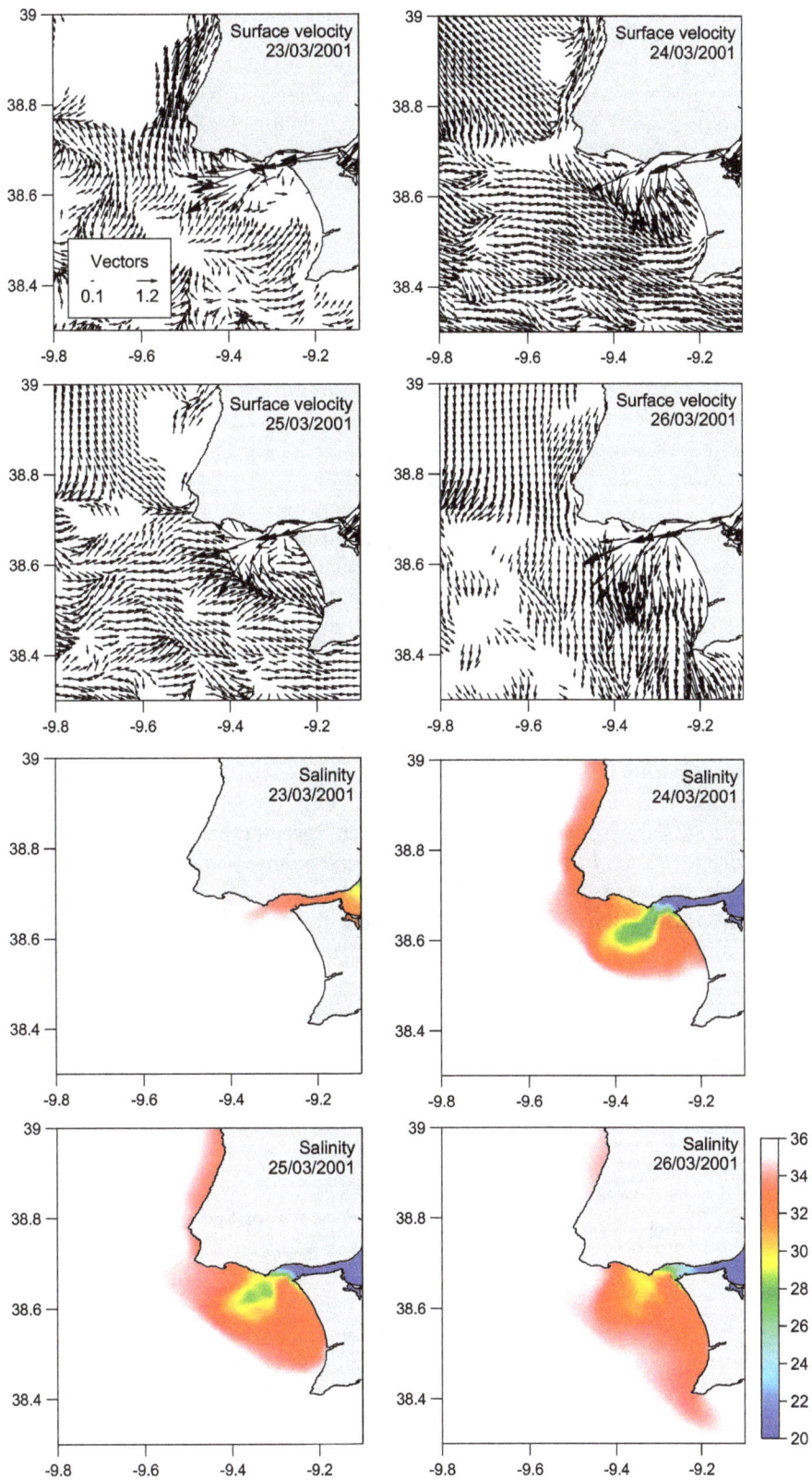

Figure 5 Model results. Model results for the surface currents and salinity in March 2001.

by the calculated DIC *versus* TA plot (Figure 6), where the separation of the two water masses (riverine and oceanic) is observed. Station 1 (with S < 30) was clearly isolated from other stations near the northern coast, displaying salinities between 30 and 34.5 (Figure 4D, F).

Simulated conditions for February 2004 show the formation of an estuary plume in the vicinity of the estuary mouth, extending westwards along the north side from the estuary as a result of geostrophic adjustment (Figure 7). This pattern is the result of the influence of moderate river flow and persistent south winds (Figure 3). Given the small variation in the forcing conditions, model results for February 2004 show little variation during the simulated period. The physical structure of the plume was consistently characterized by an offshore transport westward from the river mouth (Figure 7A), and presented a similar signature pattern in the horizontal salinity field observed in this period (Figure 7B).

The signature of the estuarine plume is also evident in the chemical and biological parameters plotted in Figure 8. The results suggest a southward transport of the plume in winter 2001 and a northward transport in winter 2004. Model results explain these patterns by providing the temporal evolution of the physical conditions of the plume in both circumstances.

The estuarine plume biogeochemistry

Contour plots show a marked estuarine plume enriched in nutrients, represented as $Si(OH)_4$, particles and calculated pCO_2 in 2001 (Figure 8A, B, E). In both sampling periods, high concentrations of suspended material (SPM) were associated with low concentrations of particulate

organic carbon (data not shown), suggesting an organically impoverished plume, similar to other estuarine plumes [6,7,28]. SPM declined outside the plume, partly due to sinking, and there was a nutrient decrease caused by the combined effects of mixing with nutrient poor offshore waters and phytoplankton uptake, as suggested by the increase in Chl *a* (Figure 8C, H) and DO (Figure 8D, I).

The different water masses observed in both scenarios are characterized by distinct environmental properties, and also reveal particular CO_2 features. Higher calculated pCO_2 occur in March 2001 (Table 1), ranging from 990 µatm outside the plume (Station 19) to 1460 µatm near Albufeira Lagoon (Station 18) in the tip of the Lisbon submarine canyon head (Figure 8E). The elevated pCO_2 values are probably a signal from the river where the most riverine station (~65 km of Station 1) values were up to 2500 µatm. In February 2004 the plume is pushed northwards (Figure 8J), and the highest value of calculated pCO_2 (654 µatm) is observed at the estuary mouth (Station 1). Both periods were characterised by CO_2 oversaturation, reaching ~400% in 2001 and ~170% in 2004. Other European estuarine plumes also presented CO_2 oversaturation, such as the Scheldt in winter [9,12], the Elbe in the spring [6], and the Loire in autumn [5]. The marked pCO_2 gradient in the buoyant plume suggests that its structure and dynamics regulates the pCO_2 property in the studied area (Figure 8E, J). Again, several other studies revealed the variability of estuarine plumes with respect to CO_2 dynamics [29].

Calculated pCO_2 values decreased from inshore to offshore (Figure 8E, J), following the decreased in salinity. A significant correlation was found between calculated pCO_2 and salinity in February 2004 ($r^2 = 0.890$, $p < 0.05$, n = 14; Figure 9A). This distribution pattern, also seen in other systems [2,9,15,25], indicates that the mixing processes influence the CO_2 pattern, which is reinforced by the proximity to the conservative mixing line (Figure 9A). However, the simultaneous calculated pCO_2 and DO decrease ($r^2 = 0.561$, $p < 0.05$, n = 14) and Chl *a* increases along the salinity gradient (Figure 9B, C), suggests the prominence of biological processes inside the plume. This is supported by the drawdown drop of calculated DIC associated with a pH increase ($r^2 = 0.704$, $p < 0.05$, n = 14) (Figure 9E, F). The non-linear relationships between calculated DIC and nutrients (only represented by NO_3 in this study) also reflect both mixing and biological processes (Figure 9G) regulating pCO_2 distribution in February 2004. These features were also found in the Mississippi River [22] and Amazon River [25] plumes.

For the outer Loire estuary [13] biological processes, namely an episodic winter phytoplankton blooms, was also pointed out as responsible for pCO_2 variability. Applying the Takahashi et al. [30] procedure to the

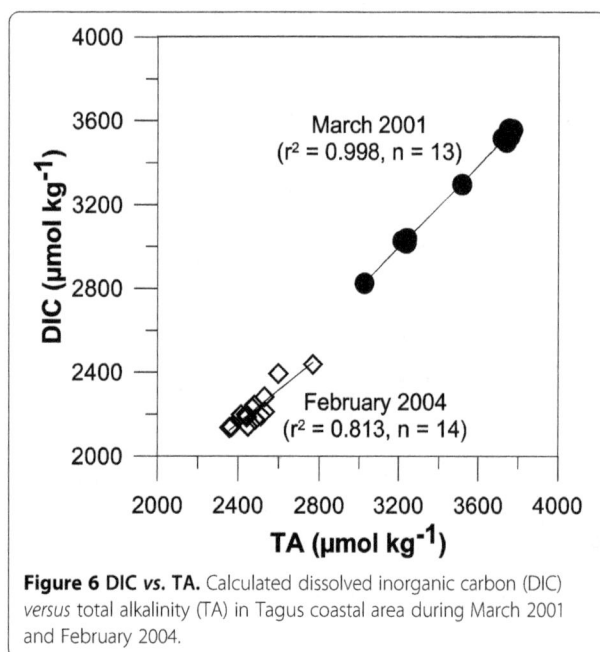

Figure 6 DIC *vs*. TA. Calculated dissolved inorganic carbon (DIC) *versus* total alkalinity (TA) in Tagus coastal area during March 2001 and February 2004.

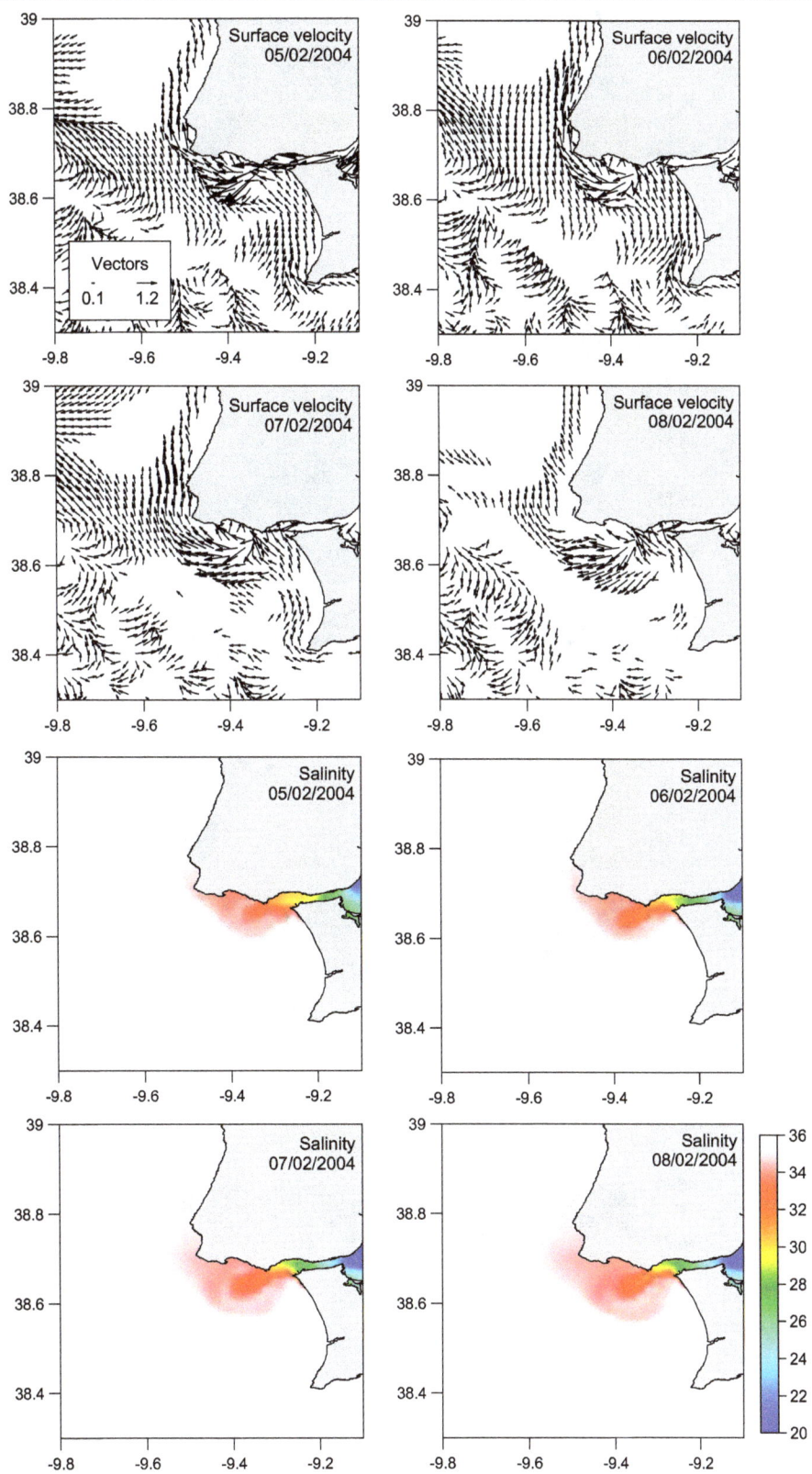

Figure 7 Model results. Model results for the surface currents and salinity in February 2004.

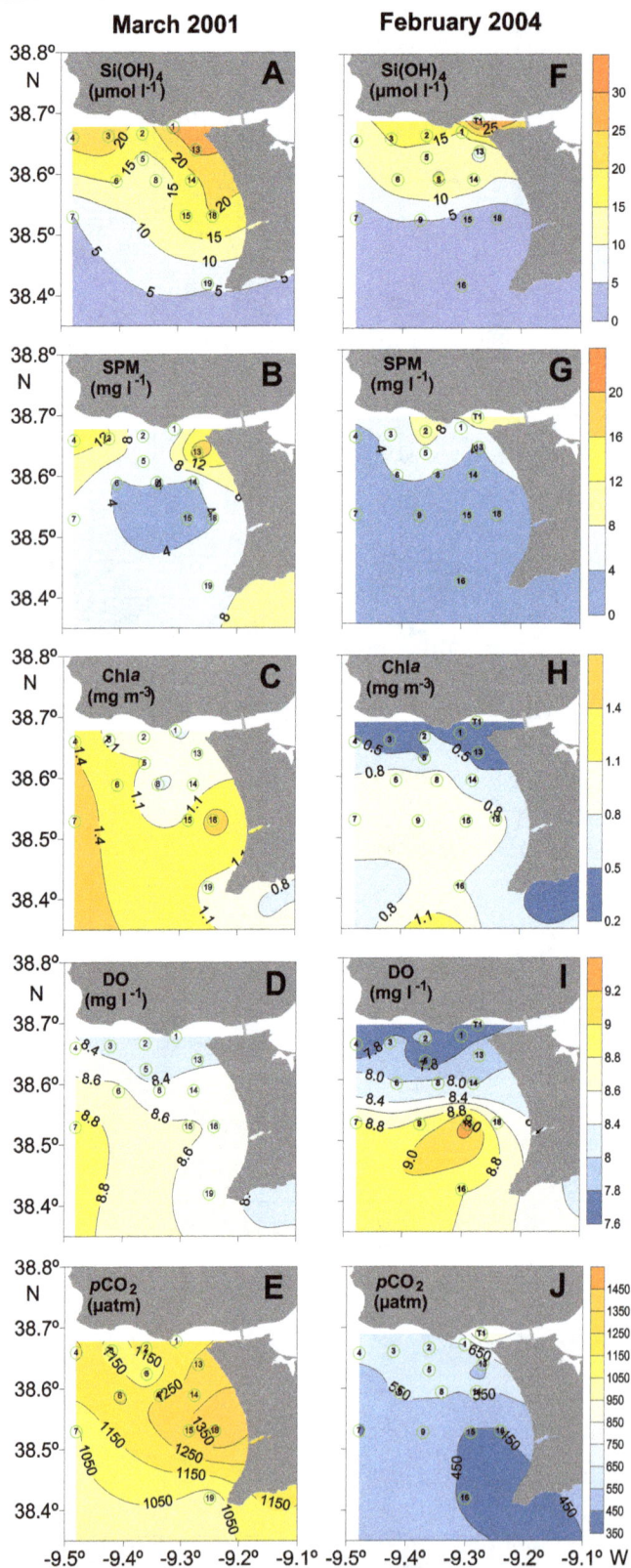

Figure 8 Surface concentration of monitored properties. Characterization of the Tagus coastal area during March 2001 and February 2004. Surface concentration of **(A,F)** silicate (Si(OH)4), **(B,G)** suspended matter (SPM), **(C,H)** chlorophyll *a* (Chl *a*), **(D,I)** dissolved oxygen (DO), and **(E,J)** calculated CO_2 partial pressure (pCO_2).

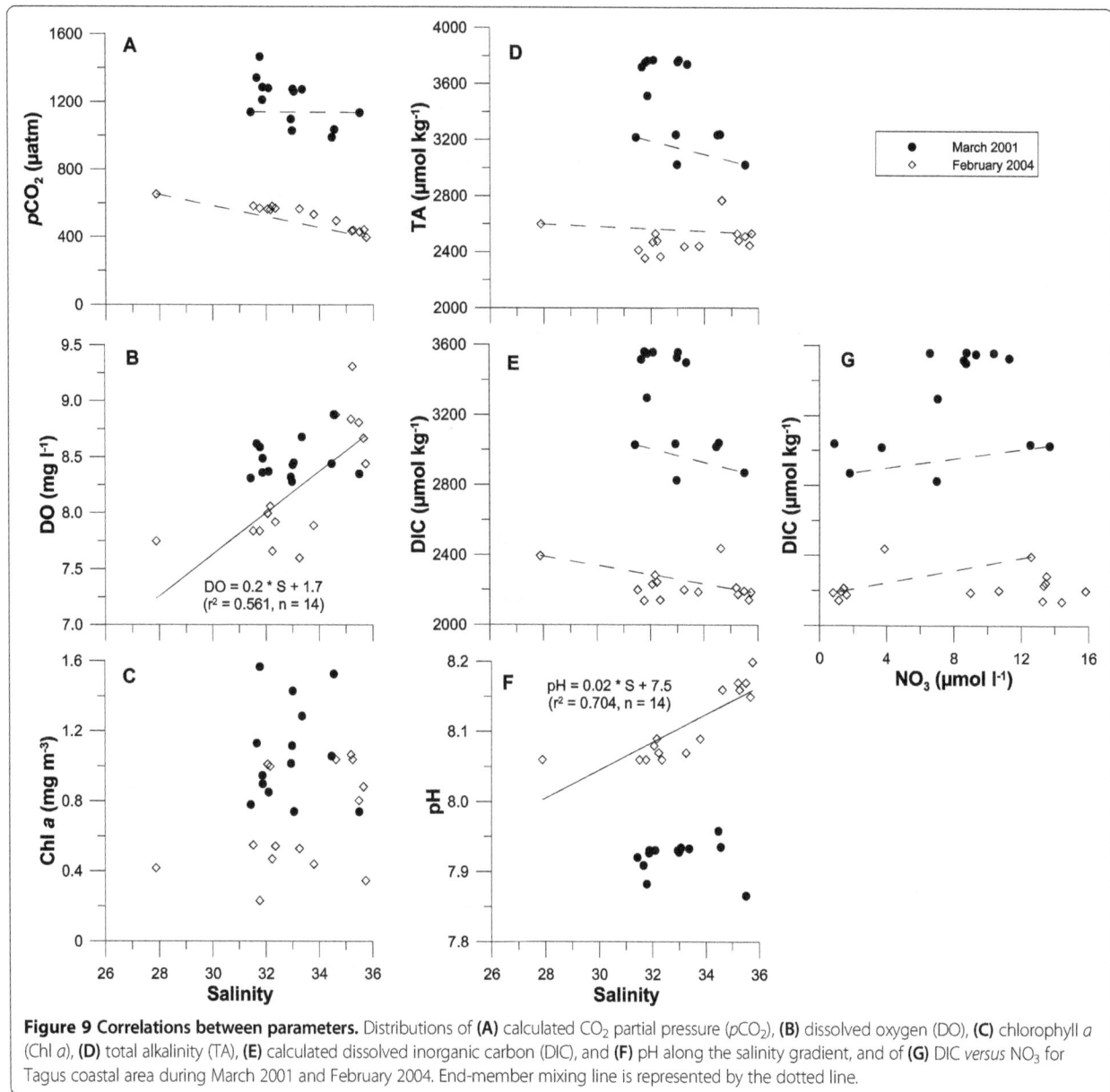

Figure 9 Correlations between parameters. Distributions of **(A)** calculated CO_2 partial pressure (pCO_2), **(B)** dissolved oxygen (DO), **(C)** chlorophyll a (Chl a), **(D)** total alkalinity (TA), **(E)** calculated dissolved inorganic carbon (DIC), and **(F)** pH along the salinity gradient, and of **(G)** DIC *versus* NO_3 for Tagus coastal area during March 2001 and February 2004. End-member mixing line is represented by the dotted line.

February 2004 data, pCO_2 variability was not affected by temperature. By contrast, in March 2001 pCO_2 variability was attributed to physical processes, such as the thermodynamic effect of temperature and the riverine/estuarine discharge [31]. Moreover, the non-linear relationships of TA and DIC with salinity (Figure 9D, E) and of calculated DIC and NO_3 (Figure 9G) reflect mixing and biological processes, inducing pCO_2 variability. However, the calculated DIC *vs.* TA plot suggests that DIC and TA resulted by the same mechanism.

For the outer Loire estuary [13] biological processes, namely an episodic winter phytoplankton blooms, was also pointed out as responsible for pCO_2 variability.

Applying the Takahashi et al. [30] procedure to the February 2004 data, pCO_2 variability was not affected by temperature. By contrast, in March 2001 pCO_2 variability was attributed to physical processes, such as the thermodynamic effect of temperature and the riverine/estuarine discharge [31]. Moreover, the non-linear relationships of TA and DIC with salinity (Figure 9D, E) and of calculated DIC and NO_3 (Figure 9G) reflect mixing and biological processes, inducing pCO_2 variability. However, the calculated DIC *vs.* TA plot suggests that DIC and TA resulted by the same mechanism.

In both winter occasions two major regions were spatially individualized in the area: a high nutrient and CO_2 enriched region inside the plume, and a warmer region characterized by higher phytoplankton biomass in the outer plume.

CO_2 fluxes across the water-air interface

The water-air CO_2 fluxes showed similar patterns in the studied periods, with lower emissions to the atmosphere outside the plume (Table 2), and the highest values coincident with high wind speeds (data not shown). A striking pattern is found in both cases, namely the reduction in CO_2 fluxes from inside the plume to outside of about 90% and 20%, in 2001 and 2004, respectively. Other authors [2,15-20] have also reported this CO_2 uptake capacity of estuarine plumes, even suggesting that other estuarine plumes might counteract inner estuary CO_2 emissions. Overall, the adjacent waters to the Tagus estuary acted as sources of CO_2 to the atmosphere, emitting 25.9 ± 4.3 mmol C m^{-2} d^{-1} in March 2001, and 2.4 ± 3.4 mmol C m^{-2} d^{-1} in February 2004 (Table 2). Thus, CO_2 emissions to the atmosphere in March 2001 were ~90% higher than in February 2004. The differences can be attributed to the variable river influence (e.g., effect of nutrients and labile organic matter, additional buoyant stability induced by freshwater fluxes), as suggested by other authors [10,25,32]. The CO_2 emissions estimated in this work are within the range of those reported for the Tagus adjacent coastal waters [31] and several other near-shore ecosystems [9].

Conclusions

Tagus estuarine plume can be traced on the shelf by gradients of salinity, but also by gradients of less conservative tracers such as water temperature, chlorophyll a, inorganic nutrients, total alkalinity and CO_2. Thus, Tagus estuary adjacent shelf exhibits a high nutrient, low chlorophyll and enriched in CO_2 estuarine plume, and a warm region impoverished in CO_2 and enriched in phytoplankton in the outer plume. Estuarine Tagus plume behaved as a weak source of CO_2 to the atmosphere, with estimated fluxes of 3.5 ± 3.7 and 27.0 ± 3.8 mmol C m^{-2} d^{-1} for February 2004 and March 2001, respectively.

Based on two winter cruises, it seems that Tagus plume significantly impacted estimates of water-air CO_2 fluxes

at a regional scale. Hence, this work emphasizes the importance of estuarine plumes on the CO_2 dynamics in coastal areas. However, due to the complexity of near shore ecosystems and processes therein the magnitude of water-air fluxes is variable from one system to another.

Also, this study reinforces the usefulness of complimentary approaches such as the application of numeric models in reproducing the physical and chemical characteristics of plumes dynamics. The model results provide the temporal evolution of the plume under varying wind and rivers discharge, providing additional information that could not be obtained otherwise and, consequently, insightful clues on the integration of field data. Still, this is a first approach to using modelling tools with field data in the Tagus estuary, and future developments will include the CO_2 dynamics in the model simulations.

Methods
Study area

The present investigation was carried out in the continental shelf offshore Tagus estuary (Figure 1) in the Portuguese coast, covering the geographic area between 38.35° – 38.80° N and 9.10° – 9.50° W. The continental shelf is ≤10 km wide south of Lisbon and presents topographic structures as prominent capes, promontories and submarine canyons. Its morphology is strongly influenced by the intense discharge of Tagus River, usually showing a pronounced dry/wet season signal as well as large inter-annual variation. The mean annual average discharge of Tagus is 350 m^3 s^{-1} [33], with monthly averages ranging from 1 to 2200 m^3 s^{-1}. The Tagus estuary is a relatively shallow mesotidal system with semi-diurnal tidal regime (1 to 4 m in amplitude range). The surface area is about 320 km^2 and the mean volume 1900×10^6 m^3. Intertidal mudflats cover an area of about 20 to 40% of the estuary.

The coastal area off Tagus estuary is characterized by the presence of upwelling plumes originated by jet-like flow extending more than 20 km seaward [34]. Advection of warmer oligotrophic oceanic waters into the shelf occurs during autumn and winter when southerly winds dominate, intensifying the poleward flow [35-37]. Episodes of reverse winds can occur during both seasons. In the absence of coastal upwelling, the surface circulation is predominantly northward [36] as a result of the geostrophic equilibrium. Also, the plume of estuarine waters is highly influenced by the coastline geometry. Intense freshwater discharge events under highly variable wind direction conditions in winter and strong upwelling episodes in spring-summer as well as fortnightly spring-neap tidal cycle, affect strongly the shape and size of Tagus plume [38]. While the plume is usually trapped close to the shore and transports estuarine water northward along the coast, under persistent northern wind conditions the plume is displaced offshore.

Table 2 Water-air CO_2 fluxes (Mean values and standard deviation)

	March 2001	February 2004
Estuarine plume (S < 34.5)	27.0 ± 3.8	3.5 ± 3.7
Outer plume (S > 34.5)	19.9 ± 1.0	0.2 ± 0.1
Overall area	25.9 ± 4.3	2.4 ± 3.4

Mean values and standard deviation of water-air CO_2 fluxes (mmol C m^{-2} d^{-1}) calculated according to [32] parameterization for stations inside and outside the Tagus estuary plume during March 2001 and February 2004.

A significant amount of phytoplankton is exported from Tagus to the estuarine plume. Field and modelling studies suggest that nutrients are not depleted by primary producers due to light limitation inside the estuary and end up by being exported, eventually enhancing primary production in the coastal area [39-43]. Tagus estuary is also a major source of nutrients [43] and suspended matter to the adjacent coastal area [38]. The transport and transformation of such materials in the area is regulated by the interplay of dynamics and structure of the Tagus plume and hydrological characteristics of the coastal area [44].

The Tagus estuary and adjacent coastal waters are sources of CO_2 to the atmosphere, with winter values ranging from 29 to 419 mmol C m^{-2} d^{-1} in the estuary, and up to 34 mmol C m^{-2} d^{-1} in the adjacent waters [45]. The carbonate system parameters have been evaluated in Tagus estuary and adjacent coastal area from 1999 to 2007 [46], and TA highest values (~4600 μmol kg^{-1}) were recorded in 2002 spring [26].

Sampling program

Surface water sampling was accomplished during ebb tide for a total of 16 stations distributed in the study area (see Figure 1), in two distinct winter periods (March 2001 and February 2004), defining winter as beginning in 21 December and ending in 21 March.

Parameters determination

Temperature (T) and salinity (S; PSS-78) parameters were determined in situ with a Seabird SBE19/CTD (Conductivity - Temperature - Depth) probe. Salinity was calibrated with an AutoSal salinometer using IAPSO standard seawater, with a variation coefficient of 0.003%.

Dissolved oxygen (DO) was analysed following the Winkler method [47] using a whole-bottle manual titration, and the coefficient of variation associated with the method ranged from 0.08 to 0.25%. pH was measured immediately after sample collection at 25°C, using a Metrohm 704 pH-meter and a combination electrode (Metrohm) standardised against 2-amino-2-hydroxymethyl-1,3-propanediol seawater buffer (ionic strength of 0.7 M), at a precision of 0.005 pH units [48]. Total alkalinity (TA) samples were filtered through Whatman GF/F (0.7 μm) filters, fixed with $HgCl_2$ and stored (refrigerated not frozen) until use. Samples were then titrated automatically with HCl (~0.25 M HCl in a solution of 0.45 M NaCl) past the endpoint of 4.5 [48], with an accuracy of ±2 μmol kg^{-1}. The respectively accuracy was controlled against certified reference material supplied by A.G. Dickson (Scripps Institution of Oceanography, San Diego, USA). Discrete water samples were also taken for nutrient determination ($NO_3^- + NO_2^-$, referred as NO3; NH_4^+ referred as NH4;

PO_4^{3-}, referred as PO4; $Si(OH)_4^-$, referred as Si(OH)4), chlorophyll a (Chl a) and suspended particulate matter (SPM). Nutrient samples were filtered through MSI Acetate Plus (0.45 μm) filters and analysed on a Traacs Autoanalyser, with a variation coefficient of ±1.0%. Chl a was measured by filtering triplicate aliquots of 250 ml water through Whatman GF/F (0.7 μm) filters under a 0.2 atm vacuum, which were immediately frozen and later extracted in 90% acetone for analysis in a fluorometer Hitachi F-7000, calibrated with commercial solutions of Chl a (Sigma Chemical Co.). The coefficient of variation associated with the method was 1.8%. For SPM measurements six aliquots of 750–1000 ml water samples were filtered through pre-combusted (2 h at 450°C) Wathman GF/F (0.7 μm) filters and determined gravimetrically (drying at 70°C). A portable Vaisala° meteorological station (Datalogger Campbell Scientific CR510) coupled with a MetOne 034A anemometer located at 11 m height was used to measure in situ wind speed and direction data at 1-minute intervals at each station. Wind speed was referenced to a height of 10 m (u_{10}) using the algorithm given by Johnson [49]. We used one standard deviation of ±2 m s^{-1} as wind speed error.

Calculated parameters

The upwelling indices (negative values indicate upwelling) were based on the northward wind stress component, and calculated according to Bakun [50]. Wind data was obtained from the meteorological weather station of Cape Carvoeiro located ~70 km north of Lisbon and supplied by the Portuguese Portuguese Institute for the Ocean and Atmosphere (IPMA, I.P.). Apparent oxygen utilisation (AOU) was calculated according to the equation:

$$AOU = O_{2sat} - DO \qquad (1)$$

where O_{2sat} is the oxygen saturation in equilibrium with atmosphere. pH values corrected to in situ temperature were calculated from total alkalinity (TA) and in situ pH and temperature following the procedure proposed by Hunter [51]. For these calculations the carbon dioxide constants of Millero et al. [52] were applied. The partial pressure of CO_2 in seawater (pCO_2) and the dissolved inorganic carbon (DIC) were calculated from the in situ temperature, TA and corrected pH, using the carbonic acid dissociation constants given by Millero et al. [52] and the CO_2 solubility coefficient of Weiss [53]. Errors associated with pCO_2 and DIC calculations were estimated to be ±10 μatm and ±5 μmol kg^{-1}, respectively (accumulated errors on TA and pH). The water-air CO_2 fluxes (CO_2 Flux) were computed according to the equation:

$$CO_2 Flux = k.K_0.\Delta pCO_2 \qquad (2)$$

where k is the gas transfer velocity (also referred to as piston velocity), K_0 is the solubility coefficient of CO_2 and ΔpCO_2 the water-air gradient of pCO_2. Positive fluxes indicate CO_2 upward water – air emission. The k value is based on the Wanninkhof [54] parameterization. Atmospheric CO_2 data were obtained from the Terceira Island's reference station (Azores, Portugal, 38°46'N 27°23'W), operated by the network of the National Oceanic and Atmospheric Administration (NOAA)/Climate Monitoring and Diagnostics Laboratory/Carbon Cycle Greenhouse Gases Group [55]. Subsequently, the observed atmospheric CO_2 content in mole fraction (in dry air) was converted into wet air values using the algorithms given by Dickson et al. [48]. Atmospheric pCO_2 data obtained from our single day shipboard were only available for some sampling periods, while Terceira data represent a readily accessible continuous thropospheric dataset for the complete study period. Significant correlations were found between Terceira data and shipboard data available ($r^2 = 0.910$, $p < 0.05$, n = 45). The discrepancies lie between 3 and 13 μatm, and the impact of using Terceira data on this study was considered negligible.

Statistical analysis

Contour plots were created using Surfer 8.0° (Golden Software, 2002) following the kriging interpolation technique considering a linear interpolation with a slope of one. Exploratory analysis and statistical procedures were implemented using the statistical software Statistica 6.0° (Statsoft Inc., 2001). Differences between sampling periods in the measured/calculated physical-chemical and biological parameters were assessed using an analysis of variance (ANOVA), and differences between means have been considered statistically significant for $p < 0.05$.

Model application

The model

The MOHID Water Modelling System (www.mohid.com) was applied to this study to simulate the circulation regime of the study area. MOHID is a three-dimensional marine model that has been implemented in several studies of estuaries and shelf circulation [56-60]. MOHID employs a 3D finite-volume approach for spatial discretization [61] using an Arakawa-C grid [62] to perform the computations. For the baroclinic force, the MOHID system uses a z-level approach with a partial step approach [63]. Temporal discretization is performed by a semi-implicit ADI (Alternating Direction Implicit) algorithm with two time levels per iteration. The hydrodynamic governing equations are the momentum and the continuity equations. The hydrodynamic model solves the primitive equations in Cartesian coordinates for incompressible flows.

The momentum and mass evolution equations are:

$$\frac{\partial u_i}{\partial t} + \frac{\partial (u_i u_j)}{\partial x_j} = -\frac{1}{\rho_0}\frac{\partial p_{atm}}{\partial x_i} - g\frac{\rho(\eta)}{\rho_0}\frac{\partial \eta}{\partial x_i}$$
$$-\frac{g}{\rho_0}\int_{x_3}^{\eta}\frac{\partial p'}{\partial x_i}dx_3 + \frac{\partial}{\partial x_j}\left(v\frac{\partial u_i}{\partial x_j}\right) - 2\varepsilon_{ijk}\Omega_j u_k \tag{3}$$

$$\frac{\partial \eta}{\partial t} = -\frac{\partial}{\partial x_1}\int_{-h}^{\eta}u_1 dx_3 - \frac{\partial}{\partial x_2}\int_{-h}^{\eta}u_2 dx_3 \tag{4}$$

where u_i is the velocity vector component in the Cartesian x_i directions, η is the free surface elevation, v is the turbulent viscosity and p_{atm} is the atmospheric pressure. ρ' is the density anomaly, ρ_0 is the reference density, g is the acceleration of gravity, t is the time, h is the depth, Ω is the Earth's velocity of rotation and ε is the alternate tensor.

The horizontal and vertical advection of momentum, heat and mass is computed using a Total Variation Diminishing (TVD) Superbee method [64]. Vertical turbulent viscosity/diffusivity coefficients are computed using a k-epsilon model coupling the MOHID system to the General Ocean Turbulence Model (GOTM) [65].

Modelled scenarios

Two distinct winter episodes were modelled: March 2001 and February 2004. Both scenarios simulate oceanic conditions based on realistic forcing for river discharge (Figure 2) and wind conditions (Figure 3). We have adopted a method using a direct initialization with values from the MERCATOR solution [66]. This methodology interpolates the initial velocity field, temperature, salinity and sea surface height from the MERCATOR solution for the D2 grid assuming geostrophic balance. A two-month period was prescribed as a spin-up period.

Model setup

The numerical model was implemented using a two level one-way nesting configuration. The first domain (D1) is a 2D barotropic tidal-driven model, forced only with the FES2004 (Finite Element Solution) tidal atlas [67,68]. This domain covers most of the Atlantic coast of Iberia and Northwest Morocco, and has variable horizontal resolution (0.02°-0.04°). The second (D2) level is a 3D baroclinic model with a 0.02° horizontal resolution and includes the Tagus Promontory area. This domain is directly coupled to D1 at the open boundaries using a one-way downscaling to impose the solution of D1. For D1 low-frequency open boundary conditions for salinity, temperature and U and V velocity components are interpolated via a downscaling of the MERCATOR operational solution for the Northeast Atlantic area (Mercator-Océan Psy2V3). A z-level vertical discretization was adopted for D2 with 33 vertical layers. In this application we have set a time step of 60 s for D1 and 15 s for D2.

Hourly values for wind, air temperature, relative humidity, barometric pressure and downward longwave and shortwave radiation, were used to calculate air-sea heat and momentum fluxes using bulk formulae. The data for atmospheric forcing was retrieved from an atmospheric modelling system based on the MM5 (Mesoscale Meteorological Model 5) model running at IST (http://meteo.ist.utl.pt). For land boundary conditions, the model uses realistic freshwater discharge and a null mass and momentum flux is imposed. River outflow was prescribed using outflow values from the Portuguese Water Institute (INAG) gauges for Tagus River.

Competing interests
The authors declare that they have no competing interests.

Authors' contributions
APO conceived this study, contributed to all sections and coordinated the main writing process. APO and MM both analyzed the results and contributed equally to the manuscript. GC and RN provided valuable input for the data analysis and discussion sections. All authors have read and approved the final manuscript.

Acknowledgements
We acknowledge the captain and the crew of RV "Mestre Costeiro" and RV "Capricórnio" for their excellent support and cooperation. We are grateful to our colleagues António Correia, António Pereira, Célia Gonçalves, Conceição Araújo, Isaura Franco, Luís Palma Oliveira, Maria Rosa Pinto and Paula Cabeçadas for their sampling, technical and analytical assistance. Thanks are due to Marta Nogueira for CTD data acquisition. This work was funded by the European Commission, Programa POpesca MARE project 22-05-01-FDR-0015 and the Portuguese Science Foundation (FCT) with which A.P. Oliveira had a Ph.D. grant, and by the Project BioPlume - Dependence of coastal ecosystems on river run-off: today & tomorrow (PTDC/AAG-REC/2139/2012).

Author details
[1]Instituto Português do Mar e da Atmosfera (IPMA), I.P., Avenida de Brasília, 1449-006 Lisboa, Portugal. [2]MARETEC, Instituto Superior Técnico, Universidade de Lisboa, Av. Rovisco Pais, 1049-001 Lisboa, Portugal.

References
1. Morris AW, Allen JI, Howland RJM, Wood RG. The estuary plume zone: source or sink for land-derived nutrient discharges? Estuar Coast Shelf Sci. 1995;40:387–402.
2. de la Paz M, Gómez-Parra A, Forja J. Inorganic carbon dynamic and air–water CO_2 exchange in the Guadalquivir Estuary (SW Iberian Peninsula). J Mar Syst. 2007;68:265–77.
3. Gaston TF, Schlacher TA, Connolly RM. Flood discharges of a small river into open coastal waters: plume traits and material fate. Estuar Coast Shelf Sci. 2006;69:4–9.
4. Lohrenz SE, Fahnenstiel GL, Redalje DG, Lang GA, Dagg MJ, Whitledge TE, et al. Nutrients, irradiance, and mixing as factors regulating primary production in coastal waters impacted by the Mississippi River plume. Cont Shelf Res. 1999;19:1113–41.
5. Sanders R, Jickells T, Mills D. Nutrients and chlorophyll at two sites in the Thames plume and southern North Sea. J Sea Res. 2001;46:13–28.
6. Dagg M, Benner R, Lohrenz S, Lawrence D. Transformation of dissolved and particulate materials on continental shelves influenced by large rivers: plume processes. Cont Shelf Res. 2004;24:833–58.
7. Dagg MJ, Bianchi T, McKee B, Powell R. Fates of dissolved and particulate materials from the Mississippi river immediately after discharge into the northern Gulf of Mexico, USA, during a period of low wind stress. Cont Shelf Res. 2008;28:1443–50.
8. Dai M, Zhai W, Cai W-J, Callahan J, Huang B, Shang S, et al. Effects of an estuarine plume-associated bloom on the carbonate system in the lower reaches of the Pearl River estuary and the coastal zone of the northern South China Sea. Cont Shelf Res. 2008;28:1416–23.
9. Borges AV, Frankignoulle M. Daily and seasonal variations of the partial pressure of CO_2 in surface seawater along Belgian and southern Dutch coastal areas. J Mar Syst. 1999;19:251–66.
10. Borges AV, Tilbrook B, Metzl N, Lenton A, Delille B. Inter-annual variability of the carbon dioxide oceanic sink south of Tasmania. Biogeosciences. 2008;5:141–55.
11. Brasse S, Nellen M, Seifert R, Michaelis W. The carbon dioxide system in the Elbe estuary. Biogeochemistry. 2002;59:25–40.
12. Schiettecatte L-S, Gazeau F, van der Zee C, Brion N, Borges AV. Time series of the partial pressure of carbon dioxide (2001–2004) and preliminary inorganic carbon budget in the Scheldt plume (Belgian coastal waters). Geochem Geophys Geosyst. 2006;7, Q06009.
13. Bozec Y, Cariou T, Mace E, Morin P, Thuillier D, Vernet M. Seasonal dynamics of air-sea CO_2 fluxes in the inner and outer Loire estuary (NW Europe). Estuar Coast Shelf Sci. 2012;100:58–71.
14. de la Paz M, Padin XA, Rios AF, Perez FF. Surface fCO_2 variability in the Loire plume and adjacent shelf waters: high spatio-temporal resolution study using ships of opportunity. Mar Chem. 2010;118:108–18.
15. Zhai W, Dai M. On the seasonal variation of air – sea CO_2 fluxes in the outer Changjiang (Yangtze River) Estuary, East China Sea. Mar Chem. 2009;117:2–10.
16. Kumar MD, Naqvi SWA, George MD, Jayakumar DA. A sink for atmospheric carbon dioxide in the northeast Indian Ocean. J Geophys Res. 1996;101:18121–5.
17. Bakker DCE, de Baar HJW, de Jong E. The dependence on temperature and salinity of dissolved inorganic carbon in East Atlantic surface waters. Mar Chem. 1999;65:263–80.
18. Chen C-TA, Wang S-L. Carbon, alkalinity and nutrient budgets on the East China Sea continental shelf. J Geophys Res. 1999;104:20675–86.
19. Cai W-J. Riverine inorganic carbon flux and rate of biological uptake in the Mississippi River plume. Geophys Res Lett. 2003;30:1032.
20. Körtzinger A. A significant CO_2 sink in the tropical Atlantic Ocean associated with the Amazon River plume. Geophys Res Lett. 2003;30:2287.
21. Lohrenz SE, Cai W-J. Satellite ocean color assessment of air-sea fluxes of CO_2 in a river-dominated coastal margin. Geophys Res Lett. 2006;33, L01601.
22. Huang WJ, Cai WJ, Powell RT, Lohrenz SE, Wang Y, Jiang LQ, et al. The stoichiometry of inorganic carbon and nutrient removal in \newline the Mississippi River plume and adjacent continental shelf. Biogeosciences. 2012;9:2781–92.
23. Borges A, Frankignoulle M. Distribution and air-water exchange of carbon dioxide in the Scheldt plume off the Belgian coast. Biogeochemistry. 2002;59:41–67.
24. Salisbury J, Vandemark D, Hunt C, Campbell J, Jonsson B, Mahadevan A, et al. Episodic riverine influence on surface DIC in the coastal Gulf of Maine. Estuar Coast Shelf Sci. 2009;82:108–18.
25. Cooley SR, Yager PL. Physical and biological contributions to the western tropical North Atlantic Ocean carbon sink formed by the Amazon River plume. J Geophys Res-Oceans. 2006;111:C08018.
26. Cabeçadas G, Oliveira AP. Impact of a Coccolithus braarudii bloom on the carbonate system of Portuguese coastal waters. J Nannoplankton Res. 2005;27:141–7.
27. Oliveira AP. Air-water CO_2 fluxes in a Portuguese estuarine system and adjacent coastal waters (in Portuguese). PhD Thesis. Lisboa: Instituto Superior Técnico; 2011.
28. Dagg MJ, Breed GA. Biological effects of Mississippi River nitrogen on the northern gulf of Mexico - a review and synthesis. J Mar Syst. 2003;43:133–52.
29. Borges AV, Delille B, Schiettecatte LS, Gazeau F, Abril G, Frankignoulle M. Gas transfer velocities of CO_2 in three European estuaries (Randers Fjord, Scheldt, and Thames). Limnol Oceanogr. 2004;49:1630–41.

30. Takahashi T, Olafsson J, Goddard JG, Chipman DW, Sutherland SC. Seasonal variation of CO_2 and nutrients in the high-latitude surface oceans - a comparative study. Global Biogeochem Cy. 1993;7:843–78.

31. Oliveira A, Fortunato AB, Rego JRL. Effect of morphological changes on the hydrodynamics and flushing properties of the Obidos lagoon (Portugal). Cont Shelf Res. 2006;26:917–42.

32. Gypens N, Lancelot C, Borges AV. Carbon dynamics and CO_2 air-sea exchanges in the eutrophied coastal waters of the Southern Bight of the North Sea: a modelling study. Biogeosciences. 2004;1:147–57.

33. Santos FD, Forbes K, Moita R. Climate change in Portugal. Scenarios, Impacts and Adaptation Measures – SIAM Project. Gradiva: Lisbon, Portugal; 2002.

34. Moita M, Oliveira P, Mendes J, Palma A. Distribution of chlorophyll a and gymnodinium catenatum associated with coastal upwelling plumes off central portugal. Acta Oecol. 2003;24:125–32.

35. Fiúza AFG, Macedo ME, Guerreiro MR. Climatological space and time variation of the Portuguese coastal upwelling. Oceanol Acta. 1982;5:31–40.

36. Haynes R, Barton ED. A poleward flow along the Atlantic coast of the Iberian peninsula. J Geophys Res-Oceans. 1990;95:11425–41.

37. Peliz A, Rosa TL, Santos AMP, Pissarra JL. Fronts, jets, and counter-flows in the Western Iberian upwelling system. J Mar Syst. 2002;35:61–77.

38. Valente AS, da Silva JCB. On the observability of the fortnightly cycle of the Tagus estuary turbid plume using MODIS ocean colour images. J Mar Syst. 2009;75:131–7.

39. Mateus M, Leitão PC, de Pablo H, Neves R. Is it relevant to explicitly parameterize chlorophyll synthesis in marine ecological models? J Mar Syst. 2012;94:23–33.

40. Mateus M, Neves R. Evaluating light and nutrient limitation in the Tagus estuary using a process-oriented ecological model. Journal of Marine Engineering and Technology. 2008;A12:43–54.

41. Mateus M, Vaz N, Neves R. A process-oriented model of pelagic biogeochemistry for marine systems. Part II: Application to a mesotidal estuary. J Mar Syst. 2012;94:90–101.

42. Saraiva S, Pina P, Martins F, Santos M, Braunschweig F, Neves R. Modelling the influence of nutrient loads on Portuguese estuaries. Hydrobiologia. 2007;587:5–18.

43. Cabeçadas L, Brogueira, Cabeçadas G. Phytoplankton spring bloom in the Tagus coastal waters: hydrological and chemical conditions. Aquatic Ecology. 1999;33:243–50.

44. Oliveira PB, Nolasco R, Dubert J, Moita T, Peliz A. Surface temperature, chlorophyll and advection patterns during a summer upwelling event off central Portugal. Cont Shelf Res. 2009;29:759–74.

45. Oliveira AP, Nogueira M, Cabecadas G. CO_2 variability in surface coastal waters adjacent to the Tagus Estuary (Portugal). Cienc Mar. 2006;32:401–11.

46. Oliveira AP, Cabeçadas G, Pilar-Fonseca T. Iberia coastal ocean in the CO_2 sink/source context: Portugal case study. J Coast Res. 2012;28:184–95.

47. Carrit DE, Carpenter JH. Comparison and evaluation of currently employed modifications of the Winkler method for determining oxygen in seawater. A NASCO Report. J Mar Res. 1966;24:286–318.

48. Dickson AG, Sabine CL, Christian JR: Guide to best practices for ocean CO_2 measurements. PICES Special Publication 3; 2007. p. 191.

49. Johnson HK. Simple expressions for correcting wind speed data for elevation. Coast Eng. 1999;36:263–9.

50. Bakun A. Coastal upwelling indices, west coast of North America. NOOA techn. Rep. NMFS-671. 1973.

51. Hunter KA. The temperature dependence of pH in surface seawater. Deep-Sea Res I Oceanogr Res Pap. 1998;45:1919–30.

52. Millero FJ, Graham TB, Huang F, Bustos-Serrano H, Pierrot D. Dissociation constants of carbonic acid in seawater as a function of salinity and temperature. Mar Chem. 2006;100:80–94.

53. Weiss RF. Carbon dioxide in water and seawater: the solubility of a non-ideal gas. Mar Chem. 1974;2:203–15.

54. Wanninkhof R. Relationship between wind speed and gas exchange over the ocean. J Geophys Res. 1992;97:7373–82.

55. Conway TJ, Lang PM, Masarie KA. Atmospheric carbon dioxide dry air mole fractions from the NOAA ESRL Carbon Cycle Cooperative Global Air Sampling Network, 1968-2011, Version: 2013-08-08, (ftp://ftp.cmdl.noaa.gov/ccg/co2/flask/event/), 2012.

56. Coelho HS, Neves RJ, Leitão PC, Martins H, A. S. The slope current along the Western European Margin: a numerical investigation. Bol Inst Esp Oceanogr. 1999;15:61–72.

57. Coelho HS, Neves RJ, White M, Leitao PC, Santos AJ. A model for ocean circulation on the Iberian coast. J Mar Syst. 2002;32:153–79.

58. Vaz N, Dias JM, Leitão PC, Martins W. Horizontal patterns of water temperature and salinity in an estuarine tidal channel: Ria de Aveiro. Ocean Dyn. 2005;55:416–29.

59. Vaz N, Dias JM, Leitão PC, Nolasco R. Application of the Mohid-2D model to a mesotidal temperate coastal lagoon. Comput Geosci-Uk. 2007;33:1204–9.

60. Mateus M, Riflet G, Chambel P, Fernandes L, Fernandes R, Juliano M, et al. An operational model for the West Iberian coast: products and services. Ocean Sci. 2012;8:713–32.

61. Martins F, Leitao P, Silva A, Neves R. 3D modelling in the Sado estuary using a new generic vertical discretization approach. Oceanol Acta. 2001;24:S51–62.

62. Arakawa A. Computational design for long-term numerical integration of the equations of fluid motion: Two-dimensional incompressible flow. Part I. J Comput Phys. 1966;1:119–43.

63. Kliem N, Pietrzak JD. On the pressure gradient error in sigma coordinate ocean models: a comparison with a laboratory experiment. J Geophys Res-Oceans. 1999;104:29781–99.

64. Vincent S, Caltagirone JP. Efficient solving method for unsteady incompressible interfacial flow problems. Int J Numer Methods Fluids. 1999;30:795–811.

65. Ruiz-Villarreal M, Bolding K, Burchard H, Demirov E. Coupling of the GOTM turbulence module to some three-dimensional ocean models. In: Baumert HZ, Simpson JH, Sundermann J, editors. Marine Turbulence: Theories, Observations, and Models Results of the CARTUM Project. Cambridge: Cambridge University Press; 2005. p. 225–37.

66. Cailleau S, Chanut J, Levier B, Maraldi C, Reffray G. The new regional generation of Mercator Ocean system in the Iberian Biscay Irish (IBI) area. Mercator Quarterly Newsletter. 2010;34:5–15.

67. Lyard F, Lefevre F, Letellier T, Francis O. Modelling the global ocean tides: modern insights from FES2004. Ocean Dyn. 2006;56:394–415.

68. Lefèvre F, Lyard FH, Le Provost C, Schrama EJO. FES99: a global tide finite element solution assimilating tide gauge and altimetric information. J Atmos Ocean Technol. 2002;19:1345–56.

## 15

Carbon storage in Ghanaian cocoa ecosystems

Askia M. Mohammed[1*], James S. Robinson[2], David Midmore[2] and Anne Verhoef[2]

Abstract

Background: The recent inclusion of the cocoa sector as an option for carbon storage necessitates the need to quantify the C stocks in cocoa systems of Ghana.

Results: Using farmers' fields, the carbon (C) stocks in shaded and unshaded cocoa systems selected from the Eastern (ER) and Western (WR) regions of Ghana were measured. Total ecosystem C (biomass C + soil C to 60 cm depth) ranged from 81.8 to 153.9 Mg C/ha. The bulk (~89 %) of the systems' C stock was stored in the soils. The total C stocks were higher in the WR (137.8 ± 8.6 Mg C/ha) than ER (95.7 ± 8.6 Mg C/ha).

Conclusion: Based on the cocoa cultivation area of 1.45 million hectares, the cocoa sector in Ghana potentially could store 118.6–223.2 Gg C in cocoa systems with cocoa systems aged within 30 years regardless of shade management. Thus, the decision to include the cocoa sector in the national carbon accounting emissions budget of Ghana is warranted.

Keywords: Carbon stocks, Cocoa ecosystem, Shaded and unshaded cocoa systems

Background

Cocoa is cultivated in the forest regions of Ghana where an estimated area of 1.45 million hectares of forest land has been displaced [1]. A substantial volume of literature is replete with evidence that the reductions in forest cover produced net sources of carbon dioxide (CO_2), the main greenhouse gas of the atmosphere [2, 3]. According to the Intergovernmental Panel on Climate Change (IPCC), global C stocks in terrestrial biomass have decreased by 25 % over the past century [3, 4]. This corresponds to an annual decline of 1.1 Gt of the global carbon stocks in forest biomass [5]. Stern [2] note that deforestation alone is responsible for 18 % of the world's greenhouse gas emissions.

Cocoa intensification for higher yields has led to a drastic reduction in shade tree density and, on many farms total elimination of the shade trees in cocoa ecosystems [6]. Essentially, cocoa expansion in Ghana has been closely linked to deforestation [7, 8]. One option

*Correspondence: mamusah@yahoo.com
[1] CSIR-Savanna Agricultural Research Institute, Nyankpala, PO Box 52, Tamale, Ghana
Full list of author information is available at the end of the article

to redress deforestation and create a carbon sink is to encourage the establishment of tree-crop farming or agroforestry systems [9–11]. Cocoa agroforestry is an age-old practice in the tropics [12]. Various recommendations have been made to farmers with regard to the number of non-cocoa trees to provide shade for cocoa during planting. However, the decision on how much shade is optimal often depends on the ecological system, social factors, biodiversity interests, ecological services and pod yields [7, 11].

With the recent inclusion of the cocoa sector in the national C emission accounting budgets of Ghana [13], the need to quantify the carbon sequestered in cocoa ecosystems is urgent. In addition to measuring the amounts of carbon stored in cocoa and shade tree biomass in the cocoa systems, the soil organic carbon content needs to be determined. Globally, the amount of C stored in soils is estimated to be 1.5–3 times more than in vegetation [9]. Thus, if Ghana is to include the C sequestered in the cocoa sector in its proposal for developing a national carbon accounting strategy, as outlined in its Readiness Plan Proposal [13], the C quantities stored both in the vegetation and the soils of the cocoa ecosystems must be included.

This paper evaluates the C storage in cocoa ecosystems from two regions of Ghana under two shade management systems and two cocoa stand age categories. It was hypothesised that; (a) the distribution of the total C stocks in the cocoa ecosystem differs between vegetation and soils, and (b) the C stocks differ between regions and shade management. The objectives were: (1) to quantify the total carbon stocks and distribution in the cocoa ecosystem, and (2) to assess the influence of shade management and the region of cocoa production on the C stocks.

Results and discussion

Selected properties of the soils under the cocoa ecosystems

The present study showed a range of 1.1–1.9 Mg/m^3 as the bulk density of the soils under the cocoa ecosystems (Table 1). As expected, the bulk density increased with soil depth from the surface. The gravimetric moisture content of the soils under the cocoa ecosystems ranged from 12.6 to 17.9 % (w/w). The soil moisture only varied with soil depth with the topsoil, 0–20 cm, being the wettest (Table 1). The ranges of the particle size fractions were: clay, 6.6–13.6 %; sand, 49–53 %, and silt, 36–41 % (Table 1). The soils are characterised as having the texture of sandy silt throughout the 0–60 cm layer (Table 1).

Biomass C concentrations

The mean carbon concentrations in above-ground components for all of the ecosystems under evaluation are presented in Fig. 1. The measured litter carbon concentration of 36.1 ± 1.1 corroborates the value of 37 % C in forest litter by Smith and Heath [14] that is currently being used as a default C concentration for litter in agroecosystems by the Intergovernmental Panel on Climate Change [3]. Similarly, the current carbon concentration value of 42.0 ± 0.4 % for the cocoa trees is in agreement with 43.7 ± 2.1 % for cocoa carbon reported by Anglaaere [15].

With the exception of litter C (a proportion of which is lost through respiration as it decomposes), the other components had a narrow range of 42.0–45.6 % C, with cocoa trees having the least C and *Persea americana* (dominant shade species in the Western region) having the highest (Fig. 1). Although few studies on agroecosystem C stocks present direct measurements of carbon with the aid of a C-analyser [16, 17], several studies have used constant values ranging from 45 to 50 % as the proportion of C for all parts of tree biomass [18, 19]. The organic carbon levels in the shade trees in the current study are not markedly different from the constant 45 % C for forest species being used by other studies [20, 21].

Soil organic carbon concentration

The soil total organic carbon concentrations differed significantly (P < 0.05) between regions, systems and soil depths (Table 1). Soil C concentration decreased with soil depth from the surface. Similar trends with depth have been noted by Cifuentes-Jara [22] and Dawoe [23]. The topsoil, 0–20 cm, contained approximately 58.8 % of the soil organic C in the 0–60 cm soil profile. This undoubtedly reflects the great mass of litter fall in cocoa ecosystems. In addition, the high C concentration in the topsoil is in accordance with the presence of 80–85 % mat of lateral roots of cocoa trees being predominantly found within the top 0–30 cm [23–25], although visible roots were excluded in sampling for the current study. The soil C concentration range of 0.6–2.0 % lies within the soil C concentration range of 0.4–2.6 %, reported by Dawoe [23] for 15 and 30 year old cocoa ecosystems in the Ashanti region, Ghana.

Above-ground carbon stocks in cocoa ecosystems

The C contribution from different cocoa ecosystem components to the total above-ground biomass C varied among regions, system, and their interactions (Table 2). On a per hectare basis, the system's biomass C components ranged as follows: cocoa trees, 11.8–16.9 Mg C/ha;

Table 1 Grand mean ± standard error of selected properties of the soils in the cocoa ecosystems for region (n = 24), system (n = 24) and depth (n = 16)

Factor	Treatment	Bulk density (Mg/m³)	Clay (%)	Sand	Silt	Moisture	C
Region	Eastern	1.5 ± 0.1	8.3 ± 0.6	51 ± 2	40 ± 1	14.7 ± 0.7	0.7 ± 0.1
	Western	1.6 ± 0.1	11.7 ± 0.6	52 ± 2	38 ± 1	14.7 ± 0.7	1.5 ± 0.1
System	Shaded	1.6 ± 0.1	10.5 ± 0.6	53 ± 2	36 ± 1	14.8 ± 0.7	1.0 ± 0.1
	Unshaded	1.5 ± 0.1	9.6 ± 0.6	49 ± 2	41 ± 1	14.7 ± 0.7	1.3 ± 0.1
Soil depth	0–20 cm	1.1 ± 0.1	6.6 ± 0.7	53 ± 2	41 ± 1	17.9 ± 0.8	2.0 ± 0.1
	20–40 cm	1.6 ± 0.1	9.9 ± 0.7	51 ± 2	39 ± 1	12.6 ± 0.8	0.8 ± 0.1
	40–60 cm	1.9 ± 0.1	13.6 ± 0.7	50 ± 2	37 ± 1	13.7 ± 0.8	0.6 ± 0.1

Age of farms appearing as covariate

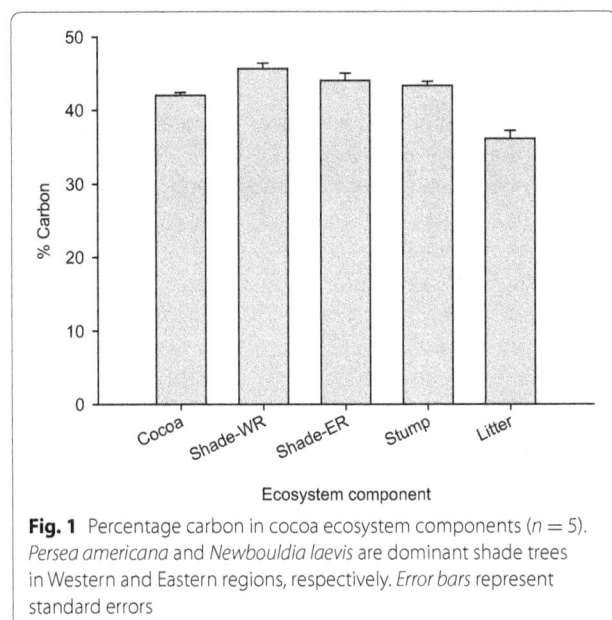

Fig. 1 Percentage carbon in cocoa ecosystem components (*n* = 5). *Persea americana* and *Newbouldia laevis* are dominant shade trees in Western and Eastern regions, respectively. *Error bars* represent standard errors

shade trees, 10.2–16.4 Mg C/ha; litter, 1.9–2.9 Mg C/ha, and stumps, 0.01–0.24 Mg C/ha (Table 2). These results compared well with the ranges of C stocks reported for cocoa trees in the literature [26, 27]. Similarly, the C stocks in shade trees of the current study agreed with the estimates for those in agro-ecosystems researched by Kürsten and Burschel [28] and Polzot [29] (3–25 and 1.9–31.8 Mg C/ha, respectively). Although the cocoa and shade trees' contributions were comparable, and together they contributed approximately 87.3–92.7 % of the total system's biomass C, only the biomass C contribution from the shade trees correlated significantly with the system's total biomass C (r = 0.9724, P < 0.001, Table 3). The lowest contribution to the system's C storage was obtained from stumps in shaded systems in the Western region (Table 2).

Overall, the mean carbon storage of cocoa trees was similar to that estimated for cocoa trees in a 30 year old cocoa system in Cameroon (14.4 Mg C/ha) reported by Norgrove and Hauser [27]. The present study estimated C stock in cocoa trees similar to those reported by Isaac et al. [30] as 10.3 Mg C/ha in an 8 year-old cocoa system in Ghana [30]. Isaac et al. [26] estimated the C storage of a 15 year-old cocoa system in Ghana as 16.8 and 15.9 Mg C/ha for a 25 year-old system, both of which agreed with the present finding that the average carbon storage of cocoa trees ranged between 11.8 and 16.9 Mg C/ha.

Soil organic carbon stocks

Understanding the effects of land use/land cover changes on ecosystem functions is often inferred from changes in soil organic carbon. However, measurements of SOC have often been excluded in many studies on land-use change because of methodological uncertainties. Jones et al. [31] reported a measurement standard error of 1000 kg/ha for SOC, due largely to wide variation in the soil C estimation at deeper soil profiles. In the current study, uncertainty was reduced in the characterization of the soil C pools from the surface to 60 cm depth by measuring C stocks in different soil layers.

Table 4 presents the measured SOC contents for different layers to 60 cm depth. There were considerable variations in SOC contents between regions and systems. The soil organic C stocks ranges in the 0–20, 20–40 and 40–60 cm depths were 35.7–70.7, 15.0–46.7 and 11.5–31.3 Mg/ha, respectively. Clearly, the bulk of the SOC was concentrated in the topsoil, 0–20 cm depth. Moreover, SOC decreased with depth under all the factors. At all depths, soils in the W had the highest mean C stocks. Significantly (P < 0.05) higher SOC stocks were measured in E than W at all depths. The system of production affected SOC storage from the surface to 40 cm depth, but not between 40 and 60 cm (Table 4).

Total cocoa ecosystem carbon stocks and accumulation

Table 5 presents the mean C stocks distributed between the biomass and soil components of cocoa ecosystems. Total above-ground C stock in ecosystem biomass was estimated as the sum of the biomass C from cocoa trees, shade trees, stumps, and litter (Table 2). The total biomass C was highly variable in the cocoa ecosystems and ranged from a minimum mean value of 16.7 ± 2.2 Mg C/ha from unshaded cocoa systems in the Western region to a maximum mean value of 31.3 ± 2.2 Mg C/ha measured in shaded cocoa systems in the Eastern region (Table 5). Statistical analysis of the total system's biomass C showed significantly higher C stocks in E than W and in shaded than unshaded systems (Table 5).

Total SOC pools from the topsoil to 60 cm depth varied considerably from a minimum of 61.7 ± 7.7 Mg C/ha in unshaded cocoa system in the Eastern region to a maximum C stock of 137.8 Mg/ha in unshaded system in the Western region (Table 5). Results from this study estimated higher SOC stocks than the mean SOC value of 60.4 Mg/ha in Dawoe [23] for 0–60 cm depth of cocoa soils in the Ashanti region, Ghana. Cumulative (0–60 cm depth) SOC indicated significant (P < 0.05) variations between regions and also between management systems (Table 5).

The total ecosystem C stock of cocoa systems was estimated as the sum of soil C within 0–60 cm depth and above-ground biomass C (trees, stump and litter C). Total ecosystem C was higher in the Western region (137.7 ± 8.6 Mg C/ha) than in the Eastern region

Table 2 Mean C stocks ± standard error (Mg/ha) in cocoa trees, shade trees, stumps and litter components as influenced by region [Eastern (E), Western (W)] and system [shaded (S), unshaded (U)] and their interactions, (n = 12)

Factor	Treatment	Cocoa	Shade	Stumps	Litter
Region	E	15.2 ± 1.0	10.2 ± 6.4	0.16 ± 0.02	2.3 ± 0.2
	W	13.5 ± 1.0	16.4 ± 6.4	0.12 ± 0.02	2.4 ± 0.2
System	S	12.7 ± 1.1	13.3 ± 4.1	0.07 ± 0.02	2.6 ± 0.2
	U	16.1 ± 1.1	n.a.	0.21 ± 0.02	2.2 ± 0.2
Region * System	E * S	13.6 ± 1.5	10.2 ± 6.4	0.12 ± 0.03	2.3 ± 0.2
	E * U	16.9 ± 1.5	n.a.	0.19 ± 0.03	2.4 ± 0.2
	W * S	11.8 ± 1.6	16.4 ± 6.4	0.01 ± 0.03	2.9 ± 0.2
	W * U	15.2 ± 1.5	n.a.	0.24 ± 0.03	1.9 ± 0.2

Age of farms appearing as covariate in the statistical model used

Not applicable

Table 3 Pearson correlation coefficients (r) for linear relationships among biomass C components in cocoa ecosystems

	Ecosystem	Cocoa	Shade	Stumps
Cocoa	0.4936			
Shade	0.9724**	0.2801		
Stump	−0.1842	0.2948	−0.2536	
Litter	0.4945	0.5397	0.3703	−0.4871

Values with '**' are significant at $P < 0.01$, and without symbol are not significant, (2—tailed test)

(95.7 ± 8.6 Mg C/ha). These C estimates are very high when compared with data from Dawoe [23]. This is attributed to the low soil C stocks (35.5–80.4 Mg C/ha) from 0–60 cm depth reported by Dawoe [23], that were equivalent to the estimated C stocks in the current study's topsoil, 0–20 cm (35.7–70.7 Mg C/ha) (see Table 4). Notably, in the current study, the soils contributed between 3 and

5 times more C than the above-ground pools of the cocoa ecosystems. Given the age range (7–28 years) of farms used in the current studies, as well as the extensive cultivation of 1.45 million hectares of cocoa in Ghana [1], it appears that approximately 118.6–223.2 Gg C could be stored in cocoa systems with stands aged within 30 years, irrespective of the shade-management system.

The relative contribution of the cocoa systems (scatter) to the overall C stocks (line) in each component varied considerably when expressed on the basis of cocoa stand age (Fig. 2). The shaded and unshaded cocoa systems appeared to contain the same biomass stocks at stand age of 10 years in age. In the above-ground biomass C stocks, both shaded and unshaded cocoa systems increase with stand age but the contribution from the shaded systems to the overall biomass C trend was much higher than the unshaded system for cocoa stands older than 10 years (Fig. 2).

With respect to the effects of cocoa systems on soil C, there appears to be a general decline of the soil C stocks as time progressed. Whereas the shaded systems indicate a slight increase, the unshaded systems showed a slight decrease in soil C (Fig. 2). The two systems have similar soil C stocks at stand age of 25 years onwards.

The primary source of soil C is from litter and so the quantity and quality of the litter inputs affect the soil C dynamics [32]. Of the systems' contribution to the total C in the ecosystems, the trend follows that of the soil C since the bulk of C (>80 %) is stored in the soil (Table 5). The trends indicate that total carbon in shaded and unshaded systems are the same at age 17 years, but the shaded system thereafter, increased in the total C higher than that of the unshaded system (Fig. 2).

Conclusions

The need to quantify the carbon stocks in cocoa systems in Ghana is necessitated by the recent inclusion of the sector as an option that could result in a net increase in

Table 4 Mean soil organic C stocks ± standard error (Mg/ha) at 0–20, 20–40 and 40–60 cm layers as influenced by region [Eastern (E), Western (W)], and system [shaded (S), unshaded (U)], (n = 12)

Factor	Treatment	0–20 cm	20–40 cm	40–60 cm
Region	E	40.2 ± 3.4	16.6 ± 3.6	14.3 ± 1.6
	W	58.4 ± 3.4	33.3 ± 3.6	25.7 ± 1.6
System	S	45.4 ± 3.4	19.0 ± 3.6	18.5 ± 1.6
	U	53.2 ± 3.4	30.9 ± 3.6	21.4 ± 1.6
Region * System	E * S	44.7 ± 4.8	18.2 ± 5.0	17.1 ± 2.2
	E * U	35.7 ± 4.8	15.0 ± 5.0	11.5 ± 2.2
	W * S	46.1 ± 5.1	19.9 ± 5.4	20.0 ± 2.4
	W * U	70.7 ± 4.8	46.7 ± 5.1	31.3 ± 2.2

Age of farms appearing as covariate in the statistical model used

Table 5 Mean cocoa ecosystem carbon stocks ± standard error, distributed between the biomass and soil (0–60 cm depth) components according to region [Eastern (E), Western (W)], and system [shaded (S), unshaded (U)], (n = 12)

Factor	Treatment	Biomass C (Mg/ha)	Soil C (Mg/ha)	Total C (Mg/ha)
Region	E	25.2 ± 1.6	70.5 ± 5.4	95.7 ± 8.6
	W	18.4 ± 1.6	113.0 ± 5.4	137.7 ± 8.6
System	S	25.8 ± 1.6	83.7 ± 5.5	115.5 ± 8.6
	U	17.8 ± 1.6	99.8 ± 5.5	117.9 ± 8.6
Region * System	E * S	31.3 ± 2.2	79.3 ± 7.7	109.5 ± 12.0
	E * U	19.0 ± 2.2	61.7 ± 7.7	81.8 ± 12.0
	W * S	20.2 ± 2.3	88.1 ± 8.2	121.5 ± 12.0
	W * U	16.7 ± 2.2	137.8 ± 7.7	153.9 ± 12.1

Age of farms appearing as covariate in the statistical model used

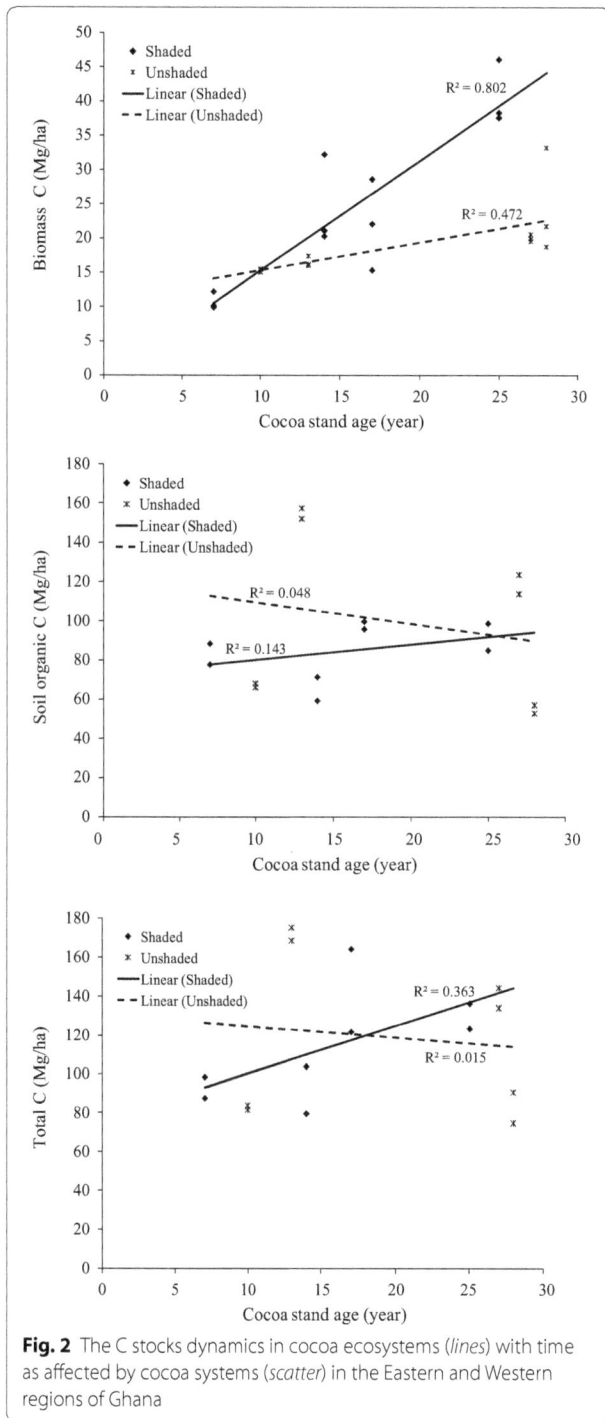

Fig. 2 The C stocks dynamics in cocoa ecosystems (*lines*) with time as affected by cocoa systems (*scatter*) in the Eastern and Western regions of Ghana

was twice that in unshaded systems, the two systems did not differ significantly with respect to total ecosystem C stocks. The bulk of the C stock was in the soil. The estimated high C stocks suggest that the cocoa sector holds a large amount of carbon and should be included in the national carbon accounting emission budget of Ghana.

Methods
Physiology of the study area
The field studies were carried out between July and October, 2011 in two regions of Ghana; the Eastern region at Duodukrom community in the Suhum district (6°2′N, 0°27′W), and the Western region at Anyinabrim in the Sefwi-Wiawso district (6°57′N, 2°35′W). Figure 3 presents the map of Ghana showing the regions and districts where the field studies were conducted.

The Eastern region covers a land area of 19,323 km^2 representing 8.1 % of the total land area of Ghana. It is located between latitude 6° and 7°N and longitude 1°30′W and 0°30′E. The region lies within the wet semi-equatorial zone which is characterized by double-maxima rainfall in June and October. The natural vegetation of the region is humid deciduous forest. Temperatures in the region are high and range between 26 °C in August and 30 °C in March. The relative humidity which is high throughout the year varies between 70 and 80 %.

The Western region occupies a land area of 23,921 km^2 which is approximately 10 % of the total land area of Ghana. The region lies in the equatorial climatic zone that is characterized by a double maxima rainfall occurring in May–July and September/October. Its vegetation is that of humid deciduous forest. The region is the wettest part of Ghana with an average rainfall of 1600 mm per annum and harbours about 24 forest reserves that account for about 40 % of the forest reserves in Ghana. The climate creates much moisture culminating in high relative humidity, ranging from 70 to 90 % in most part of the region. Temperatures range between 22 °C at nightfall and 34 °C during the day.

Thus, the two regions experience similar climate and vegetation. The major soils found in both regions are mostly well drained *Ochrosols* or *Oxisols* suitable for the production of industrial crops such as cocoa, pineapple, pawpaw cola nut and oil palm. However, the Eastern region has been producing cocoa long before cultivations started in the Western region.

Selection of farms
Eight farms, comprising four from the Duodukrom community in the Suhum district of the Eastern region, and four from the Anyinabrim community in the Sefwi-Wiawso district of Western region were selected for sampling cocoa stands on the basis of shade management

terrestrial carbon stocks. Hence, this paper estimated the carbon stocks in shaded and unshaded cocoa systems at different age categories; the fields were selected from the Eastern region (E) and Western region (W) of Ghana. Total ecosystem carbon was higher in the W than E. While the biomass C stock from shaded systems

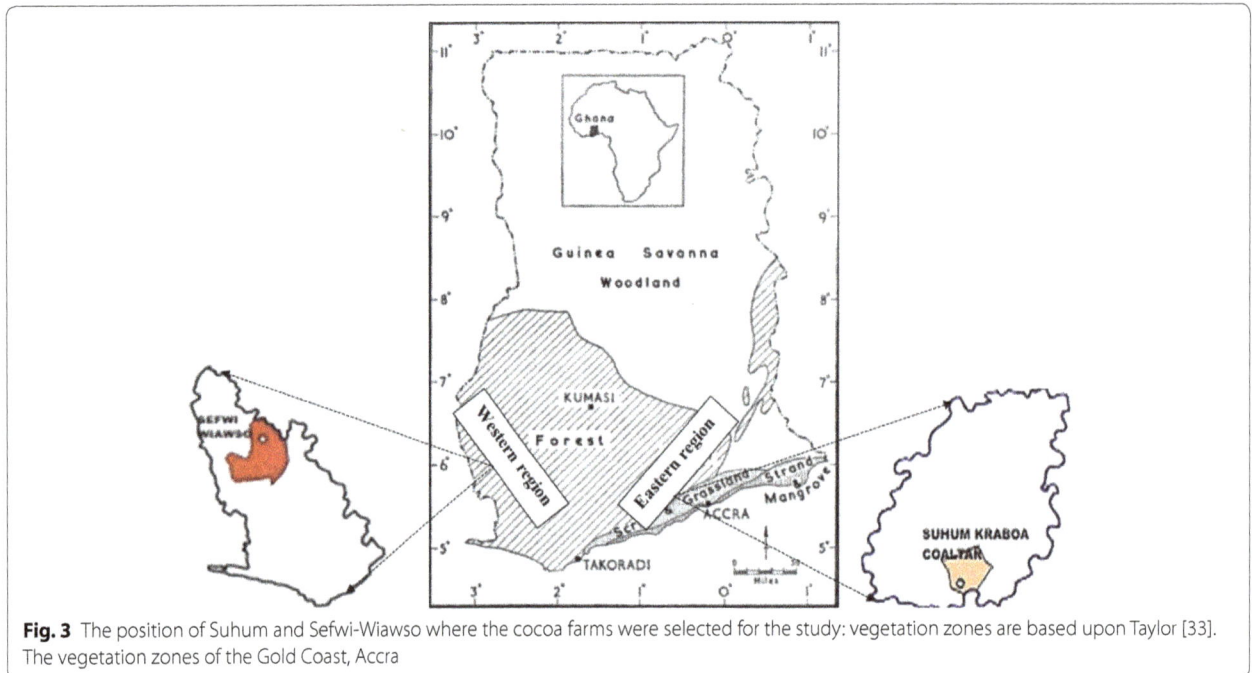

Fig. 3 The position of Suhum and Sefwi-Wiawso where the cocoa farms were selected for the study: vegetation zones are based upon Taylor [33]. The vegetation zones of the Gold Coast, Accra

(shaded, unshaded). Selected farms had cocoa stand ages of 10, 14, 25 and 28 years in the Eastern region (E) and 7, 13, 17 and 27 years in the Western region (W).

At each farm, plot sizes of 30 × 90 m were demarcated for sampling. Two 30-m transects dividing the plot into three of 30 × 30 m (~0.23 acre or 0.09 ha) subplots were demarcated to give three pseudo-replications of each farm. The common shade tree species identified on the cocoa farms included *Terminalia ivorensis*, *Terminalia superba*, *Entandrophragma cylindricum*, *Entandrophragma angolense*, *Newbouldia laevis*, *Persea americana*, *Celtis mildbraedii*, *Cola nitida*, *Carica papaya*, *Palmae* sp., *Spondia smombin*, *Ficus exasperate*, *Citrus sinensis* (L.) Osbeck, *Acacia mangium*, and other forest tree species. Avocado (*Persea americana*) was the dominant shade tree in cocoa farms found in the Western region whilst *Newbouldia laevis* was the dominant shade tree in the Eastern region's cocoa farms.

All trees were counted, and their diameters at breast height (DBH) measured, sorted and grouped into three diameter class sizes (upper, middle, and lower) relative to the DBH range of cocoa trees on the farms; 16 cocoa trees, comprising two cocoa trees per farm were randomly selected such that the diameter of one tree lay within the upper class and the other in the lower class for destructive sampling. The felled trees were each separated into trunks, branches and foliage (leaves, fruits); these parts were cut to smaller pieces, weighed in batches and then summed to give total component weight. Fresh leaf samples of the dominant shade trees found in each

region were also taken. Based on the measured DBH and the biomass per tree of the 16 cocoa trees that were destructively sampled across all the study sites, an allometric relation was developed using regression techniques to estimate standing cocoa tree biomass. The general equation from FAO [34], recommended by UNFCCC [35], was used to estimate the above-ground biomass of the shade tree species.

$$AgB = \exp\left[-2.134 + 2.530 \ln (DBH)\right] \qquad (1)$$

where AgB denotes above-ground biomass, kg tree^{-1}, and DBH = diameter at breast height, cm.

Soil moisture and bulk density

Soil samples at 0–20, 20–40 and 40–60 cm depths were taken from a total of 16 plots comprising 2 micro-plots of (50 × 50 cm) that were established at random within the eight cocoa farms. Two core soil samples per depth were taken randomly at each micro plot using an auger after removing visible litter from the soil surface. Soil bulk density gives an indication of the level of soil compaction [36]. Soil bulk density and moisture contents at each sampling depth were determined on the undisturbed core samples, as outlined in Blake and Hartge [37].

Texture

Another set of soil samples from the same micro-plots was air-dried for 72 h, and ground to pass through a 2-mm mesh sieve to yield the fine earth fraction for chemical analysis. The soil particle size distribution was determined by laser granulometry, using a Coulter LS230 particle size analyser connected to a Windows-based computer [38–40].

Carbon concentration

Weights between 0.9–1.1, and 8.0–12.0 mg were taken respectively, from plant and soil samples the determination of C concentrations. The organic C concentration in the samples was determined using the Europa Roboprep connected to a VG 622 Mass Spectrophotometer.

Carbon stocks

There are various C pools, or compartments, within cocoa ecosystems. These include the soil C pool, the litter C pool and the woody biomass C pool in trees. The quantity of C stored in each pool is reported as the C stock, and the sum of the C stocks from the different pools constitutes the total ecosystems C stocks. On each farm, the total biomass-C stock was estimated as the sum of the C stocks in cocoa tree components (root, stem, branch, and leaf litter), floor litter and shade trees (if any) as expressed in Eq. (2). The cocoa tree component-C stocks were calculated as the product of the mean C concentration and the biomass per hectare [41]. The mean C concentration of leaves of shade trees was used as the average C concentration of the whole shade tree in estimating the C stock of the shade trees.

$$
\begin{aligned}
TotalC_{biomass} = \{&[(\%C_{root} \times root_{biomass}) \\
&+ (\%C_{stem} \times stem_{biomass}) \\
&+ (\%C_{branch} \times branch_{biomass}) \\
&+ (\%C_{leaf} \times leaf_{biomass})]_{cocoa} \\
&+ (\%C_{litter} \times litter_{biomass}) \\
&+ (\%C_{shade} \times Shade_{biomass})\}
\end{aligned} \tag{2}
$$

Soil C stocks were also calculated using the formula:

$$
SOC = \sum \rho_i \times d_i \times \%C_i \tag{3}
$$

where SOC denotes soil organic carbon stock (Mg/ha); ρ = soil bulk density (g/cm^3); i = 0–20, 20–40, and 40–60 cm sampling depth; d = depth over which the sample was taken (cm); and $\%C$ = soil carbon concentration (%). The total cocoa ecosystem carbon stock for each farm/system was then estimated as the sum of Eqs. (2) and (3).

Data analyses

The data were tested for normality using q–q plot with Anderson–Darling P values in MINITAB v16. Where the tested component C was found to be non-normal,

the appropriate transformation was determined with the help of Box-Cox transformation and optimal or rounded lambda that suggested one of the following transformational method as appropriate: square root, reciprocal square root, natural logarithm or inverse transformation method, according to the skewness of the data [42]. Specifically, litter, stumps and total ecosystem C data were normal (P > 0.05) without transformation; biomass and soil C were inversely transformed; cocoa tree C was transformed with square root and shade tree C was normalized using natural logarithms. The transformed data were analysed by the Linear MIXED Model of IBM SPSS statistics 20th edition to determine significant differences between Eastern and Western regions and between shaded and unshaded systems as well as the interactions on carbon stocks controlling for the ages (covariate) of cocoa farms. The means were then estimated by restricted maximum likelihood (REML) and back-transformed to maintain the original form of the measurement. Correlation analyses by Pearson's rank matrix were also carried out to determine any relationships among some of the ecosystem variables.

Authors' contributions

All authors (AMM, JSR, DM and AV) took part in the development of the main idea and the writing of the manuscript. The field data and laboratory analysis were done by AMM who also performed the majority of the statistical analysis. The data and analytical results were discussed among all four authors. All authors read and approved the final manuscript.

Author details

[1] CSIR-Savanna Agricultural Research Institute, Nyankpala, PO Box 52, Tamale, Ghana. [2] School of Archaeology, Geography and Environmental Science, University of Reading, Reading, UK.

Acknowledgements

This paper is an output from a Ph.D. study in the University of Reading, UK, with funds from the Ghana Education Trust Fund, and the authors are grateful for the permission granted to publish this work. The support by cocoa farmers in Ghana to use their fields for data collection is also gratefully acknowledged.

Competing interests

The authors declare that they have no competing interests.

References

1. Anim-Kwapong G, Frimpong E. Vulnerability of agriculture to climate change-impact of climate change on cocoa production. Final Report submitted to the Netherlands climate change studies assistance programme: NCAP; 2005. 2.
2. Stern NH. Stern review: the economics of climate change, vol. 30. London: HM treasury London; 2006.
3. IPCC. Climate change 2007 the physical science basis. Agenda. 2007;6:07.
4. IPCC. Climate change: the scientific basis. Cambridge: Cambridge University Press; 2001.
5. Marland G, Boden TA, Andres RJ, Brenkert A, Johnston C. Global, regional, and national fossil fuel CO$_2$ emissions. Trends: a compendium of data on global change; 2003. p. 34–43.

6. Seeberg-Elverfeldt C, Schwarze S, Zeller M. Payments for environmental services: incentives through carbon sequestration compensation for cocoa-based agroforestry systems in Central Sulawesi. Grauer; 2008.

7. Wade ASI, Asase A, Hadley P, Mason J, Ofori-Frimpong K, Preece D, Spring N, Norris K. Management strategies for maximizing carbon storage and tree species diversity in cocoa-growing landscapes. Agric Ecosyst Environ. 2010;138(3–4):324–34.

8. Gockowski J, Sonwa D. Cocoa intensification scenarios and their predicted impact on CO_2 emissions, biodiversity conservation, and rural livelihoods in the Guinea Rain Forest of West Africa. Environ Manage. 2011;48(2):307–21.

9. Dixon R. Agroforestry systems: sources of sinks of greenhouse gases? Agrofor Syst. 1995;31(2):99–116.

10. UNFCCC. Calculation of the number of sample plots for measurements with A/R CDM project activities version 02; 2009.

11. Oke D, Olatiilu A. Carbon storage in agroecosystems: a case study of the cocoa based agroforestry in Ogbese forest reserve, Ekiti State,Nigeria. J Environ Prot. 2011;2(8):1069–75.

12. Oke D, Odebiyi K. Traditional cocoa-based agroforestry and forest species conservation in Ondo State, Nigeria. Agric Ecosyst Environ. 2007;122(3):305–11.

13. Chagas T, O'Sullivan R, Bracer C, Streck C. Consolidating national REDD + accounting and subnational activities in Ghana. UNDP, Global Environment Facility, NORAD, and Gordon and Betty Moore Foundation; 2010. p. 36.

14. Smith JE, Heath LS. Identifying influences on model uncertainty: an application using a forest carbon budget model. Environ Manage. 2001;27(2):253–67.

15. Anglaaere LCN. Improving the sustainability of cocoa farms in Ghana through utilization of native forest trees in agroforestry systems, Ph.D. Thesis, University of Wales, Bangor; 2005. p. 340.

16. Kraenzel M, Castillo A, Moore T, Potvin C. Carbon storage of harvest-age teak (Tectona grandis) plantations, Panama. For Ecol Manage. 2003;173(1–3):213–25.

17. Losi CJ, Siccama TG, Condit R, Morales JE. Analysis of alternative methods for estimating carbon stock in young tropical plantations. For Ecol Manage. 2003;184(1–3):355–68.

18. Snowdon P, Raison J, Keith H, Ritson P, Grierson P, Adams M, Montagu K, Bi H-Q, Burrows W, Eamus D. Protocol for sampling tree and stand biomass. Canberra: Australian Greenhouse Office; 2002. p. 76.

19. IPCC. Good practice guidelines for National Greenhouse gas inventories. Hayama: Institute for Global Environmental Strategies (IGES); 2006.

20. Noordwijk MV, Rahayu S, Hairiah K, Wulan Y, Farida A, Verbist B. Carbon stock assessment for a forest-to-coffee conversion landscape in Sumber-Jaya (Lampung, Indonesia): from allometric equations to land use change analysis; 2002. p. 1–12.

21. Smiley G, Kroschel J. Temporal change in carbon stocks of cocoa–gliricidia agroforests in Central Sulawesi, Indonesia. Agrofor Syst. 2008;73(3):219–31.

22. Cifuentes-Jara M. Aboveground biomass and ecosystem carbon pools in tropical secondary forests growing in six life zones of Costa Rica, Ph.D. Thesis, Oregon State University; 2008. p. 195.

23. Dawoe E. Conversion of natural forest to cocoa agroforest in lowland humid Ghana: impact on plant biomass production, organic carbon and nutrient dynamics, Ph.D. Thesis, Kwame Nkrumah University of Science and Technology; 2009. p. 279.

24. Wood GAR, Lass R. Cocoa. Hoboken: Wiley-Blackwell; 2008.

25. de Oliveira Leite J, Valle RR. Nutrient cycling in the cacao ecosystem: rain and throughfall as nutrient sources for the soil and the cacao tree. Agric Ecosyst Environ. 1990;32(1):143–54.

26. Isaac M, Gordon A, Thevathasan N, Oppong S, Quashie-Sam J. Temporal changes in soil carbon and nitrogen in west African multistrata agroforestry systems: a chronosequence of pools and fluxes. Agrofor Syst. 2005;65(1):23–31.

27. Norgrove L, Hauser S. Carbon stocks in shaded Theobroma cacao farms and adjacent secondary forests of similar age in Cameroon. Trop Ecol. 2013;54(1):15–22.

28. Kürsten E, Burschel P. CO_2-mitigation by agroforestry. Water Air Soil Pollut. 1993;70(1):533–44.

29. Polzot CL. Carbon storage in coffee agroecosystems of southern Costa Rica: Potential applications for the clean development mechanism, M.Sc. Thesis, York University, Toronto; 2004. p. 162.

30. Isaac ME, Timmer VR, Quashie-Sam SJ. Shade tree effects in an 8-year-old cocoa agroforestry system: biomass and nutrient diagnosis of Theobroma cacao by vector analysis. Nutr Cycl Agroecosyst. 2007;78(2):155–65.

31. Jones J, Graham W, Wallach D, Bostick W, Koo J. Estimating soil carbon levels using an ensemble Kalman filter. Trans Am Soc Agric Eng. 2004;47(1):331–42.

32. Post WM, Kwon KC. Soil carbon sequestration and land-use change: processes and potential. Glob Change Biol. 2000;6(3):317–27.

33. Taylor CJ. The vegetation zones of the Gold Coast. Forestry Department Bulletin 4, Accra. 1952; pp. 48–51.

34. FAO. Estimating biomass and biomass change of tropical forests: a primer, in FAO Forestry Paper 134. Rome: FAO; 1997. p. 55.

35. UNFCCC. Monitoring methodologies for selected small-scale afforestation and reforestation project activities under the Clean Development Mechanism. Bonn; 2006.

36. Hillel D. Environmental soil physics: fundamentals, applications, and environmental considerations. New York: Academic Press; 1998. p. 19–201.

37. Blake G, Hartage K. Bulk density. Methods of soil analysis: part 1—physical and mineralogical methods, 1986(methodsofsoilan1). p. 363–75.

38. Buurman P, Pape T, Muggler C. Laser grain-size determination in soil genetic studies 1. Practical problems. Soil Sci. 1997;162(3):211–8.

39. Muggler C, Pape T, Buurman P. Laser grain-size determination in soil genetic studies 2. Clay content, clay formation, and aggregation in some brazilian oxisols. Soil Sci. 1997;162(3):219–28.

40. Arriaga FJ, Lowery B, Mays MD. A fast method for determining soil particle size distribution using a laser instrument. Soil Sci. 2006;171(9):663–74.

41. Subedi BP, Pandey SS, Pandey A, Rana EB, Bhattarai S, Banskota TR, Charmakar S, Tamrakar R. Forest carbon stock measurement: guidelines for measuring carbon stocks in community-managed forests; 2010. p. 69.

42. Hamilton LC. Modern data analysis: a first course in applied statistics. California: Brooks/Cole Pacific Grove; 1990.

## 16

A multi-tiered approach for assessing the forestry and wood products industries' impact on the carbon balance

Marcus Knauf

Abstract

Background: The forestry and wood products industries play a significant role in CO_2 emissions reduction by increasing carbon stocks in living forest biomass and wood products. Moreover, wood can substitute for fossil fuels. Different methods can be used to assess the impact of regional forestry and wood products industries on regional CO_2 emissions. This article considers three of those methods and combines them into a multi-tiered approach.

Results: The multi-tiered approach proposed in this article combines: 1) a Kyoto-Protocol-oriented method focused on changes in CO_2 emissions resulting from regional industrial production, 2) a consumer-oriented method focused on changes in CO_2 emissions resulting from regional consumption, and 3) a value-creation-oriented method focused on changes in CO_2 emissions resulting from forest management and wood usage strategies. North Rhine-Westphalia is both a typical German state and an example of a region where each of these three methods yields different results. It serves as a test case with which to illustrate the advantages of the proposed approach.

Conclusions: This case study argues that the choice of assessment methods is essential when developing and evaluating a strategy for reducing CO_2 emissions. Emissions can be reduced through various social and economic processes. Since none of the assessment methods considered above is suitable for all of these processes, only a multi-tiered approach may ensure that strategy development results in an optimal emissions reduction strategy.

Keywords: Forestry and wood products industry; Multi-tiered approach for assessing; Kyoto-protocol-oriented approach; Consumer-oriented approach; Carbon footprint; Value-creation-oriented approach; C-sink; Substitution; North Rhine-Westphalia

Background

The forestry and wood products industries' impact on climate protection and how it is reflected in climate reporting

Forests have a significant impact on the global carbon cycle and therefore on the climate [1]. On the one hand, they help improve local climates by moderating temperature and humidity, while on the other, they absorb atmospheric carbon (CO_2) through photosynthesis and forest growth and store it for the long term (sequestration). When wood is used in furniture or construction, for example, the sequestered carbon remains stored in the wood product (wood carbon stock) [2-4]. Wood can substitute for fossil fuels such as oil, gas or coal. Not

only can wood be burned for fuel (fuel substitution) [5-7], production and disposal of wood products typically require less energy than products made from other materials (material substitution) [8-15].

In the context of climate reporting, article 3.4 of the Kyoto Protocol recognizes an increase in the forest's carbon stocks as a CO_2 emissions reduction measure. Forest carbon sequestration is assessed within the framework of the "land use, land use change and forestry" (LULUCF) sector, cf. [16,17]. According to the IPCC Good Practice Guidance for Land Use, Land Use Change and Forestry [16,17] accounting must include the five carbon pools: (1) above-ground biomass, (2) below-ground biomass, (3) deadwood, (4) litter and (5) organic soil carbon [18]. These reports are included in national greenhouse gas inventories [19]. In the first commitment period of the Kyoto Protocol, from 2008 to 2012, the

Correspondence: m@knauf-consulting.de
Knauf Consulting, Dorotheenstraße 7, D-33615 Bielefeld, Germany

assumption was that "all carbon removed in wood and other biomass from forests is oxidized in the year of removal" [20]. However, this assumption did not take into consideration that wood removal does not result in an immediate release of CO_2 [21,22]. The decisions made at the Conferences of the Parties in Copenhagen 2009, Durban 2011 and Doha 2012, necessitated a follow-up agreement to the Kyoto Protocol to address the sink function of harvested wood products (HWP) [23]. A forest management reference level (FMRL) [24,25] accounts for carbon stored in HWP. In climate reporting, the effects of substitution are recorded as CO_2 reductions within the industry and energy sectors and are, therefore, not recognized as a contribution of the forestry and wood products industries.

To fully assess and increase the forestry and wood products industries' contribution to climate protection, an integrated study of the forestry and wood products industries which accounts for all storage and substitution effects [1,14,26,27] is needed. Such a study could be conducted either at the national [14,28] or regional level [29].

The Kyoto Protocol's geographical approach and its limitations

Climate reporting under the Kyoto Protocol follows a geographical approach [30], assigning territorial responsibility for CO_2 emissions to the producers (e.g. industry CO_2 emissions). It allocates emissions to the emitter. This is also the case for the forestry and wood products industries. Studies which assess the overall impact of CO_2 emissions on climate protection (see previous paragraph), follow the same geographical approach by observing a specific forest area (whether national or regional) and analyzing all of its associated carbon reduction effects (sequestration and substitution). Consumers are not included in this analysis. Many authors, however, have proposed including consumer effects in general, i.e. not specifically related to the forestry and wood products industries [31-36]. This argument is based on the premise that "the responsibility for carbon dioxide (CO_2) emissions from economic activities lies with people's attempts to satisfy certain functional needs and desires" [31]. This concept is generally referred to as the carbon footprint (CF) [37]. Current studies on the forestry and wood products industries' impact on climate protection do not include the life cycle effects of consumption. Instead, the CF is only calculated at the individual products level [38]. However, following the premise that the forestry and wood products industries do in fact contribute to the reduction of CO_2 emissions and thereby to climate protection, and also assuming that promoting wood use (e.g. increasing wood-based construction) is a sensible climate policy measure, then there is a deficit in the data needed to evaluate such measures in terms of consumer effects (e.g. CF). The CF is normally associated with CO_2

emissions and thus accounted for as a debit. However, in the case of the forestry and wood products industries, the CF is typically associated with a reduction in CO_2 and expressed as a credit. The following paper treats the CF of the forestry and wood products industries as a negative CF.

The purpose of this paper is to present an analytical model combining three approaches

This paper presents a tool that accounts for the CO_2-responsibility of consumers [32] in assessing the forestry and wood products industries' impact on climate protection. It proposes an analytical model with three distinct approaches: 1) an approach based on the principles of the Kyoto Protocol and focused on emitters, 2) a consumer-oriented approach (carbon footprint) and 3) an approach based on the value chain of the forestry and wood products industries. A particular aspect of this analytical model with its three approaches is that the effects of wood usage (storage and substitution) are allocated differently. The paper presents its analytical model in the form of a generally accepted matrix that incorporates the four CO_2 reducing effects of the forestry and wood products industries (forest sink, HWP sink, fuel substitution, material substitution). The model is then tested in a case study assessing the German state of North Rhine-Westphalia.

Results and discussion
Development of a universal model for assessing the forestry and wood products industries' impact on the carbon balance

The three approaches developed for this study uniformly assess sequestration in the forest according to the IPCC Good Practice Guidance for Land Use, Land Use Change and Forestry [16-18] and in conformity with the international conventions of climate reporting. The approaches differ with respect to the assessment of wood usage and its related C-effects. The following three questions characterize the different approaches:

- Approach I (Kyoto-Protocol-oriented): "How do the forestry and wood products industries in Area x impact the CO_2 balance in Area x (and the global CO_2 balance)?"
- Approach II (consumer-oriented): "How do the forest and consumers in Area x impact the global CO_2 balance?"
- Approach III (value creation-oriented): "How do the forest and harvested wood products from Area x impact the global CO_2 balance?"

Approach I applies the Kyoto Protocol approach and concentrates on the emitter. In addition to sequestration in the forest, this model takes into account the HWP from Area x. The carbon stock levels in harvested wood products are calculated according to the IPCC classification [25]; this assessment model provides a country-specific method of calculation. It also takes CO_2 emissions reduction through the substitution of fossil fuels into account. This model includes wood in Area x used as fuel (fuel substitution), as well as wood in Area x used to manufacture products (material substitution). The emissions reductions are calculated based on the volumes of wood utilized and manufactured into finished products by multiplying the mass C in wood (expressed as t C) with the underlying substitution factors. To assess fuel substitution, for example, the substitution factor is calculated from the difference in emissions between fossil fuels (e.g. defined mix) and wood, based on the carbon content of the wood utilized [14,39,40]. Excluding the fossil fuel consumption inherent in forest management, timber harvesting and transport, which makes up less than 10% of the total emissions profile, the use of wood for fuel is considered CO_2 neutral, cf. [39,41]. Material substitution is assessed using the general substitution factors developed by [15]. Here the difference between CO_2 emissions (expressed as C) of competing products (wood versus non-wood) with the same functionality are set in relation to their carbon content. In addition to general substitution factors, product or product-specific substitution factors can also be used, cf. [14,27]. The substitution effects are not visible per se but are reflected in the greenhouse gas inventory of Area x (if a regional greenhouse gas inventory has been conducted). They are expressed as CO_2 emissions reductions in the energy and industrial sectors, but not attributed to the forest-based industries. The accounting method of Approach I makes it possible to identify the substitution effects as contributions of the forestry and wood products industries and evaluate them accordingly.

In contrast to Approach I, Approach II does not concentrate on CO_2 emissions (e.g. of the industrial sector) or the CO_2 emissions avoided through substitution, but instead focuses on the consumption of wood products in Area x. Approach II shows the CO_2 effects associated with the wood products used in Area x (over their entire life cycle). The approach concentrates on the carbon footprint and offers a way to incorporate consumer responsibility into the analysis of the forestry and wood product industries' impact on CO_2 emissions reduction. It also prevents so-called leakage effects [42]. An assessment method that disregards the location of wood usage ignores any climate-conscious consumption and investment decisions impacting the CO_2 footprint of a region or a country, rendering them irrelevant. The impact from fuel substitution can be measured the same way as in Approach I. The evaluation of the CO_2 effects of wood products, however, differs in that Approach II links

the wood products used to Area x. These wood products are taken into account when determining the carbon stock in wood products and assessing material substitution. The volume of utilized wood products is calculated using input-output analyses based on either official statistics [43], or empirical studies [44]. The material-substitution calculations are based on the utilized wood products and the substitution factor for the material substitution [15].

Approach III can be used as a basis for determining which forest management measures and wood usage strategies in Area x have the greatest impact on CO_2 reduction. This approach is referred to as the value-creation approach because it emphasizes assessment of the effects of wood harvested in a region's forests as part of the value chain in the forestry and wood product industries. Approach III provides useful information for designing the forest-wood value chain to optimize climate benefits. For example, this approach can be used to develop forest management scenarios for future forest development [29,45]. In Approach III carbon stocks in the forest and harvested wood are evaluated using the same methodology as in Approach I [18,25]. Calculating the fuel and material substitution levels requires a harvested wood products utilization model that takes into account current material flows (related to any fuel and material use). Fuel and material substitution levels are calculated based on this utilization model and substitution factors (see explanations under Approach I).

These three approaches are equally valid but can be used to draw different conclusions. While Approach III is favored for the development of climate-optimal strategies for the forestry and wood products industries, Approach II is primarily used when assessing of wood usage and Approach I is helpful in evaluating the local forest-based industry and its specific impacts. The paper sets the three approaches in relation to the four effects associated with the forestry and wood products industries' contribution to climate protection (forest sink, HWP sink, fuel substitution, material substitution) to create a 12-field matrix and adds a totals column to form a 15-field matrix (Table 1).

In a country without any international trade in timber or wood-based products, the differences in the three different approaches would be redundant, as all three approaches would yield the same results. A similar situation would arise if there were no foreign trade surplus in the trade of wood products. However, a country without forestry and wood product industries that only imports wood products would show a zero value in Approaches I and III (assuming it would not use any wood for fuel), while Approach II would yield a value based on the amount of imported wood (carbon stock in the wood and material substitution; the disposal of the wood products would have to be taken into account, however). A heavily forested country with strong

Table 1 Matrix for assessing the impact of the forestry and wood products industries on the CO_2 balance (Area x)

Approaches for assessing the forestry and wood products industries' impact on the CO_2 balance (Area x)	Category of emission reduction/stock/sink				TOTAL
	Forest sink	HWP sink	Fuel substitution	Material substitution	
	Changes in forest carbon stock	Changes in HWP carbon stock	CO_2 impact of wood fuel	CO_2 impact of wood product usage	
	[Mt CO_2]	[Mt CO_2]	[Mt CO_2]	[Mt CO_2]	[Mt CO_2]
How do the forestry and wood product industries in Area x impact the CO_2 balance in Area x (and the global CO_2 balance)? **Approach I: Kyoto Protocol-oriented approach**	Forest Area x	HWP from Forest Area x	All wood used as fuel in Area x	All wood processed/ manufactured in Area x	Overall impact Approach I
How do the forest and consumers in Area x impact the global CO_2 balance? **Approach II: Consumer-oriented approach**	Forest Area x	All HWP used in Area x	All wood used as fuel in Area x	All HWP used in Area x	Overall impact Approach II
How do the forest and the harvested wood products from Area x impact the global CO_2 balance? **Approach III: Value-creation approach**	Forest Area x	HWP from Forest Area x	Wood from Area x used as fuel	HWP from Forest Area x	Overall impact Approach III

forestry and wood products industries, a small population and high timber exports would exhibit a significantly higher value in Approaches I and III than in Approach II. Looking at all countries combined, the sum of all of the values under Approach I equals the sum of all of the values under Approach II.

Applying the model to the German state of North Rhine-Westphalia

A model combining these three assessment approaches makes it possible to calculate the CO_2 effects for any area based on data pertaining to its forestry and wood products industries and wood usage levels. In most cases such data is readily available at the national level or can be extrapolated to supplement existing data (see above; or below, under Methods). In addition to national commitments, a growing number of states (e.g. the German state of North Rhine-Westphalia, NRW) have incorporated climate protection targets into their own statutes [46]. NRW has created greenhouse gas inventories based on the principles of the IPCC [47]. So far, these greenhouse gas inventories do not include forestry in the LULUCF sector and have also excluded any substitution effects of the forestry and wood products industries. Consequently, they fail to provide information on the forestry and wood product industries' impact on the CO_2 balance and climate protection in North Rhine-Westphalia. Using North Rhine-Westphalia as an example, the following paper shows how effects at the state level can be evaluated by applying a model combining three approaches. The paper discusses methodological difficulties. North Rhine-Westphalia serves as a case study; the results are exemplary and methodically transferable to any region.

North Rhine-Westphalia is characterized by its urban centers (Ruhr area, Rhineland) and with nearly 18 million consumers is Germany's most populous state. Considering the number of inhabitants, it has relatively little forested area (915,800 hectares) and low harvest levels (approximately 6 million m^3 per year) [48]. However, the forest-based sector's annual turnover of some EUR 35 billion (2010) [49] attests to the strong timber processing and wood manufacturing sector in NRW, which includes Eastern Westphalia, an important center for the German and European furniture industry and in particular kitchen manufacturing.

Table 2 shows the model results for North Rhine-Westphalia based on the matrix in Table 1. The values refer to the period from 2005 to 2010 and are expressed in units of Mt CO_2 (conversion from Mt C using the conversion factor of $44/12 = 3.67$). Table 2 shows CO_2 reductions as negative values, as stipulated in international climate reporting guidelines. The Methods section specifies the exact calculation methods used. The values shown in Table 2 were determined on the basis of official statistics and empirical studies. The field "All HWP used in Area x" (Approach II), uses nation-wide data [44] because no regional statistics were available. The proportion of wood used was derived as a percentage of total wood use equal to the population of NWR expressed as a percentage of the total population. This assumption is justified because North Rhine-Westphalia is considered an average German state, with regard to relevant structural parameters for construction and wood usage (e.g., home ownership rates, income levels, building structures) [50]. For example, residents of NRW fall exactly in the middle of the range of purchasing power in Germany. Basing the analysis on national statistics presents a methodological weakness given the current lack of available data. This

Table 2 Matrix for assessing forestry and wood products industries' impact on the CO_2 balance (North Rhine-Westphalia, Germany for the period from 2005 to 2010)

Approaches for assessing forestry and wood products industries' impact on the CO_2 balance (North Rhine-Westphalia, NRW)	Category of emission reduction/stock/sink				TOTAL
	Forest sink	HWP sink	Fuel substitution	Material substitution	
	Changes in forest carbon stock	Changes in HWP carbon stock	CO_2 impact of wood fuel	CO_2 impact of wood product usage	
	[Mt CO_2]	[Mt CO_2]	[Mt CO_2]	[Mt CO_2]	[Mt CO_2]
How do the forestry and wood product industries in NRW impact the CO_2 balance in NRW (and the global CO_2 balance)?	Forest in NRW	HWP from forests in NRW	All wood used as fuel in NRW	All wood processed/ manufactured in NRW	
Approach I: Kyoto Protocol-oriented approach	(−4)	−1.1	−5.0	−7.9	−18.0
How do the forest and consumers in NRW impact the global CO_2 balance?	Forest in NRW	All HWP used in NRW	All wood used as fuel in NRW	All HWP used in NRW	
Approach II: Consumer-oriented approach	(−4)	−3.3	−5.0	−9.1	−21.4
How do the forest and the harvested wood products from NRW impact the global CO_2 balance?	Forest in NRW	HWP from forests in NRW	Wood from NRW used as fuel (of which ca. 0.5 Mt is not statistically recorded as harvested wood)	HWP from forests in NRW	
Approach III: Value-creation approach	(−4)	−1.1	−2.5	−3.6	−11.2

flaw makes it impossible to track individual developments in NRW in relation to the national average (for instance, to evaluate a program to increase wood use in construction or for heating with wood/pellets at the state level). However, it does not represent a basic or fundamental methodological weakness; it can be overcome if data from future empirical studies such as [44] become available not only for the national but also for the regional level.

Currently no regional data are available with which to calculate forest carbon storage levels. Therefore, a value was inferred based on the most recent national forest inventory in 2002 [51] (see Methods). This is a conservative estimate; the value is considered a placeholder until regional forest inventory data become available (BWI III or national forest inventory; 2015 or 2016). The value is given in parentheses to emphasize its statistical uncertainty.

North Rhine-Westphalia is a region with no congruency between forest areas, timber processing, wood product manufacturing and demand for wood products. That it is the most populous state in Germany is reflected in the value shown in Approach II of −21.4 Mt C, which is almost twice as high as the value determined for its own forest area and timber production (−11.2 Mt C, Approach III). The value in Approach I is relatively high, at −18 Mt C, when compared with the amount of timber sourced from its own forests. This value can be explained on the one hand by the fact that North Rhine-Westphalia has an extensive wood-products and furniture manufacturing industry and therefore allocates a value of −7.9 Mt C to material substitution. On the other hand, Approach I includes fuel substitution, to a large

extent through wood burning in biomass incineration plants. Since this bioenergy use is partially coupled with the consumer markets through the recycled wood market or takes place within the wood products industry, Approach I yields a higher value than Approach III.

The values shown in Table 2 can be used as a benchmark for comparison with other areas (both states and countries) but the values need to be normalized. For Approach II it is suitable to refer to population levels and for Approach III to the forest area. For Approach I, it may make more sense, rather than using population as the reference, to draw a connection to a different economic indicator (e.g. economic output).

In principle, the values determined in Approaches I and II can also be set in relation to CO_2 emissions in the corresponding area. The value of −18 Mt CO_2 annually determined by Approach I can be compared, in accordance with the model, to the greenhouse gas emissions of NRW of 307 Mt CO_2 (2010) [47]. The annual value of −18 Mt CO_2 (about −1 t CO_2 per capita) can therefore be interpreted such that the total CO_2 emissions of North Rhine-Westphalia would be 18 million tons higher without the impact of the forestry and wood-products industries. This would represent a 5.9% increase in total greenhouse gas emissions for North Rhine-Westphalia in 2010. The value of −21.4 Mt CO_2 in Approach II means that (global) CO_2 emissions would be 21.4 million tons higher if the forest in North Rhine-Westphalia and the NRW consumers, by opting for wood products and wood energy, made no impact on the CO_2 balance. The carbon footprint for NRW consumers would increase by 21.4 Mt CO_2 (about 1.2 tons CO_2 per capita) if the impact of the forestry and wood products industries were not considered. Assuming the current carbon footprint for NRW consumers of approximately 200 Mt CO_2 (derived from the nationwide figures according to [42]), the result would be a 10.7% larger carbon footprint.

The value of 5.9% in Approach I is about half as high as the 10.7% in Approach II. Since in the past analysis of the forestry and wood products industries' impact on the CO_2 balance was assigned to national GHG emissions, it was only used as a reference in Approach I [52]. For a state such as North Rhine-Westphalia, however, such an approach is not appropriate. On the one hand, it does not adequately address the impact of wood usage; on the other hand, it is necessary to adjust the reference framework. While Approach I systematically uses the total GHG emissions as a reference, Approach II uses the carbon footprint. Approach II provides a consistent basis with which to integrate the overall carbon footprint of NRW consumers. This observation is particularly relevant for a state such as North Rhine-Westphalia. Nearly 22% of the population of Germany lives in North Rhine-Westphalia, and yet the state accounts for approximately 30% of German greenhouse gas emissions. The heavy industrial sector in North Rhine-Westphalia and the production of energy (mainly from fossil fuels) which provide products and energy for both the German and international markets create higher than average emissions levels for NRW [47]. Against this background, it is important to keep in mind that a resident of North Rhine-Westphalia with statistical CO_2 emissions of 15 to 16 t CO_2 per year is not 1.5 times more detrimental to the climate than the average German citizen (with CO_2 emissions of approximately 10 t CO_2 per year). The inclusion of these very important parameters is only possible if, as is done in Approach II, the carbon footprint is systematically evaluated.

Conclusions

The multi-tiered approach presented in this paper with three distinct assessment approaches enables a more comprehensive analysis of the forestry and wood products industries' impact on the CO_2 balance. The model can be modified to meet specific objectives (e.g. to evaluate climate policy measures or to develop a basis for optimizing the forestry and wood products industries' contribution to climate protection). Models like the one presented here, which allow for differentiation, pose a risk of being manipulated as a means to support a specific position. A region could, for instance, choose the model that yields the highest values in order to present its climate impact in the best possible light. The highly populated NRW region may select Approach II while the heavily forested region B may opt for Approach III, etc. This practice is an abuse of the model. However, given

the considerations presented in this article, this abuse can be discovered and evaluated accordingly. Therefore, instead of promoting random and embellished conclusions (as is often the case when certain assessments are cherry picked to fit a respective view), the methodology in this study aims to explicate its results. The approaches presented in this paper aim to avoid obfuscated results by ensuring that the analysis of the forestry and wood product industries' impact on climate change is carried out in a logical and consistent manner as proposed in Table 1.

The proposed methodology may at times be limited due to a lack of statistical data. At the national level these data are usually available (depending on the quality of official statistics) or can be derived from existing empirical studies. Regional statistics, however, are often lacking, and as a result the flow of goods between different regions (e.g. between different German states) within a larger territory (e.g. Germany) are poorly documented. This means that studies are left to rely on empirical data or national statistics, as is the case for all three assessment approaches described in this paper.

The underlying assumptions presented in this publication can be improved through in-depth analysis, such as: a) studies on wood use over the wood's entire life span, including disposal; b) more in-depth studies on wood fuel (i.e., thermal heat, thermal heat extraction in power plants, domestic woodstoves, industrial and commercial wood fuel usage, efficiency); c) achieving greater accuracy in determining a region's fuel substitution by defining a regional-specific fossil energy mix and a specific energy recovery key; d) achieving greater accuracy in assessing material substitution through more in-depth research (with regard to wood use, but also in terms of the comparison between products or product groups with the same functionality). The extent and depth of such additional studies should depend on the desired level of accuracy weighed against the significantly higher amount of time and effort they require.

Further studies are needed on how to allocate avoided CO_2 emissions in the process chain in cases where semi-finished wood products are initially processed in region A and their end product is finished in region B. Approach I assesses material substitution proportionally in relation to the process chain. At the same time, the simplifying assumption is made that the substitution factors used are universally based on [15]; in this case, more factors specific to product groups could be used in the future [14,27].

Methods

In addition to the explanation presented above under "Results and discussion," the following illustrates in a structured manner how an analysis can be conducted at the national level (or for an area with well-defined statistics, e.g. the EU). Germany is used as an example and the assessment data is derived from official statistics and empirical studies available in Germany. The explanation is intended only as an overview. Finally, a detailed description is given of how the model was applied to the case study (NRW) presented in this paper. Possible problems arising from the processing of the data are discussed.

Basic analysis

The matrix in Table 1 has eight different fields, which can be defined as areas of study; they are shown in Table 3.

To begin with, a substitution factor is determined for material substitution SF_{Ma} and fuel substitution SF_{Fuel}. SF_{Fuel} is used as a multiplier in calculating fields IV-V and SF_{Ma} is used as a multiplier in calculating the fields VI-VIII, for example a universal substitution factor SF_{Ma} can be taken from [14,27] and SF_{Fuel} can be taken from [14,27]. For the purpose of the study, universal means that the total wood volume was not assigned to different products or product groups, but rather that the same substitution factor can be used for all of the wood included in the study. For Germany, the universal substitution factors used are SF_{Ma} = 1.50 tC/tC and SF_{Fuel} = 0.67 tC/tC because the timber market and the distribution of products and product groups has already been taken into account in their calculation [27]. The substitution factor multiples for the C content of wood products or products used for fuel show the carbon mitigation of the wood use. In this sense the substitution factor can be described as a multiplier for calculation on the basis of C content in wood products.

Overview of the calculations for the eight areas of study:

I. Forest Area x:
Based on national forest inventory studies [51], or on the basis of the LULUCF values reported in national greenhouse inventories [19].

II. HWP from Forest Area x:
Based on a country-specific model according to [25] (see also [53]). Alternatively, an empirical input-output model can be applied, as for example by [44] (empirical determination of the inputs and outputs via analysis of disposal statistics).

III. All HWP used in Area x:
Based on official statistics (net balance between data on production statistics and international trade statistics) or aggregated in wood balances, e.g. [54], or from empirical studies, e.g. [44].

IV. All wood used as fuel in Area x:
Volumes are calculated based on empirical studies [55,56] or aggregated [57], or supplementary data e.g. from [58]; SF_{Fuel} = 0.67 tC/tC is used as multiplier.

V. Wood from Forest Area x used as fuel:
Volumes are calculated based on a material-flow model, tracking the flow from timber harvesting (e.g. based on official harvest statistics) to end product [59]; supplementary data on harvested wood not included in official statistics, usually harvested for fuel (e.g. [60]) or as a balance of raw timber balances (e.g. [61]) and wood balances, (e.g. [54]); SF_{Fuel} = 0.67 tC/tC is used as multiplier; alternatively, using additional evaluations of the energy balance [62].

VI. All wood processed and manufactured in Area x:
To determine wood volumes, see II.; SF_{Ma} = 1.50 tC/tC used as multiplier.

VII. All HWP used in Area x:
To determine wood volumes, see III.; SF_{Ma} = 1.50 tC/tC used as multiplier.

VIII. All HWP from Forest Area x:
To determine amounts, see V.; SF_{Ma} = 1.50 tC/tC used as multiplier.

Table 3 Differentiation of the areas of study in the matrix for assessing the forestry and wood-products industries' impact on the CO_2 balance

Approaches for assessing the forestry and wood products industries' impact on the CO_2 balance (Area x)	Category of emission reduction/stock/sink			
	Forest sink	HWP sink	Fuel substitution	Material substitution
	Changes in forest carbon stock	Changes in HWP carbon stock	CO_2 impact of wood fuel	CO_2 impact of wood product usage
	[Mt CO_2]	[Mt CO_2]	[Mt CO_2]	[Mt CO_2]
How do the forestry and wood product industries in Area x impact the CO_2 balance in Area x (and the global CO_2 balance)?	Forest Area x	HWP from Forest Area x	All wood used as fuel in Area x	All wood processed/ manufactured in Area x
Approach I: Kyoto Protocol-oriented approach	I	II	IV	VI
How do the forest and consumers in Area x impact the global CO_2 balance?	Forest Area x	All HWP used in Area x	All wood used as fuel in Area x	All HWP used in Area x
Approach II: Consumer-oriented approach	I	III	IV	VII
How do the forest and the harvested wood products from Area x impact the global CO_2 balance?	Forest Area x	HWP from Forest Area x	Wood from Area x used as fuel	HWP from Forest Area x
Approach III: Value-creation approach	I	II	V	VIII

Analysis: NRW

The same eight fields are used as those in Table 3. They are also defined as areas of study and calculated as follows:

I. Forest Area NRW:

No current statistical data exists for calculating the carbon stock levels in the forests of North Rhine-Westphalia. The forest carbon stock is determined based on forest inventories carried out every 10 to 15 years in Germany [51]. The last national forest inventory was conducted in 2002. The data collected retrospectively for the regional analysis is older, also at the time of the inventory due to the methodology used (for the period 1987-2002). A 2008 interim inventory [63] with reduced sample size provides only national and no regional data. Data are expected to be available for North Rhine-Westphalia in 2015 following the 3rd National Forest Inventory (BWI III). A provisional (and thus given in parentheses) value for the year 2010 was determined on the basis of a simulation for the period of 2002-2010 [64]. The modeling of stock development since 2002 is subject to significant uncertainty, due in part to storm damage in 2007 (storm Kyrill). The recorded value is consistent with the calculations made by [65]. With the publication of the 3rd National Forest Inventory in 2015 and a statewide forest inventory, probably in 2016, data will become available that can be used to accurately calculate stock levels for the period 2002-2012. There is discussion about conducting less expensive, modified inventories covering shorter periods to ensure timely and consistent monitoring of the forest stock (for example, with a tool used in North Rhine-Westphalia called "virtual forest" [66]).

II. HWP from forests in North Rhine-Westphalia (sink):

A wood usage key was created using official statistics on the wood harvested from North Rhine-Westphalia. In addition, a material flow model was developed (based on impact statistics) that tracks wood processing from harvest to the end product. This study was carried out in accordance with [59] by distinguishing between softwood and hardwood varieties (because they have different timber yield levels and uses as industrial wood). The state forestry service made a distinction between logs and industrial wood based on consumer demand, which deviated from the national figures. An average value for the years 2005 to 2010 was calculated. The value provides a more accurate record of annual forest stock levels by taking in to account values distorted by storm damage in 2007 (storm Kyrill). Carbon storage levels in HWP were calculated based on the IPCC approach [25]. This approach

determines the input of wood products into the HWP carbon storage pool and calculates the withdrawal from the product store (products with long, medium and short life-span (e.g. paper) as well as fuel wood) on the basis of a mathematical function (exponential decomposition curve).

III. All HWP used in NRW (sink):

In contrast to II, the carbon storage in wood products used in NRW is calculated using an input-output measurement. The measurement was carried out using a modified model based on the nationwide material-flow model from [44] for the year 2007; the share attributed to NRW was derived based on the state's share of the overall German population (21.8%); cf. "Results and discussion".

IV. All wood used as fuel in NRW:

Fuel substitution was quantified based on official statistics and various studies on bio-energy and renewable energy. The reporting recognizes the use of solid biomass (wood) for generating electricity. Since its intent is to promote bioenergy through renewable energy regulations (EEG) [67], the reporting has a relatively high level of transparency [39,58]. The much more extensive use of bioenergy for generating heat is significantly less transparent [68]. For the electricity sector, data are available at the state level [58], and thus for NRW (1.3 billion kWh). For the heating sector, these data were derived from nationwide data and supplementary studies. On the basis of [68], the calculations were structured into the following three areas: a) biogenic solid fuels used in households (share of NRW 8.5%; based on [56]: 6.8 billion kWh); b) biogenic solid fuels used in industry (share of NRW 19.5%; based on a comparison between [69] and [70]): 4.6 billion kWh); c) biogenic solid fuels used in heating plants and from heat extraction (percentage NRW 15.2% [58]: 1.0 billion kWh). The total value of 12.3 billion kWh is derived from the 2010 calculations of biomass heat generation in NRW. Based on the conversion numbers of the energy content in the CO_2 emissions (GHG emissions) set by [68], a CO_2 reduction (energy substitution) of 5.0 Mt CO_2 was calculated for 2010.

V. Wood from forests in NRW used as fuel:

Emissions reduction levels resulting from the use of NRW wood for fuel are quantified for the period of 2005-2010 (according to II.). The analysis is based on the following assumptions: a) 100% of the fuel wood reported in the official harvest statistics is used as fuel; b) 100% of the by-products from wood harvesting and processing that are not used as material for product manufacturing are used as fuel wood; c) 50% of the bark is used as fuel; d) 85% of industrial wood used in the paper industry is used as fuel energy at the end of the paper's life cycle (according to experts from the pulp and paper industry). Given the steady paper consumption since 2000, it was assumed that the use of fuel wood is consistent over time; e) 68% of the recycled wood taken from the HWP sink is used as fuel (according to experts, taking into account the losses in the utilization phase and the recycling ratio, cf. [61]); f) based on expert estimates 700,000 to 800,000 m^3 of wood is removed each year from NRW forests and not included in the official harvest statistics; this wood is presumably used as fuel. Emissions reduction was calculated using the substitution factor SF_{Fuel}=0.67 tC/tC.

VI. All wood processed and manufactured in NRW: The calculation was carried out in three steps:

1. Comparison of German and North Rhine-Westphalia energy balances [69,70]. Comparing energy usage for two relevant product groups from the forest-based industry – wood product (WZ 16) and furniture (WZ 31) manufacturing. The proportion of energy usage by the wood products and furniture manufacturing industries in NRW compared to the industry total in Germany is determined to be 22.1%. The calculated proportion roughly corresponds to the North Rhine-Westphalian population's share of the German population (21.8%).

2. The proportion of wood processed and manufactured in NRW is quantified based on finished product statistics, such as those found in [44] (proportionate to the population; cf.) III).

3. Emissions reduction is calculated using the substitution factor SF_{Ma} =1.50 tC/tC. Calculating material substitution in this study area poses a methodological problem. Currently, a scientifically accurate assessment of material substitution is only possible at the end-product level. The manufacturing process leading to these end products includes many stages, and only the sum of these stages can be used to determine material substitution values. Therefore, assessing only the end products manufactured in North Rhine-Westphalia (e.g. according to official manufacturing statistics) does not yield an accurate figure, since many of the manufacturing stages leading to the finished products take place outside of North Rhine-Westphalia (for both wood and non-wood products). At the same time, steps in the manufacturing process performed within the North Rhine-Westphalian wood products industry contribute to the production of end products outside North Rhine-Westphalia. Therefore, the energy balance calculation relies

on official statistics. The application of this procedure must meet the following conditions: (1) the energy consumption statistics are properly recorded. (2) the share of energy consumed in specific sectors (e.g. woodworking or carpentry) is the same as in the industrial sectors. (3) the plants and equipment used by companies in North Rhine-Westphalia and Germany are equal in terms of energy consumption. (4) the production methods in the wood product and furniture manufacturing industries are comparable; that is, they include similar production steps and products. These four conditions only partially apply. Since the substitution factors for specific wood product groups in the construction industry [27] are roughly the same as those for product groups in the furniture manufacturing sector, using this methodological approach to generate estimates is justified. This approach is currently best suited to making statements about the material substitution performance of a region's wood products industry.

VII. All HWP used in NRW: For quantification of wood volumes, see III; emissions reduction was calculated using the substitution factor SF_{Ma} =1.50 tC/tC.

VIII. HWP from NRW forests: For determining the amounts, see V.; emissions reduction was calculated using the substitution factor SF_{Ma} =1.50 tC/tC.

Competing interests

The author declares that he has no competing interests.

Acknowledgements

This research was part of the study "Beitrag des NRW Clusters ForstHolz zum Klimaschutz" [NRW cluster ForstHolz's contribution to climate change mitigation] [26]. The study was conducted on behalf of the Ministry for Climate Protection, Environment, Agriculture, Nature Conservation and Consumer Protection (MKULNV) of the German State of North Rhine-Westphalia and the Landesbetrieb Wald und Holz NRW [North Rhine-Westphalia Agency for Forestry and Timber Management]. I thank all of the participants from the State Forest Administration of North Rhine-Westphalia. A special thanks to Volker Holtkämper for helping make this study possible and Rainer Joosten for overseeing the study through the ministry. For scientific co-operation, I am thankful to Arno Frühwald (Department of Wood Science and Technology, University of Hamburg), Michael Köhl, Volker Mues, and Konstantin Olschofsky (Institute of World Forestry, University of Hamburg).

References

1. Burschel P, Kürsten E, Larson BC. Die Rolle von Wald und Forstwirtschaft im Kohlenstoffhaushalt: Eine Betrachtung für die Bundesrepublik Deutschland, vol. 126. München: Schriftenreihe der Forstwissenschaftlichen Fakultät der Universität München und Bayerischen Forstlichen Versuchs- und Forschungsanstalt; 1993.
2. Frühwald A, Wegener G. Energiekreislauf Holz – ein Vorbild für die Zukunft. Holz-Zentralblatt. 1993;119:1949ff.
3. Perez-Garcia J, Lippke B, Comnick J. An assessment of carbon pools, storage, and wood products market substitution using life-cycle analysis results. Wood Fiber Sci. 2005;37(CORRIM Special Issue):140–8.

4. Skog KE. Sequestration of carbon in harvested wood products for the United States. For Prod J. 2008;56:56–72.

5. Reijnders L. Conditions for the sustainability of biomass based fuel use. Energy Policy. 2006;34:863–76.

6. Gustavsson L, Holmberg J, Dornburg V, Sathre R, Eggers T, Mahapatra K, et al. Using biomass for climate change mitigation and oil use reduction. Energy Policy. 2007;35:5671–91.

7. Sathre R, Gustavsson L. A State-of-the-Art Review of Energy and Climate Effects of Wood Product Substitution. Växjö (Sweden): School of Technology and Design Reports 57, University Växjö; 2009.

8. Frühwald A, Solberg B. LCA – a challenge for forestry and forest products industry. In: Frühwald A, Solberg B, editors. EFI Proceedings No 8. Joensuu; 1995.

9. Puettmann ME, Wilson JB. Life-cycle analysis of wood products: Cradle-to-gate LCI of residential wood building materials. Wood Fiber Sci. 2005;37(CORRIM Special Issue):18–29.

10. Wilson JB, Sakimoto ET. Gate-to-gate life-cycle inventory of softwood plywood production. Wood Fiber Sci. 2005;37(CORRIM Special Issue):58–73.

11. Winistorfer P, Chen Z, Lippke B, Stevens N. Energy Consumption and Greenhouse Gas Emissions Related to the use, Maintenance, and Disposal of a Residential Structure. Wood Fiber Sci. 2005;37:128–39.

12. Karjalainen T, Zimmer B, Berg S, Welling J, Schwaiger H, Finér L, et al. Energy, Carbon and Other Material Flows in the Life Cycle Assessment of Forestry and Forest Products. Joensuu: European Forest Institute Discussion Paper 10; 2001.

13. Lippke B, Wilson J, Perez-Garcia J, Bowyer J, Meil J. CORRIM: Life-cycle environmental performance of renewable building materials. For Prod J. 2004;54:8–19.

14. Taverna R, Hofer P, Werner F, Kaufmann E, Thürig E. The CO_2 Effects of the Swiss Forestry and Timber Industry. Bern: Scenarios of future potential for climate-change mitigation; 2007.

15. Sathre R, O'Connor J. A Synthesis of Research on Wood Products and Greenhouse Gas Impacts. 2nd ed. Vancouver: FPInnovations; 2010.

16. UNFCCC. Report of the Conference of the Parties on Its Seventh Session, Held at Marrakesh From 29 October to 10 November 2001. FCCC/CP/2001/13/Add.1. 2002.

17. IPCC. Good Practice Guidance for Land Use, Land-Use Change and Forestry. Hayama: Institute for Global Environmental Strategies (IGES); 2003.

18. IPCC. IPCC Guidelines for National Greenhouse Gas Inventories; Reference Manual. Hayama: Institute for Global Environmental Strategies (IGES); 2006.

19. Umweltbundesamt. Berichterstattung unter der Klimarahmenkonvention der Vereinten Nationen und dem Kyoto-Protokoll 2013. Nationaler Inventarbericht zum deutschen Treibhausgasinventar 1990 – 2011. 2013.

20. IPCC. 2006 IPCC Guidelines for National Greenhouse Gas Inventories Volume 4 Agriculture, Forestry and Other Land Use. Hayama: Intergovernmental Panel on Climate Change; 1997.

21. Mackensen J, Bauhus J, Webber E. Decomposition rates of coarse woody debris – A review with particular emphasis on Australian tree species. Aust J Bot. 2003;51:27–37.

22. Köhl M, Stümer W, Kenter B, Riedel T. Effect of the estimation of forest management and decay of dead woody material on the reliability of carbon stock and carbon stock changes—A simulation study. For Ecol Manag. 2008;256:229–36.

23. UNFCCC. Ad Hoc Working Group on Further Commitments for Annex I Parties Under the Kyoto Protocol, Consideration of Further Commitments for Annex I Parties Under the Kyoto Protocol. Revised Proposal by the Chair. FCCC/KP/AWG/2010/CRP.4/Rev.4. 2010.

24. UNFCCC. Synthesis Report of the Technical Assessments of the Forest Management Reference Level Submissions. Note by the Secretariat. 2011.

25. IPCC. 2013 Revised Supplementary Methods and Good Practice Guidance Arising From the Kyoto Protocol. Hayama: Intergovernmental Panel on Climate Change; 2014.

26. Knauf M, Frühwald A. Beitrag des NRW Clusters ForstHolz zum Klimaschutz. Münster: Landesbetrieb Wald und Holz Nordrhein-Westfalen; 2013.

27. Frühwald A, Knauf M. Carbon Aspects Promote Building with Wood. In: World Conference on Timber Engineering WCTE. 2014.

28. Rüter S, Rock J, Koethke M, Dieter M. Wie viel Holznutzung ist gut fürs Klima? AFZ, der Wald. 2011;15:19–21.

29. Wördehoff R, Spellmann H, Evers J, Nagel J. Kohlenstoffstudie Forst und Holz Niedersachsen. Göttingen: Beiträge aus der Nordwestdeutschen Forstlichen Versuchsanstalt Band 6; 2011.

30. Bastianoni S, Pulselli FM, Tiezzi E. The problem of assigning responsibility for greenhouse gas emissions. Ecol Econ. 2004;49:253–7.

31. Druckman A, Jackson T. The carbon footprint of UK households 1990–2004: A socio-economically disaggregated, quasi-multi-regional input–output model. Ecol Econ. 2009;68:2066–77.

32. Ipek Tunç G, Türüt-Aşık S, Akbostancı E. CO_2 emissions vs. CO_2 responsibility: An input–output approach for the Turkish economy. Energy Policy. 2007;35:855–68.

33. Dong H, Geng Y, Fujita T, Jacques DA. Three accounts for regional carbon emissions from both fossil energy consumption and industrial process. Energy. 2014;67:276–83.

34. Lenzen M, Murray J, Sack F, Wiedmann T. Shared producer and consumer responsibility — Theory and practice. Ecol Econ. 2007;61:27–42.

35. Springmann M. Integrating emissions transfers into policy-making. Nat Clim Chang. 2014;4:177–81.

36. Peters GP. From production-based to consumption-based national emission inventories. Ecol Econ. 2008;65:13–23.

37. Salazar J, Meil J. Prospects for carbon-neutral housing: the influence of greater wood use on the carbon footprint of a single-family residence. J Clean Prod. 2009;17:1563–71.

38. Garcia R, Freire F. Carbon footprint of particleboard: a comparison between ISO/TS 14067, GHG Protocol, PAS 2050 and Climate Declaration. J Clean Prod. 2014;66:199–209.

39. Umweltbundesamt. Emissionsbilanz Erneuerbarer Energieträger. Durch Einsatz Erneuerbarer Energien vermiedene Emissionen im Jahr 2010. Aktualisierte Anhänge 2 und 4 der Veröffentlichung „Climate Change 12/2009". Dezember 2011, Korrigiert März 2012. Dessau; 2012.

40. Rüter S. Welchen Beitrag leisten Holzprodukte zur CO_2-Bilanz? AFZ, der Wald. 2011;15:15–8.

41. Frühwald A, Wegener G, Krüger S, Beudert M. Forst- und Holzwirtschaft unter dem Aspekt der CO_2-Problematik. Forstabsatzfonds Bonn: Forschungsbericht; 1994.

42. Aichele R, Felbermayr G. Carbon Footprints. München: ifo Schnelldienst 21/2011; 2011.

43. Thoroe C, Dieter M. Forst- und Holzwirtschaft in der Bundesrepublik Deutschland nach neuer europäischer Sektorenabgrenzung. Forstw Cbl. 2003;122:138–51.

44. Mantau U, Bilitewski B. Stoffstrom-Modell-Holz. Bestimmung des Aufkommens, der Verwendung und des Verbleibs von Holzprodukten. Celle: Forschungsbericht für den Verband Deutscher Papierfabriken e.V. (VDP); 2010.

45. Lundmark T, Bergh J, Hofer P, Lundström A, Nordin A, Poudel B, et al. Potential Roles of Swedish Forestry in the Context of Climate Change Mitigation. Forests. 2014;5:557–78.

46. Klimaschutzgesetz NRW. Gesetz zur Förderung des Klimaschutzes in Nordrhein-Westfalen. 2013.

47. Hoffmann V, Opitz S, Hoppe D. Treibhausgas-Emissionsinventar Nordrhein-Westfalen 2011. Recklinghausen: LANUV–Fachbericht 51; 2014.

48. MKULNV Ministerium für Klimaschutz Umwelt Landwirtschaft Natur- und Verbraucherschutz des Landes Nordrhein-Westfalen. Landeswaldbericht 2012. Düsseldorf; 2012.

49. Landesbetrieb Wald und Holz Nordrhein-Westfalen. Nachhaltig wachsen. Wald und Holz NRW, Nachhaltigkeitsbericht 2010/2011. Münster; 2012.

50. Riemhofer H. Aspekte des Kohlenstoffmanagements in der Holzverwendung Nordrhein-Westfalens. Holzwirtschaft: Diplomarbeit. Universität Hamburg; 2012.

51. BMELV Bundesministerium für Ernährung Landwirtschaft und Verbraucherschutz. Die Bundeswaldinventur. http://www.bundeswaldinventur.de. Bonn; 2004.

52. Heuer E. Kohlenstoffbilanzen – Schlüssel zur forstlichen Klimapolitik. AFZ, der Wald. 2011;17:16–8.

53. Rüter S. Projections of Net-Emissions From Harvested Wood Products in European Countries. Hamburg: Johann Heinrich von Thünen-Institut (vTI). Work Report of the Institute of Wood Technology and Wood Biology, Report No: 2011/1; 2011.

54. Seintsch B, Weimar H. Holzbilanzen 2010 bis 2012 für die Bundesrepublik Deutschland. Hamburg: Thünen-Institut (TI); 2013.

55. Weimar H, Döring P, Mantau U. Standorte der Holzwirtschaft – Holzrohstoffmonitoring. Einsatz von Holz in Biomasse-Großfeuerungsanlagen 2011. Abschlussbericht. Hamburg: Universität Hamburg, Zentrum Holzwirtschaft; 2012.

56. Mantau U. Energieholzverwendung in privaten Haushalten. Marktvolumen und verwendete Holzsortimente. Hamburg: Universität Hamburg, Zentrum Holzwirtschaft/Infro; 2012.

57. FNR. Basisdaten Bioenergie Deutschland, August 2013. Gülzow; 2013.

58. DBFZ Deutsches Biomasseforschungszentrum. Monitoring zur Wirkung des Erneuerbare-Energien-Gesetz (EEG) auf die Entwicklung der Stromerzeugung aus Biomasse. Final Report, March 2012. Leipzig; 2012.

59. Seintsch B. Entwicklungen des Clusters Forst und Holz: Studie „Volkswirtschaftliche Bedeutung des Clusters Forst und Holz"im Rahmen der „Bundesweiten Clusterstudie Forst und Holz. Holz-Zentralblatt. 2008;134:1390–1.

60. Weimar H. Der Holzfluss in der Bundesrepublik Deutschland 2009. Hamburg: Methode und Ergebnis der Modellierung des Stoffflusses von Holz; 2011.

61. Mantau U. Holzrohstoffbilanz Deutschland, Entwicklungen und Szenarien des Holzaufkommens und der Holzverwendung 1987 bis 2015. Hamburg: Infro/Universität Hamburg, Zentrum Holzwirtschaft; 2012.

62. Arbeitsgemeinschaft Energiebilanzen e.V. Energiebilanz der Bundesrepublik Deutschland 2012. Berlin; 2014.

63. Oehmichen K, Demant B, Dunger K, Gruneberg E, Hennig P, Kroiher F, et al. Inventurstudie 2008 und Treibhausgasinventar Wald. Braunschweig: Landbauforschung vTI Agriculture and Forestry Research; 2011.

64. Köhl M, Mues V, Olschofsky K. Szenarien/Simulation der potenziellen Waldentwicklung Nordrhein-Westfalens bis 2100. Münster. In: Landesbetrieb Wald und Holz NRW, editor. Beitrag des NRW Clusters ForstHolz zum Klimaschutz. 2013.

65. Niesar M, Zúbrik M, Kunca A. Waldschutz im Klimawandel. Münster: Landesbetrieb Wald und Holz NRW; 2013.

66. Staub M. Digitale Bäume wachsen in den Himmel. Schweizer Holzzeitung. 2012;124:4–6.

67. EEG 2014. Gesetz für den Ausbau Erneuerbarer Energien (Erneuerbare-Energien-Gesetz). 2014.

68. BMU (Bundesministerium für Umwelt, Naturschutz und Reaktorsicherheit). Erneuerbare Energien in Zahlen. Nationale und internationale Entwicklung. Juli 2012. Berlin; 2012.

69. IT NRW (Information und Technik Nordrhein-Westfalen Geschäftsbereich Statistik). Energiebilanz und CO_2-Bilanz in Nordrhein-Westfalen 2009. Düsseldorf; 2011.

70. Statistisches Bundesamt. Erhebung über die Energieverwendung – Energieverbrauch nach Energieträgern – Sonderauswertung (Berichtzeitraum 2009/2010). Wiesbaden; 2012.

PERMISSIONS

All chapters in this book were first published in CBM, by Springer; hereby published with permission under the Creative Commons Attribution License or equivalent. Every chapter published in this book has been scrutinized by our experts. Their significance has been extensively debated. The topics covered herein carry significant findings which will fuel the growth of the discipline. They may even be implemented as practical applications or may be referred to as a beginning point for another development.

The contributors of this book come from diverse backgrounds, making this book a truly international effort. This book will bring forth new frontiers with its revolutionizing research information and detailed analysis of the nascent developments around the world.

We would like to thank all the contributing authors for lending their expertise to make the book truly unique. They have played a crucial role in the development of this book. Without their invaluable contributions this book wouldn't have been possible. They have made vital efforts to compile up to date information on the varied aspects of this subject to make this book a valuable addition to the collection of many professionals and students.

This book was conceptualized with the vision of imparting up-to-date information and advanced data in this field. To ensure the same, a matchless editorial board was set up. Every individual on the board went through rigorous rounds of assessment to prove their worth. After which they invested a large part of their time researching and compiling the most relevant data for our readers.

The editorial board has been involved in producing this book since its inception. They have spent rigorous hours researching and exploring the diverse topics which have resulted in the successful publishing of this book. They have passed on their knowledge of decades through this book. To expedite this challenging task, the publisher supported the team at every step. A small team of assistant editors was also appointed to further simplify the editing procedure and attain best results for the readers.

Apart from the editorial board, the designing team has also invested a significant amount of their time in understanding the subject and creating the most relevant covers. They scrutinized every image to scout for the most suitable representation of the subject and create an appropriate cover for the book.

The publishing team has been an ardent support to the editorial, designing and production team. Their endless efforts to recruit the best for this project, has resulted in the accomplishment of this book. They are a veteran in the field of academics and their pool of knowledge is as vast as their experience in printing. Their expertise and guidance has proved useful at every step. Their uncompromising quality standards have made this book an exceptional effort. Their encouragement from time to time has been an inspiration for everyone.

The publisher and the editorial board hope that this book will prove to be a valuable piece of knowledge for researchers, students, practitioners and scholars across the globe.

LIST OF CONTRIBUTORS

Fabián B Gálvez, Andrew T Hudak, John C Byrne and Nicholas L Crookston
USDA Forest Service, Rocky Mountain Research Station, 1221 South Main St., Moscow, ID 83843, USA

Robert F Keefe
Department of Forest, Rangeland, and Fire Sciences, University of Idaho, 975 West 6th St., Moscow, ID 83844-1133, USA

David C. Marvin and Gregory P. Asner
Department of Global Ecology, Carnegie Institution for Science, 260 Panama St., Stanford, CA 94305, USA

Tarquinio Mateus Magalhães
Departamento de Engenharia Florestal, Universidade Eduardo Mondlane, Campus Universitário Principal, Edifício no 1, Maputo, Mozambique
Department of Forest and Wood Science, University of Stellenbosch, Private Bag X1 Matieland, 7602 Stellenbosch, South Africa

Thomas Seifert
Department of Forest and Wood Science, University of Stellenbosch, Private Bag X1 Matieland, 7602 Stellenbosch, South Africa

David V. D'Amore and Paul E. Hennon
U.S. Department of Agriculture, Forest Service, Pacific Northwest Research Station, Juneau Forestry Sciences Laboratory, 11175 Auke Lake Way, Juneau, AK 99801, USA

Kiva L. Oken
Quantitative Ecology and Resource Management, University of Washington, Box 355020, Seattle, WA 98195, USA

Paul A. Herendeen
Graduate Degree Program in Ecology, Colorado State University, Fort Collins, CO 80523, USA

E. Ashley Steel
U.S. Department of Agriculture, Forest Service, Pacific Northwest Research Station, 400 N 34th Street, Suite 201, Seattle, WA 98103, USA

Coeli M. Hoover and James E. Smith
USDA Forest Service, Northern Research Station, Durham, NH, USA

Ahmed A Balogun
Federal University of Technology, Akure, Nigeria

Matthias Mauder
Institute of Meteorology and Climate Research, Karlsruhe Institute of Technology, Garmisch-Partenkirchen, Germany

Emmanuel Quansah
Federal University of Technology, Akure, Nigeria
Kwame Nkrumah University of Science and Technology, Kumasi, Ghana

Leonard K Amekudzi
Kwame Nkrumah University of Science and Technology, Kumasi, Ghana

Luitpold Hingerl and Jan Bliefernicht
Chair for Regional Climate and Hydrology, University of Augsburg, Augsburg, Germany

Harald Kunstmann
Institute of Meteorology and Climate Research, Karlsruhe Institute of Technology, Garmisch-Partenkirchen, Germany
Head of Chair for Regional Climate and Hydrology, University of Augsburg, Augsburg, Germany

Yosio E Shimabukuro
Remote Sensing Division, National Institute for Space Research (INPE), São José dos Campos, SP CEP 12201-970, Brazil

Veronika Leitold
Remote Sensing Division, National Institute for Space Research (INPE), São José dos Campos, SP CEP 12201-970, Brazil
Biospheric Sciences Laboratory, NASA Goddard Space Flight Center, Greenbelt, MD 20771, USA

Bruce D Cook and Douglas C Morton
Biospheric Sciences Laboratory, NASA Goddard Space Flight Center, Greenbelt, MD 20771, USA

Michael Keller
International Institute of Tropical Forestry, USDA Forest Service, San Juan 00926, Puerto Rico
EMBRAPA Satellite Monitoring, Campinas SP CEP 13070-115, Brazil

Virpi Junttila and Ekaterina Nikolaeva
Department of Mathematics and Physics, Lappeenranta University of Technology, P.O. Box 20, 53851 Lappeenranta, Finland

Basanta Gautam, Katri Tegel, Katja Gunia, Jarno Hämäläinen, Petri Latva-Käyrä and Jussi Peuhkurinen
Arbonaut Ltd, Kaislakatu 2, 80130 Joensuu, Finland

Tuomo Kauranne
Department of Mathematics and Physics, Lappeenranta University of Technology, P.O. Box 20, 53851 Lappeenranta, Finland
Arbonaut Ltd, Kaislakatu 2, 80130 Joensuu, Finland

Bhaskar Singh Karky
International Centre for Integrated Mountain Development (ICIMOD), G.P.O. Box 3226, Khumaltar, Lalitpur, Kathmandu, Nepal

Almasi Maguya
Department of Mathematics and Physics, Lappeenranta University of Technology, P.O.Box 20, 53851 Lappeenranta, Finland
Mzumbe University, P.O. Box 1, Mzumbe, Morogoro, Tanzania

Wenli Huang, Anu Swatantran, Laura Duncanson, Hao Tang, George Hurtt and Ralph Dubayah
Department of Geographical Sciences, University of Maryland, College Park, USA

Kristofer Johnson
USDA Forest Service, Northern Research Station, Newtown Square, PA, USA

Jarlath O'Neil Dunne
Rubenstein School of the Environment and Natural Resources, University of Vermont, Burlington, USA

Ernest William Mauya, Endre Hofstad Hansen, Terje Gobakken, Ole Martin Bollandsås and Erik Næsset
Department of Ecology and Natural Resource Management, Norwegian University of Life Sciences, P.O. Box 5003, Oslo, NO 1432, Ås, Norway

Rogers Ernest Malimbwi
Department of Forest Mensuration and Management, Sokoine University of Agriculture, P.O. Box 3013, Morogoro Tanzania

Svein Solberg
Norwegian Forest and Landscape Institute, P.O. Box 115, 1431 Ås, Norway

Erik Næsset, Terje Gobakken and Ole-Martin Bollandsås
Norwegian University of Life Sciences, P.O. Box 5003, 1432 Ås, Norway

Steven Jay, Robert Crabtree and Maggi Kraft
Yellowstone Ecological Research Center, 2048 Analysis Dr. Ste. B, Bozeman, MT 59718, USA

Christopher Potter
CASA Systems 2100, LLC, PO Box 1631, Los Gatos, CA 95030, USA

Vanessa Genovese
Science and Environmental Policy, California State University, Monterey Bay, 100 Campus Center, Seaside, CA 93955, USA

Daniel J. Weiss
Yellowstone Ecological Research Center, 2048 Analysis Dr. Ste. B, Bozeman, MT 59718, USA
Department of Zoology, University of Oxford, The Tinbergen Building, S Parks Rd, Oxford, OX1 3PS, UK.
Fig. 9 Monthly NPP comparisons at the Donaldson tower site in Florida

Ida Theilade
Faculty of Science, Institute of Food and Resource Economics, University of Copenhagen, Rolighedsvej 25, 1958 Frederiksberg C, Denmark

Ervan Rutishauser
Carbo-ForExpert, 1248 Hermance, Switzerland

Michael K Poulsen
Nordic Agency for Development and Ecology (NORDECO), Skindergade 23, 1159 Copenhagen K, Denmark

Ana P Oliveira and Graça Cabeçadas
Instituto Português do Mar e da Atmosfera (IPMA), I.P., Avenida de Brasília, 1449-006 Lisboa, Portugal

Marcos D Mateus and Ramiro Neves
MARETEC, Instituto Superior Técnico, Universidade de Lisboa, Av. Rovisco Pais, 1049-001 Lisboa, Portugal

Askia M. Mohammed
CSIR-Savanna Agricultural Research Institute, Nyankpala, PO Box 52, Tamale, Ghana

James S. Robinson David Midmore and Anne Verhoef
School of Archaeology, Geography and Environmental Science, University of Reading, Reading, UK

Marcus Knauf
Knauf Consulting, Dorotheenstraße 7, D-33615 Bielefeld, Germany

Index